Springer Laboratory

Springer Laboratory Manuals in Polymer Science

Pasch, Trathnigg: HPLC of Polymers
ISBN: 3-540-61689-6 (hardcover)
ISBN: 3-540-65551-4 (softcover)

Mori, Barth: Size Exclusion Chromatography
ISBN: 3-540-65635-9

Pasch, Schrepp: MALDI-TOF Mass Spectrometry of Synthetic Polymers
ISBN: 3-540-44259-6

Kulicke, Clasen: Viscosimetry of Polymers and Polyelectrolytes
ISBN: 3-540-40760-X

Hatada, Kitayama: NMR Spectroscopy of Polymers
ISBN: 3-540-40220-9

Brummer, R.: Rheology Essentials of Cosmetics and Food Emulsions
ISBN: 3-540-25553-2

Mächtle, W., Börger, L.: Analytical Ultracentrifugation of Polymers and Nanoparticles
ISBN: 3-540-23432-2

Walter Mächtle • Lars Börger

Analytical Ultracentrifugation of Polymers and Nanoparticles

125 Figures and 5 Tables

Dr. Walter Mächtle
BASF AG
Kunststofflaboratorium
67056 Ludwigshafen
Germany
e-mail: walter.maechtle@t-online.de

Dr. Lars Börger
BASF AG
Abt. Polymerphysik
GKP - G201
67056 Ludwigshafen
Germany
e-mail: lars.boerger@basf-ag.de

Library of Congress Control Number: 2005933893

ISBN-10 3-540-23432-2 **Springer Berlin Heidelberg New York**
ISBN-13 978-3-540-23432-6 **Springer Berlin Heidelberg New York**

Springer is a part of Springer Science+Business Media
springer.com

Printed in Germany

Cover design: *design&production*, Heidelberg, Germany
Typesetting and production: LE-TEX Jelonek, Schmidt & Vöckler GbR, Leipzig, Germany

2/3141 YL 5 4 3 2 1 0 - Printed on acid-free paper

Springer Laboratory Manuals in Polymer Science

Preface

In 1924, the Swedish chemist The(odor) Svedberg invented the analytical ultracentrifuge (AUC) to characterize nanoparticles. He used it, for instance, to measure the particle size distribution of very small ($d = 2\,\text{nm}$) gold colloid particles. Already in 1926, the Nobel Prize in Chemistry was awarded to Svedberg for his work. Later, he expanded his investigations to biochemistry, and determined the first molar masses of biopolymers, especially proteins, via AUC. In the following years, the AUC became the most important instrument for the characterization of biopolymers, culminating in the famous density gradient experiment of Meselson and Stahl in 1957, which proved that the DNA replication mechanism, proposed by the Nobel Prize winners Watson and Crick, was correct. Later, with the appearance of other new methods, the AUC lost this prominent position, also because there was no further instrumental development. Around 1980, only a few laboratories were still dealing with AUC. This changed with (1) the launch of a redesigned, fully computerized AUC, the Optima XL-A in 1991, and especially with the XL-A/I in 1997 by Beckman Instruments, Inc., Palo Alto, USA, and (2) some important instrumental developments in specialized AUC laboratories, in particular in the new field of synthetic polymers and colloids. A renaissance of AUC was starting.

In general, analytical ultracentrifugation is a powerful method for the characterization of polymers, biopolymers, polyelectrolytes, nanoparticles, dispersions, emulsions, and other colloid systems. The method is suited to determine the molar mass, particle size, particle density, and interaction parameters such as virial coefficients and association constants. Because AUC is also a fractionation method, the determination of the molar mass distribution, particle size distribution, and particle density distribution is possible as well. A special AUC technique, the density gradient method, allows us to fractionate heterogeneous samples according to their chemical nature, which means that the chemical heterogeneity of a sample can be detected.

The latest textbooks on AUC were published in 1992 and 1994. They deal mainly with biopolymers, theoretical considerations, and do not describe the new possibilities of the Optima XL-A/I. It is the intention of our book to fill this gap, and to demonstrate by means of carefully selected application examples that, especially in the field of synthetic polymers and organic and inorganic nanoparticles, the AUC is an excellent characterization tool for such species. Our book is written for beginners as well as for experienced chemists, physicists and material scientists. It allows the reader to become familiar with the actual status of instrumentation, which means the latest state of the art and the different AUC techniques. All these techniques are described in a simple manner and by means of examples. Detailed instructions for conducting experiments and for their evaluation are given, including explanations of the theoretical background. In this laboratory manual, emphasis is laid more on practical aspects, rather than on details of centrifugation theory. The book is subdivided into seven chapters, concerning the history and basic theory, instrumentation, sedimentation velocity experiments, density gradient experiments, sedimentation equilibrium experiments, application examples, and possible future developments. In particular, the detailed application chapter demonstrates the versatility and power of AUC by means of many interesting and important practical industrial examples. Most of these examples stem from an industrial AUC research laboratory of a leading chemical company, where both authors have gained many years of experience in AUC instrumentation, and AUC characterization of complex polymer and nanoparticle systems.

The authors wish to express their gratitude and appreciation to all colleagues who provided experimental details and data, in particular M.D. Lechner (Osnabrück), T.M. Laue (Durham, NH, USA) and H. Cölfen (Potsdam). The support of many colleagues from the Polymer Research Laboratory of BASF Aktiengesellschaft, Ludwigshafen, Germany, in providing devices, samples and advice, is gratefully acknowledged, above all U. Klodwig. We also would like to thank our coworkers M. Page, U. Gonnermann and M. Stadler for their great commitment in preparing the manuscript and the figures, and H. Roth, M. Kaiser, K. Vilsmeier and K.H. Zimmermann, too, who carried out nearly all measurements presented in this book in an accurate manner.

Carefully reviewing a book means lots of work but not much appreciation for the reviewers. Therefore, the authors wish to express their deep gratitude to Helmut Cölfen (Potsdam) and Karl-Clemens Peters (Bad Dürkheim) for taking on this difficult job.

Furthermore, the authors would like to thank the management of BASF Aktiengesellschaft for supporting the publication of this book.

Last but not least, we thank our families, in particular our spouses, for their indulgence and understanding during the long time of preparing the manuscript of this book, when we often did not adequately take part in family life.

Ludwigshafen, November 2005

Walter Mächtle
Lars Börger

List of Symbols and Abbreviations

β	Density gradient constant
ε	Specific decadic absorption coefficient
η_s	Viscosity of solvent
ϕ	Angle of sector-shaped AUC cell
λ	Wavelength
φ	Volume fraction
ϱ_p	Particle or polymer density
ϱ_s	Solvent density
σ	Standard deviation
τ	Turbidity
$\overline{v}$	Partial specific volume
ω	Angular velocity
a	Optical path length
a_{centr}	Gravitational field
A	Absorption
A_2	Second virial coefficient
AUC	Analytical ultracentrifuge
c	Sample concentration
c_0	Initial sample concentration
d_p	Particle or polymer diameter
D	Diffusion coefficient
DLS	Dynamic light scattering
EM	Electron microscopy
f	Frictional coefficient
F_b	Buoyant force
F_f	Frictional force
F_s	Gravitational force
FFF	Field flow fractionation
g	Acceleration due to gravity
$g(s)$	Differential sedimentation coefficient distribution
$G(s)$	Integral sedimentation coefficient distribution
I	Intensity of light
J	Interference fringe displacement
k_s	Concentration dependence coefficient
K	Kelvin

m	Mass of sample
M	Molar mass
M_{n}	Number-average molar mass
M_{w}	Weight-average molar mass
M_{z}	z-average molar mass
MMD	Molar mass distribution
n	Refractive index
n_{p}	Refractive index of particle or polymer
N	Rotor speed
N_{A}	Avogadro's number
P	Skewness parameter of the *MMD*
PSD	Particle size distribution
r	Radial distance from axis of rotation
r_{b}	Radial distance of the bottom of cell from axis of rotation
r_{m}	Radial distance of the meniscus from axis of rotation
R	Universal gas constant
RCF	Relative centrifugal field
rpm	Revolutions per minute
s	Sedimentation coefficient
s_0	Sedimentation coefficient at infinite dilution
S	Svedberg unit
SEC	Size exclusion chromatography
t	Experimental time
u	Sedimentation velocity
UV	Ultraviolet
V	Volume
$w(M)$	Differential molar mass distribution
$W(M)$	Integral molar mass distribution
X	Relative squared radial distance, $\left(r^2 - r_{\mathrm{m}}^2\right) / \left(r_{\mathrm{b}}^2 - r_{\mathrm{m}}^2\right)$

Table of Contents

1 Introduction

Various questions have to be answered before writing a book about analytical ultracentrifugation, a topic that is nowadays not of wide interest, but more a specialized recess. These questions may be summarized as follows: is it worth putting a lot of work into a book on just one, not widely spread technique, and will there be any readers?

To answer the latter question first: as you, the reader, hold this book in your hands, there is obviously at least one interested reader. It is more difficult to answer the first question. If we had not answered this question with a clear yes, you would not be reading this introduction now. The motivation to invest this huge amount of time in our book arises mainly from three aspects:

Firstly, the authors simply do not understand why the powerful technique we are talking about, the analytical ultracentrifuge (AUC), is widely used in the field of biology and adjacent areas, but to our knowledge is not, or almost never applied to colloid and synthetic polymers (especially not in the measurement of particle size distribution in the range 1 – 5000 nm). One reason might be that simply nobody knows about AUC? Here, our book may be helpful.

Secondly, also the authors believe that there is a need for a book that takes into account the latest developments of the last decade, since the most recent books on AUC were published in 1992 [1] and 1994 [2]. There are some other well-known, older books dealing with analytical ultracentrifugation [3–10], starting with the first in 1940 by Svedberg and Pedersen. However, the focus of nearly all of these books lies on biological systems. In contrast, we would like to remind scientists of a technique they may know but may have forgotten, and put the focus of this book on how powerful the AUC can be, applied on synthetic polymers and colloids.

Thirdly, we would like to emphasize that in times where mega-trends such as nanotechnology, soft materials and biotechnology are en vogue, the need for accompanying analytical methods is increasing. By the end of this book, the reader should be convinced, if necessary, that AUC can be a helpful tool in these modern scientific fields.

The power of AUC is often underestimated. The reasons for this are not easy to address. Looking back to the very beginning of this technique, one is automatically confronted with the work of the Nobel Prize winner The(odor) Svedberg [11,12]. He invented the first practical, usable *analytical* ultracentrifuge (that is, an ultracentrifuge with an optical detecting system) in 1924, together with his coworker Rinde, with the motivation to learn about colloidal systems, and especially about the size and size distribution of colloidal systems [13]; the first centrifuge that was equipped with an optical detecting system was built by Svedberg and Nichols in 1923 [14]. Hence, writing a book focusing more on the use of AUC in colloidal science takes us back to the origin of ultracentrifugation. In the years following his invention, Svedberg turned his attention more from colloidal toward biological questions, such as the determination of the molar mass of proteins [15]. The focus of the analytical ultracentrifugation community stayed on these biological and biochemical questions during the next decades. Still today, most publications containing AUC investigations deal with this field of science.

The AUC was the first instrument delivering reliable values for molar masses of biopolymers, and therefore had its outstanding place in biochemistry. To a certain degree, AUC lost this place with the upcoming of new methods such as the laser technique (allowing light scattering measurements), the development of electron microscopes (EM), the polyacrylamide gel electrophoresis (PAGE), and the size exclusion chromatography (SEC). SEC is today the dominating method to measure average molar masses M and molar mass distributions (MMD) of synthetic polymers. All this went along with a lack of improved instrumentation after the most successful AUC apparatus, the Model E by Beckman, became obsolete. Also several other companies that built ultracentrifuges turned their interest away from the AUC (see Chap. 2). The field of analytical ultracentrifugation was "starving" around 1980, with just a very few laboratories still dealing with the technique. This changed with the launch of a redesigned AUC, the Optima XL-A by Beckman in 1991, and nowadays there is a trend reversal.

Certainly, at present there is a demand for a *fractionating* measurement tool such as the AUC, which provides physicochemical information on a wide choice of topics. And this demand may increase due to the recent scientific mega-trends described above: nanotechnology and biotechnology.

1.1 Historic Examples of Ultracentrifugation

With respect to these mega-trends, two highlights from scientific history may be given in this introduction to illustrate the importance of AUC (and to serve as an appetizer to read the rest of the book that contains a lot of modern examples):

(i) Investigations on gold colloids in 1924, and
(ii) Investigations on the structure of DNA in 1957.

These two historical examples have also been selected because they illustrate two major principles of centrifugation: *sedimentation* velocity runs, and (density) *equilibrium* runs. Both examples reflect the variety of fields covered by analytical ultracentrifugation: while sedimentation velocity runs on colloids, first done by Svedberg, are representative for the field of inorganic nanoparticles and colloids in general (investigations on synthetic polymers may be implied here as well), the Meselson–Stahl density gradient experiment stands for biochemical or pure biological questions.

1.1.1 Investigations on Gold Colloids in 1924

In 1926, Svedberg won the Nobel Prize in chemistry for "his work on disperse systems", just one year after the German chemist Zsigmondy received the prize "for his demonstration of the heterogeneous nature of colloid solutions and for the methods he used, which have since become fundamental in modern colloid chemistry". Zsigmondy invented the *ultra*microscope, and used it to prove the particle nature of colloids with particle diameters in the nanometer range. Later in his Nobel Prize lecture [16], Svedberg pointed out that in his opinion the ultramicroscope of Zsigmondy had a big disadvantage: "The *distribution* of the particle size cannot be determined". And, in fact, Svedberg invented the analytical ultracentrifuge with the intention to determine particle size distributions of colloids by fractionation. Later, its value for the analysis of polymeric systems, both biopolymers and synthetic polymers, was discovered. In fact, Svedberg chose the name *ultra*centrifuge in analogy to Zsigmondy's *ultra*microscope.

The heart of any AUC is a rotor that contains parts called analytical cells (see Chap. 2). These cells house the samples to be investigated. By centrifuging the rotor at high speed, a centrifugal field is generated, and the reaction of the sample on the field can be studied with analytical detectors. As we will see below, one of the possible reactions of the sample to the centrifugal field is the sedimentation of the dispersed or dissolved particles with a characteristic velocity. The underlying principle that allows us to learn about particle size distributions from the ultracentrifuge is that the sedimentation velocity is in general well correlated to particle size, in that the larger a particle, the faster it sediments. This is a major subject of this book, and it will be discussed in detail below. Figure 1.1 shows the original data of the sedimentation experiment that Rinde and Svedberg performed

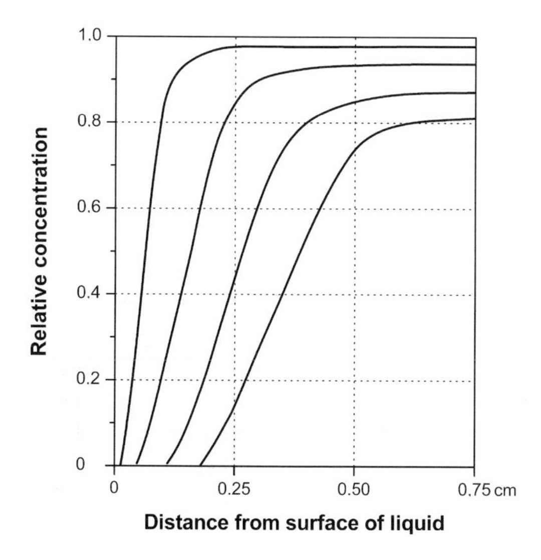

Fig. 1.1. Radial concentration distribution of a highly disperse gold colloid in an AUC cell, recorded 5, 10, 15, and 20 min after beginning of centrifugation (centrifugal field 28 800 times gravity; original work of The Svedberg in 1924; reprinted with permission from [13])

on a gold colloid sample in 1924. The presented radial concentration profile of the sedimenting colloidal gold particles inside the measuring cell, recorded every 5 min, was measured in the first AUC apparatus by Rinde and Svedberg (the details of these results will be the subject of Chaps. 2 and 3).

Each of these radial concentration profiles of gold colloids in the cell is a measure for the sedimentation velocity, or more precisely, for the sedimentation velocity distribution of the differently sized gold particles. In contrast to microscopic methods, not only a few but *all* particles of the sample contribute to the measuring signal in the example shown. Thus, a high statistical relevance is guaranteed. Each of the different radial concentration profiles in Fig. 1.1, recorded at different, well-defined times, can be converted by means of Stokes' law into a particle size distribution , abbreviated *PSD* (for details, see Chap. 3). The resulting (differential) *PSD* is given in Fig. 1.2. Within the errors of measurement, all these radial concentration profiles yield the same *PSD*.

The historical unit μμ on the axis of abscissas in Fig. 1.2 stands for millimicron (also mμ), and that is what we call today a nanometer (nm). Hence, the maximum of the *PSD* given in Fig. 1.2 is close to 1.5 nm, and the whole diameter range lies between 0.7 and 2.2 nm. By means of these, and comparable AUC measurements, Rinde and Svedberg were able to demonstrate that the gold colloids observed by Zsigmondy in his "classic" work were in fact not as narrowly distributed as thought before. Obviously, Svedberg and coworkers were able to characterize colloids that would be named *nanoparticles* in today's terminology. In Chap. 3, we will demonstrate that the advantages of a nanoparticle analysis done by applying AUC, first performed by Svedberg, are still true today.

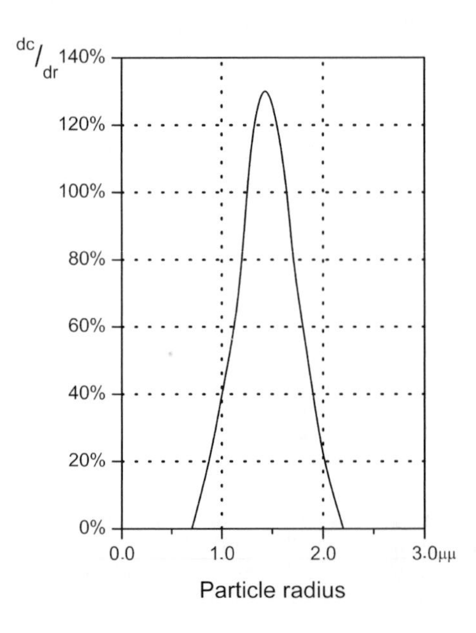

Fig. 1.2. Differential particle size distribution of a highly disperse gold colloid, calculated from Fig. 1.1, taken from the original Svedberg paper (reprinted with permission from [13])

1.1.2 Investigations on the Structure of DNA in 1957

In 1953, the Noble Prize winners James D. Watson and Francis H.C. Crick published their famous paper [17] on the structural model of deoxyribonucleic acid (DNA). The importance of this paper for biology, and for science in general, can surely not be overestimated, and has been described in innumerable publications.

Watson and Crick suggested the DNA to be shaped like a double helix. Their model opened the possibility to answer one of the most important questions in biology in general, namely, how the genetic information contained in an organism is replicated. According to the double helix model, two different possibilities for the replication mechanism can be considered, these being the semi-conservative and the conservative (see Fig. 1.3; for comprehensibility's sake, here we leave out the third possible mechanism, i.e., random disperse; for details, the reader can consult the references cited).

Each of these possibilities predicts different DNA molecules in the descended generation. In the semi-conservative case, the double-strand DNA of the progeny in the first generation after replication consists of one strand from the parent and one new strand. In the conservative case, the parent DNA will be still intact, and the progeny DNA will consist of two new strands.

At this point, two other scientists came into play: Matthew Meselson and Frank Stahl [18, 19]. Stimulated by the question indicated above, they thought of experiments to rule out possibilities and/or to prove one of the mechanisms to be true. The details of the history that led to the "most beautiful experiment in biology" are excellently described in a book on AUC technique recently published by F.L. Holmes [20]: the *equilibrium density gradient*. The principle of this new kind of experiment invented by Meselson and Stahl will be pointed out in detail in Chap. 4. Therefore, here we concentrate on the essential aspects needed to understand the brilliance of Meselson and Stahl's experimental design. The experiment makes use of the fact that an aqueous CsCl solution forms a density gradient when exposed to a high gravitational field in an AUC measuring cell. These heavy cesium chloride molecules (density in the solid state $\rho = 4.1\,\mathrm{g/cm^3}$) are subject to sedimentation as well as diffusion. Due to the first dominating sedimentation, the concentration of the salt is increased at the bottom of the cell. The resulting radial CsCl concentration gradient creates a back-transport of salt under diffusion, according

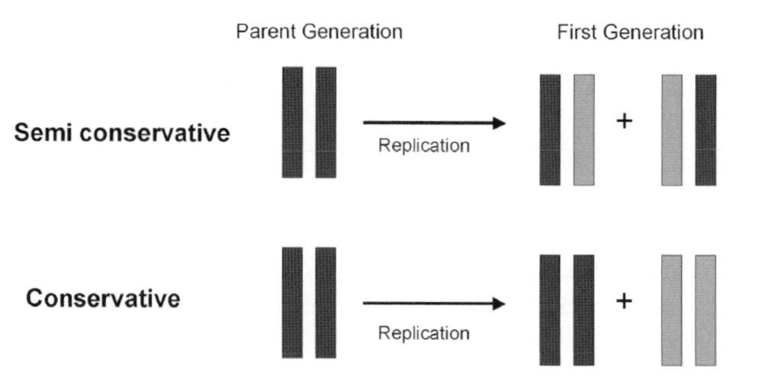

Fig. 1.3. Possible mechanisms of the double-strand DNA replication

to Fick's first law. After appreciable time, this leads to equilibrium between sedimentation and back-diffusion, resulting in a time-independent radial exponential concentration gradient, $c(r)$, of CsCl in the centrifuge cell. The different radial CsCl concentrations are correlated to different corresponding local densities. Therefore, not only a concentration, but also a radial exponential density gradient, $\varrho(r)$, is created when an aqueous CsCl solution is centrifuged.

Particles or macromolecules of a sample (e.g., DNA) added to such an equilibrium density gradient will sediment or float toward the isopycnic point in the density gradient, where the sample particle density is matched by the local gradient density. At this radial point, the sample particle will rest because no net force acts. If the sample consists of components exhibiting different particle densities, then these particular components will be found at different radial positions in the cell. Effectively, the sample has then been fractionated according to the density of its components.

Meselson and Stahl made use of this new type of experiment in order to differentiate between the different replication mechanisms. The design of their experiment, which became famous as the Meselson–Stahl experiment, may be described as follows.

The basic idea they had (stimulated by the work of renowned scientists such as Linus Pauling, Howard Schachman und Max Dellbrück) was to distinguish between the parent and daughter DNA generations by marking the parent generation with heavy isotopes. They choose the ^{15}N isotope, and incorporated it into the DNA of *Escherichia coli* by letting *E. coli* grow in a "heavy" medium that contained solely ^{15}N-marked nutrients for many generations. Eventually, the *E. coli* bacterium was likely to contain almost only the heavy isotope. Therefore, the density of the DNA of this *E. coli* modification would have been shifted to higher density values, as ^{15}N is, of course, heavier than ^{14}N.

As proof of the concept, Meselson and Stahl tested if they were able to distinguish between pure ^{15}N DNA and pure ^{14}N DNA. Figure 1.4 shows their results from an AUC CsCl/water–density gradient experiment in which, as sample, a mixture of ^{15}N DNA and ^{14}N DNA was added. Two signals at different radial positions

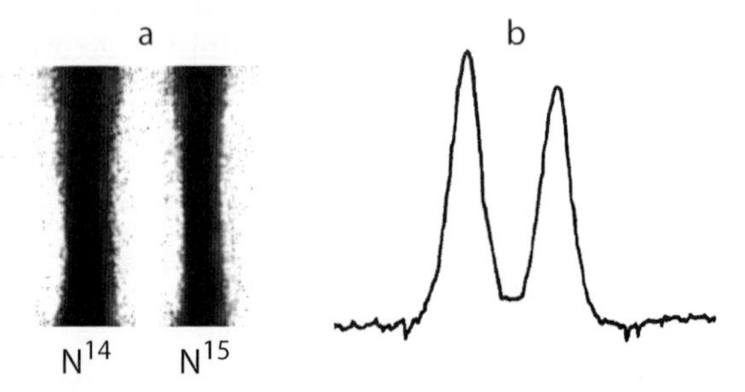

Fig. 1.4a,b. The proof of fractionation of ^{14}N- and ^{15}N-DNA in a CsCl/water–AUC density gradient, taken from the original paper [19]. **a** Photograph taken after 24 h of centrifugation at 44 700 rpm. **b** Microdensitometer trace of that photograph showing the DNA distribution (reprinted with permission from [19])

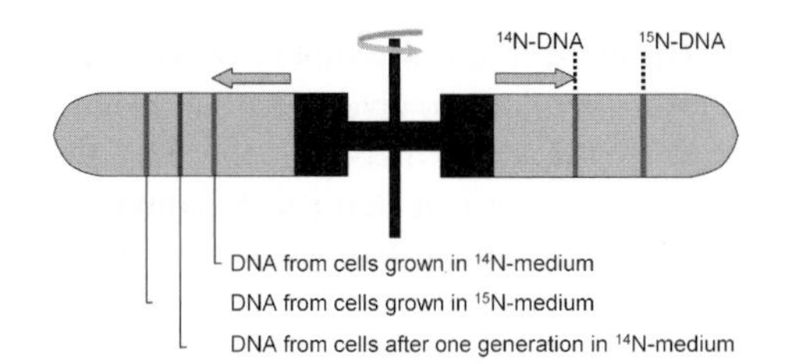

Fig. 1.5. Schematic results from the Meselson–Stahl AUC density gradient experiment

are recorded, one resulting from the ^{15}N DNA, the other from the ^{14}N DNA. These different radial band positions represent a density difference between marked and unmarked DNA of merely $\Delta\varrho = 0.014\,\mathrm{g/cm^3}$ (a modern example concerning non-deuterated and deuterated samples, so-called nanogels, is presented in Sect. 6.2 and Fig. 6.13).

The actual replication experiment [20] was the following: Meselson and Stahl took the ^{15}N *E. coli* DNA they grew as described above and placed it into a normal ^{14}N nutrient medium. The *E. coli* DNA was then allowed to reproduce. At different times, samples from this "growing" solution were taken and were examined in an AUC density gradient experiment. According to Fig. 1.3, in the conservative case two signals are expected for the first generation of DNA, whereas the semi-conservative mechanism predicts only one signal of a DNA species with a density between the those of pure ^{14}N DNA and pure ^{15}N DNA.

The results of Meselson and Stahl were clear (see Fig. 1.5): they found only *one* signal in the AUC density gradient. This means the conservative mechanism could be ruled out. Furthermore, the observed species exhibited a density slightly lower than the parent density, and between the ^{15}N DNA and the ^{14}N DNA densities. This second finding strongly suggested the replication mechanism to be semi-conservative.

All in all, this AUC experiment proved the proposal made already by Crick and Watson that the replication of DNA follows the semi-conservative mechanism. This is briefly the story of the Meselson–Stahl experiment, which was not arbitrarily named the "most beautiful experiment" in biology.

1.2 Basic Theory of Ultracentrifugation

In analytical ultracentrifugation experiments, dissolved or dispersed samples inside AUC measuring cells are exposed to high gravitational fields induced by the spinning of the centrifuge rotor (see Chap. 2). The reaction of the sample to the gravitational field is followed by optical detection systems that basically measure (in the general case of sedimentation velocity runs) the concentration change of the sample with time and radius, $c(r, t)$. In the case of equilibrium runs, only $c(r)$ is measured.

In this Sect. 1.2, we will give a short introduction into the theory of AUC. In this overview, we will concentrate on the basic equations to give a general insight into how physicochemical parameters such as the molar mass M, the particle density ϱ_p, the particle diameter d_p, the diffusion coefficient D, the sedimentation coefficient s, and the frictional coefficient f are connected by AUC theory, and thus are measurable by means of AUC. It must be emphasized that, inside a rotating AUC cell, fractionation of the dissolved particles takes place according to their size and density. Thus, not only the average of physicochemical parameters can be measured, but also their distributions, such as the molar mass distribution *MMD*, and the *PSD*.

The introduction of the AUC theory will be given in two steps. First, a simplified theory based on mechanics according to Svedberg is given. In the second step, the famous general Lamm equation based on first principles from thermodynamics will be derived.

1.2.1 Svedberg's Simplified Theory

For the first step, we consider a particle (or macromolecule) of mass m_p and density ϱ_p, suspended (or dissolved) in a solvent with density ϱ_s and viscosity η_s. When this particle is exposed to a gravitational field $\omega^2 r$ at a radial distance r from the center of rotation, there are basically three forces acting upon it (see Fig. 1.6):

(i) The gravitational force F_s, induced by the centrifugal acceleration $\omega^2 r$ inside the spinning AUC rotor:

$$F_s = m_p \omega^2 r = \frac{M}{N} \omega^2 r \,, \tag{1.1}$$

where M is the molar mass of the solute, and N the Avogadro's number.

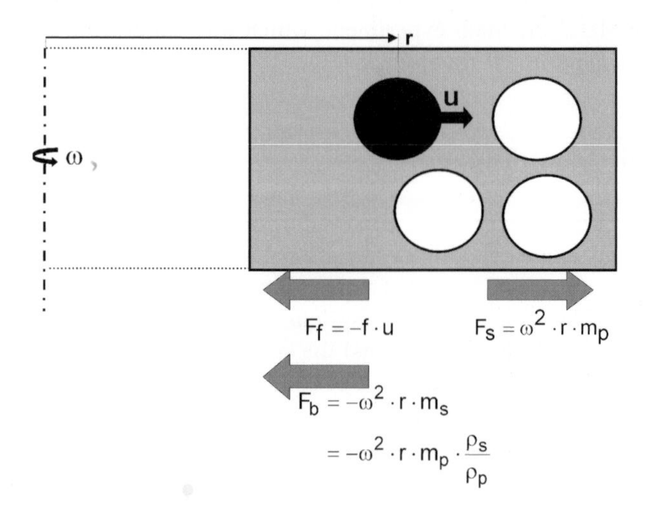

Fig. 1.6. Schematic diagram of the forces acting on a suspended particle sedimenting in a gravitational field in an AUC

(ii) The buoyant force F_b, which is – according to the principle of Archimedes – proportional to the mass of the displaced solvent m_s and $\omega^2 r$ (in opposite direction to F_s):

$$F_b = -m_s \cdot \omega^2 r = -m_p \cdot \overline{v} \cdot \varrho_s \cdot \omega^2 r = -\frac{M}{N} \cdot \overline{v} \cdot \varrho_s \cdot \omega^2 r \,, \tag{1.2}$$

with the partial specific volume of the solute $\overline{v} = \left(\varrho_p\right)^{-1}$, which is the reciprocal of the particle density ϱ_p.

(iii) The frictional force F_f (also in opposite direction to F_s), induced by the movement of the particle through the solvent:

$$F_f = -f \cdot u \,, \tag{1.3}$$

with the frictional coefficient f and the sedimentation velocity u of the solute.

The three forces shown above come into balance immediately, causing the particle to move with a constant sedimentation velocity u. Therefore, the forces can be summed up to give (1.4):

$$F_s + F_b + F_f = 0 = \frac{M}{N} \cdot \omega^2 r - \frac{M}{N} \cdot \overline{v} \cdot \varrho_s \cdot \omega^2 r - f \cdot u \,. \tag{1.4}$$

Rearrangement of (1.4) leads to (1.5):

$$\frac{M\left(1 - \overline{v} \cdot \varrho_s\right)}{N \cdot f} = \frac{u}{\omega^2 r} \equiv s \,. \tag{1.5}$$

Equation (1.5) contains the definition of the sedimentation coefficient $s \equiv u/\left(\omega^2 r\right)$. The frictional coefficient f is well known from hydrodynamic theory. It depends on the shape and size of the moving particle. For the simplest case of a sedimenting (or floating) spherical particle, the Stokes–Einstein Eq. (1.6) and the Stokes Eq. (1.7) can be assumed to hold:

$$f = \frac{kT}{D} = \frac{RT}{ND} \,, \tag{1.6}$$

with the Boltzmann number k, and the gas constant R, and

$$f = 3\pi \cdot \eta_s \cdot d_p \,, \tag{1.7}$$

with the particle diameter d_p of the spherical particle.

Insertion of (1.6) into (1.5) leads directly to the famous Svedberg equation (1.8), which allows us to determine M by measuring s, D and $\overline{v}$ (see Sect. 5.5):

$$M = \frac{s \cdot RT}{D \cdot \left(1 - \overline{v} \cdot \varrho_s\right)} \,. \tag{1.8}$$

Insertion of (1.6) and (1.7) into (1.5) leads to (1.9). This equation is the basis for measuring particle sizes d_p (or, more precisely, Stokes-equivalent (sphere) diameters) by AUC via measurement of s and u, respectively:

$$d_p = \sqrt{\frac{18 \cdot \eta_s \cdot s}{(\varrho_p - \varrho_s)}} = \sqrt{\frac{18 \cdot \eta_s}{(\varrho_p - \varrho_s)} \cdot \frac{u}{\omega^2 r}} \,. \tag{1.9}$$

This section demonstrated Svedberg's derivation of the basic AUC equations, by applying simply a mechanical force consideration.

1.2.2 Derivation of Lamm's Equation

As a second step to introduce AUC theory, in this section a more accurate thermodynamic approach is presented that will lead us to the general Lamm equation. As mentioned above, the basic information obtained in an AUC experiment is the change of sample concentration as a function of radius and time, $c(r,t)$. This change of sample concentration, as described below, basically holds for all different types of AUC experiments.

The change of the sample concentration with radius and time during an ultracentrifugation experiment is given by the so-called Lamm equation [6,21]:

$$\frac{\partial c}{\partial t} = \underbrace{D\left(\frac{\partial^2 c}{\partial r^2} + \frac{1}{r}\frac{\partial c}{\partial r}\right)}_{\text{diffusion term}} - \underbrace{\omega^2 s\left(r\frac{\partial c}{\partial r} + 2c\right)}_{\text{sedimentation term}} \,, \tag{1.10}$$

with the sample concentration $c(r,t)$, the radial distance from axis of rotation r, the diffusion coefficient D, the angular velocity ω, and the sedimentation coefficient s.

Obviously, this equation consists of two terms: The first one describes the diffusion, and the second term the sedimentation of the sample particles. This easily leads to the distinction of different basic types of AUC experiments according to the dominating term, as we will see below (see Sect. 1.3 and Table 1.1).

Equation (1.10) can be derived as follows (for details, see [6]). If the experiment is well performed (no convection, etc.), only two different mass transport mechanisms for the sample particles occur in an ultracentrifuge cell: transport via sedimentation, and transport via diffusion.

For a start, the mass transport due to sedimentation is addressed. Consider a volume element $\mathrm{d}V$ in a sector-shaped ultracentrifugation cell (see Fig. 1.7) rotating at given angular velocity ω (in rad/s). The volume element may reach from r to $r + \mathrm{d}r$.

The mass of solute transported via sedimentation across a given surface per unit time ($\mathrm{d}m_s/\mathrm{d}t$) is given by the product of the concentration c of the solute at the surface, the surface area, and the sedimentation velocity u. The magnitude of the gravitational field at given radial distance from the axis of rotation r is $\omega^2 r$. The area of the sector-shaped cell is $\phi \cdot a \cdot r$, with the angle of the sector ϕ (in rad) and the thickness of the cell along the optical path a. The sedimentation velocity

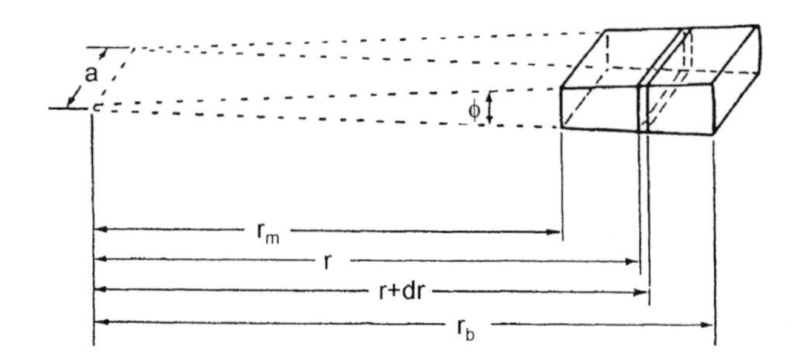

Fig. 1.7. Schematic diagram of a sector-shaped ultracentrifuge cell

$u = s \cdot \omega^2 r$ follows from the definition of the sedimentation coefficient s in (1.5). Thus, the final equation for dm_s/dt is:

$$\frac{dm_s}{dt} = c \cdot \phi\, a\, r \cdot u = c \cdot \phi\, a\, r \cdot s \cdot \omega^2 r \,. \tag{1.11}$$

Now, the second mechanism of mass transport due to diffusion (in the opposite direction of sedimentation) is addressed. The transported mass across the given surface per unit time (dm_D/dt) is given by Fick's first law, in (1.12):

$$\frac{dm_D}{dt} = -D \cdot \phi\, a\, r \cdot \frac{\partial c}{\partial r} \,, \tag{1.12}$$

with the diffusion coefficient D, and the concentration gradient along the radius $\partial c/\partial r$. The combination of (1.11) and (1.12) leads to the following expression for the net mass transport across the surface at the given radial distance r per unit time:

$$\frac{dm}{dt} = \phi\, a\, r \left[cs\omega^2 r - D\frac{\partial c}{\partial r} \right] . \tag{1.13}$$

A similar equation can be written for the net transport across the surface at the radial distance $r + dr$. Subtraction of these two equations gives the net accumulation of mass in the volume element per unit time. The quotient of the change of mass per unit time dm/dt and volume element dV gives the concentration change in the volume element per unit time, in (1.14):

$$\frac{\partial c}{\partial t} = \frac{dm/dV}{dt} = \frac{dm}{dt} \cdot \frac{1}{\phi\, a\, r \partial r} \,. \tag{1.14}$$

This leads to the following expression for the concentration change per unit time at the radial position r:

$$\frac{\partial c}{\partial t} = \frac{1}{r}\frac{\partial}{\partial r} \left[\left(D\frac{\partial c}{\partial r} - \omega^2 s r \right) r \right] . \tag{1.15}$$

If it is assumed that s and D do not depend on the concentration, this equation can be transformed into (1.10). This general differential Lamm equation can in principle describe any experiment performed in an analytical ultracentrifuge. In practice, it is often difficult to make use of this equation, and therefore very often the simpler equations derived in the first part of the basic theory paragraph are used instead.

1.3 Basic Experiment Types of Ultracentrifugation

There are five basic types of experiments that can be performed with an analytical ultracentrifuge (see Table 1.1). Each of these can deliver its own range of physicochemical information on samples. In this chapter, a simple overview of the different experimental procedures is given. The most relevant types of experiments will be described in detail in following chapters.

A short summary of each of these five different experimental approaches will be given in the following five sections.

1.3.1 Sedimentation Velocity Experiment

The sedimentation velocity experiment carried out at high centrifugal fields is the most important AUC technique for macromolecule and nanoparticle characterization, especially for the measurement of particle size distributions. An

Table 1.1. The five basic experiment types of analytical ultracentrifugation

Experiment	Operative term in the Lamm equation (1.10)	Characteristics of experiment	Main accessible physicochemical parameter
(1) Sedimentation velocity	Sedimentation term dominates diffusion term	High rotational speed	s, s distribution, M, MMD, d_p, PSD
(2) Synthetic boundary	Both terms operative, diffusion dominates	Low and high rotational speed, special synthetic boundary cells required	D, c, small s values, dn/dc
(3) Sedimentation equilibrium	Both terms operative, equilibrium between sedimentation and diffusion	Low to moderate rotational speed, time consuming	M, MMD, equilibrium constants and stoichiometrics of interacting systems, A_2
(4) Density gradient	Both terms operative, equilibrium between sedimentation and diffusion	High rotational speed, either light/heavy solvent mixture or heavy additive as support agent	Density ϱ_p, relative $\Delta\varrho$, chemical heterogeneity
(5) Approach to equilibrium (Archibald method)	Both terms operative, sedimentation dominates	Low to moderate rotational speed	M

example of this type of experiment has already been presented in this chapter. Here, the molecules/particles sediment according to their mass/size, density, and shape, without significant back-diffusion according to the simultaneously generated concentration gradient. Under such conditions, a fractionation of mixture components, mainly according to size, takes place, and one can detect this fractionation as a step-like or broadly distributed radial concentration profile $c(r, t)$ in the ultracentrifuge cell. These profiles usually exhibit an upper and a lower plateau, as shown in Fig. 1.1. In most cases, these measured sedimentation profiles can be transferred into average values of s, M or d_p, or into their corresponding distributions, such as *MMD* and *PSD*. The sedimentation velocity experiment will be discussed exhaustively in Chap. 3.

1.3.2 Synthetic Boundary Experiment

In a synthetic boundary experiment, time-dependant radial concentration changes within a boundary between solution and solvent are observed. At low centrifugal fields, where no, or nearly no sedimentation of the sample occurs, diffusion dominates, and thus it is possible to precisely measure the diffusion coefficient D. At high centrifugal speeds, it is possible to measure the sedimentation coefficient s of samples sedimenting very slowly ($0.3 < s < 2S$), which cannot be measured in a standard sedimentation run because the sample moves away too slowly from the meniscus.

Such experiments require special cells where it is possible to layer the solvent upon the solution column, under the action of a certain centrifugal field during the run. This is achieved either by capillaries that connect the solvent compartment with the solution compartment in a double-sector cell, or in an overlaying cell of the valve type (see Chap. 2 and Fig. 2.7). Both types of synthetic boundary cells (see Sect. 2.3) allow solvent superimposition to occur at a certain hydrostatic pressure. At the beginning of the solvent superimposition, $t = 0$, a sharp boundary between pure solvent and pure solution is obtained, comparable to the meniscus. Within minutes, this boundary spreads by diffusion of the sample particles according to Fick's first law (1.12), due to the high sample concentration gradient (dc/dr) within the boundary. The spreading of the boundary with radius and time is monitored.

The following information is available by this method: the plateau concentration of the upper plateau can be determined, which is a measure for the loading concentration c_0 and the specific refractive index increment (dn/dc). Furthermore, the radial refractive index increment (dn/dr) at the boundary can be measured. This allows the determination of the diffusion coefficient D. A possible radial movement of the boundary allows us to determine (very small) sedimentation coefficients s. The synthetic boundary experiment will be discussed in Chap. 3.

1.3.3 Sedimentation Equilibrium

In contrast to the sedimentation velocity run, a sedimentation equilibrium experiment is carried out at moderate centrifugal fields. Here, the sedimentation of the dissolved sample is balanced by back-diffusion, according to Fick's first law, caused by the established concentration gradient. This means that in Lamm's Eq. (1.10), $\partial c/\partial t = 0$ is valid. After the equilibrium between these two transport processes is achieved, a radial exponential concentration gradient $c(r)$ has formed in the ultracentrifuge cell. Therefore, the sedimentation equilibrium analysis is based on thermodynamics. The time of equilibrium attainment depends considerably on the column height of the solution [22], so that short-column techniques with solution columns of about $(r_b - r_m) = 1$ mm are common practice nowadays wherever a rapid equilibrium within 1–2 h or less is desired [23].

The radial concentration gradient $c(r)$ contains information about the absolute molar mass M and the MMD of the dissolved sample, the second osmotic virial coefficient A_2 or interaction constants in case of interacting systems, independently of the shape of the dissolved macromolecules. An advantage is that the detection of the concentration gradient is possible without disturbing the chemical equilibrium even of weak interactions. This experiment type will be described more closely in Chap. 5.

1.3.4 Density Gradient

In the introduction of this chapter, we have described the very first density gradient experiment performed in an AUC. A density gradient experiment is the second principal possibility of separation/fractionation in an analytical ultracentrifuge. It is the separation due to the chemical structure expressed in different solute densities ρ_p in a density gradient. In order to generate a continuous radial density gradient $\rho(r)$ in the ultracentrifuge cell, either a high-density salt (CsCl, etc.) or organic substances such as sucrose are dissolved in water, or a mixture of two organic solvents with very different density is applied. Under the action of the centrifugal field, the "heavy" salt, or the solvent component showing a higher density will sediment toward the cell bottom, and thus change the radial density distribution of the two-component density gradient solution continuously toward the cell bottom. If sample particles are placed into this density gradient solution, they will sediment/float to the isopycnic radial position where their density matches that of the gradient medium. In the case of a chemically heterogeneous sample, this leads to a banding of the different components due to their chemical structure/density at different radial positions. The range of densities that can be covered using density gradients is wide (0.8–2.0 g/cm^3; [24]), sufficient for the separation of nearly all polymeric substances. However, inorganic or organic–inorganic hybrid colloids have a density range that is in most cases too high for the successful application of a density gradient. Furthermore, these so-called *static* equilibrium density gradient experiments take a long time, usually in the order of 1–5 days, due to the slow formation of the density gradient $\rho(r)$. This disadvantage can be overcome to a certain degree by applying a technique called *dynamic* density gradients.

In this very fast type of experiment, a time-dependent radial density gradient $\rho(r,t)$ is created by overlaying H_2O on a D_2O solution of the sample to be examined. The H_2O and the D_2O diffuse into each other. Therefore, the density in the cell is inhomogeneous: At the beginning of the experiment, the bottom region of the cell is enriched with D_2O, whereas the meniscus region is enriched with H_2O. All possible densities, from 1.0 g/cm^3 (H_2O) to 1.1 g/cm^3 (D_2O) can be found in between, depending on the radial position in the cell. At the end of the experiment, there is a homogenous medium density all over the cell. The sample's density can be measured within minutes (if the sample moves fast enough) by recording the radial position of the sample in the dynamic density gradient. The main drawback of this dynamic density gradient method is the limited density range covered. Nevertheless, density gradients of both types, static ones and dynamic ones, are an excellent tool for the investigation of structural differences or chemical heterogeneities in sample mixtures. The density gradient experiments will be presented exhaustively in Chap. 4.

1.3.5 Approach-to-Equilibrium (or Archibald) Method

This procedure is often referred to as the Archibald method [25], in honor of the man who first described it. In principle, the approach-to-equilibrium method is a method to measure molar masses M. It depends on the fact that the conditions for sedimentation equilibrium are fulfilled at both ends of the solution column at all times during every kind of centrifugation experiment. This means that no solute can pass through the air–solution meniscus or out through the bottom of the cell. The net flow of the solute is zero at r_m and r_b at all times. Thus, the Archibald equation is similar to the equilibrium equation for this transient state. From a measurement of both the concentration c and the concentration gradient dc/dr, at either r_m or r_b, the molar mass M of the solute can be calculated. The Archibald method is described in Sect. 5.5.

1.4 Closing Remarks

The authors of this book are working in the world's leading chemical company producing polymer materials and colloids. Within our company, the AUC is intensively used in research, development and production. Thus, it is not surprising that most of the measuring examples presented in this book to illustrate the five basic types of AUC experiments stem from these areas and are, as goes without saying, of major industrial importance.

As mentioned at the beginning of this chapter, the most recent book summarizing the latest advances in AUC instrumentation was published in 1992 [1]. Thus, Chap. 2 of this AUC book deals with new instrumentation developed to date (e.g., the new Optima XL-A/I, Schlieren optics, detectors, multiplexer).

While the instrumentation and the most important types of AUC experiments (and what can be learned from them) will be introduced in Chaps. 2–5, Chap. 6 will give interesting examples of application to display the full power of the AUC as an analytical tool, especially for complex mixtures. Samples from both industrial and scientific fields tend to become more complicated and complex at the nano-scale.

In many cases, one analytical technique applied solely cannot reveal sufficient information. Therefore, we will also show in Chap. 6 how different AUC methods combined with each other, or with other analytical techniques such as field flow fractionation (FFF), dynamic light scattering (DLS) or electron microscopy (EM), will yield information that goes well beyond that provided by a single method.

In the final Chap. 7, we dare to give an outlook on what might be the future of analytical ultracentrifugation. One of these future trends (also discussed in Chap. 6) is the so-called global analysis, i.e., the analysis of very complex samples via combination of different physical methods and powerful, fast data analysis.

As textbooks for a deeper study of analytical ultracentrifugation, we recommend the references [1–10]. Furthermore, we recommend becoming part of the various internet AUC user groups [26–28]. These are offering a permanent and actual discussion platform for all questions concerning AUC. In particular, they are a source for a lot of interesting, free evaluation AUC software.

References

1. Harding SE, Rowe AJ, Horton JC (1992) (eds) Analytical ultracentrifugation in biochemistry and polymer science. The Royal Society of Chemistry, Cambridge
2. Schuster TM, Laue TM (1994) (eds) Modern analytical ultracentrifugation. Birkhäuser, Boston
3. Svedberg T, Pedersen KO (1940) Die Ultrazentrifuge. Steinkopff, Dresden
4. Schachman HK (1959) Ultracentrifugation in biochemistry. Academic Press, New York
5. Fujita H (1962) Mathematical theory of sedimentation analysis. Academic Press, New York
6. Fujita H (1975) Foundations of ultracentrifugal analysis. Wiley, London
7. Williams JW (1963) (ed) Ultracentrifugal analysis. Academic Press, New York
8. Bowen TJ, Rowe AJ (1970) An introduction to ultracentrifugation. Wiley, London
9. Williams JW (1972) Ultracentrifugation of macromolecules. Academic Press, New York
10. Rickwood D (1984) (ed) Centrifugation. IRL Press, Oxford
11. Svedberg T, Rinde H (1923) J Am Chem Soc 45:943
12. Svedberg T (1925) Kolloid-Z Zsigmondy Festschr Erg-Bd Zu 36:53
13. Svedberg T, Rinde H (1924) J Am Chem Soc 46:2677
14. Svedberg T, Nichols JB (1923) J Am Chem Soc 45:2910
15. Svedberg T, Fahraeus R (1926) J Am Chem Soc 48:430, and Svedberg T, Nichols JB (1926) J Am Chem Soc 48:3081
16. Svedberg T (1927) Nobel lecture
17. Watson JD, Crick FHC (1953) Nature 171:737
18. Meselson M, Stahl FW, Vinograd J (1957) Proc Natl Acad Sci 43:581
19. Meselson M, Stahl FW (1958) Proc Natl Acad Sci 44:671
20. Holmes FL (2001) Meselson, Stahl, and the replication of DNA. Yale University Press, New Haven
21. Lamm O (1929) Z Phys Chem A143:177
22. van Holde KE, Baldwin RL (1958) J Phys Chem 62:734
23. Yphantis DA (1960) Ann N Y Acad Sci 88:586
24. Mächtle W (1992) In: Harding SE, Rowe AJ, Horton JC (eds) Analytical ultracentrifugation in biochemistry and polymer science. The Royal Society of Chemistry, Cambridge, p. 147
25. Archibald WJ (1947) J Phys Colloid Chem 51:1204
26. http://www.bbri.org/RASMB/rasmb.html
27. http://www.cauma.uthsca.edu/
28. http://www.nanolytics.de/e/auz/auz.htm

2 Analytical Ultracentrifugation, Instrumentation

Centrifugation is a common technical process for the separation of materials consisting of two (or more) compounds having different sizes and/or showing different densities, such as dissolved macromolecules in a solvent or dispersed particles in a liquid. A centrifuge is basically an apparatus to create centrifugal fields by fast rotation of a rotor. The special design of centrifuges allows us letting these fields act on samples inside the rotor, which leads to a sedimentation (or flotation) of the dissolved/dispersed macromolecules/particles and also to their fractionation, if the macromolecules/particles are different in size and/or density. There is no clear definition from which magnitude of the centrifugal field on a centrifuge is called an *ultra*centrifuge. Usually, centrifuges that create fields higher than 5000 times the acceleration due to the Earth's gravitational field (g) are called ultracentrifuges [1].

An *analytical* ultracentrifuge, in turn, is an ultracentrifuge with one or several optical detection systems, which allow the observation of the fractionation process while the sample is centrifuged.

An analytical ultracentrifuge consists of different main components, namely, the centrifuge housing (see Sect. 2.1 and part 2.1 in Fig. 2.1) itself (including motor bearing, rotor axle/drive, safety vacuum chamber, temperature control, velocity control), the analytical rotor (see 2.2), one or several sample cells (including housing) inside the rotor (see 2.3), the optical detector unit (see 2.4), and a multiplexer unit if several measuring cells have to be detected simultaneously in the same experimental run (see 2.5). Figure 2.1 serves as guide to the structure of this Chap. 2.

While writing about instrumentation, it must be emphasized that the AUC bears considerable danger. The involved powers are enormous (for illustrative pictures of an accident with a preparative ultracentrifuge in 1998, see [2]). A lot of engineering has therefore been put into the development of proper safety elements. For example, the centrifuges are equipped with a system to prevent the rotor from

(i) oscillations created by imbalance at the start of the run or during the run, e.g., if a window breaks, and
(ii) being run with overspeed.

This overspeed control is particularly important, as the typical rotor would explode if speeded up to about 120 000 rpm, or more. For the case of explosion, the rotor itself is surrounded by a safety chamber of heavy steel.

In this chapter, the most important components of analytical ultracentrifuges are described. We will concentrate on state-of-the-art instrumentation, and we

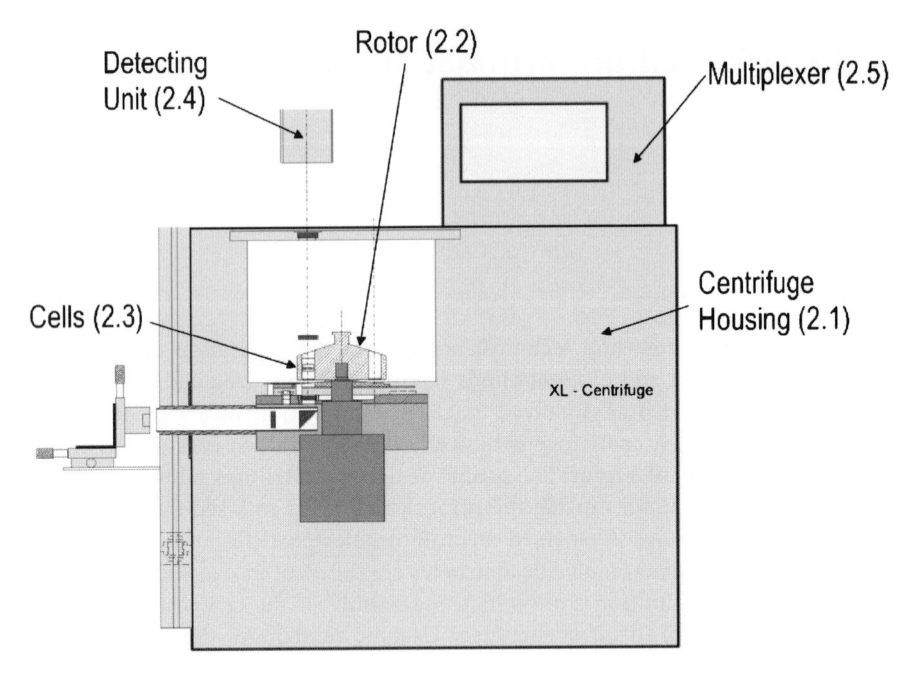

Fig. 2.1. Schematic picture of an analytical ultracentrifuge and its most important components

will therefore leave out most of the historically relevant aspects. For the latter, the reader is referred to [3] or [4].

2.1 Ultracentrifuges

It has been a long way from the room-filling, user-made, really efficient first analytical ultracentrifuge in Uppsala developed by Svedberg and coworkers in the 1930s, to the desk-sized compact, ready-to-use machines of the early 1990s, such as the Beckman Optima XL-A.

Starting with mainly homemade individual items in the early days of analytical ultracentrifugation, the development has subsequently been driven by different companies including Beckman, Phywe, Christ (Heraeus), Metrimpex (MOM), Hitachi, MSE, and Sorvall. Still, all these companies, with the exception of Beckman-Coulter, gave up the production of AUCs and are producing only preparative ultracentrifuges today.

So, nowadays, basically two types of analytical ultracentrifuges are playing a role in the scientific community:

(i) the commercially available Beckman-Coulter AUC (Optima XL-A/I), and
(ii) user-made analytical ultracentrifuges, created mostly as modifications of the commercially available preparative ultracentrifuges.

The present instrument of Beckman-Coulter is the Optima XL-A/I. This is the follow-up model of the perhaps most successful AUC apparatus (in terms of numbers of distributed machines), the famous Model E, which was already equipped with a UV scanner, six-cell multiplexer, interference and Schlieren optics with photographic detection. The modern Optima XL-A/I series was released in 1990/1997, inducing a renaissance of the analytical ultracentrifugation technique in general. The integration of modern computer technology, the more compact design, a higher safety standard, a better electrical drive, a better temperature control of the apparatus, the introduction of a CCD camera, and the online digitization of the measuring data transformed the whole measurement procedure into a more user-friendly and easy-forward process, allowing even non-specialists to perform experiments to a certain degree. However, analytical ultracentrifugation is still not a black-box method, and will probably never become one.

2.1.1 The Beckman-Coulter Optima XL-A/I

The Optima XL-A/I series (see Fig. 2.2) offers two modern, independent optical systems – the UV/Vis absorption optics (since 1990), and the Rayleigh interference optics (since 1997). In contrast to the precursor Model E, unfortunately no Schlieren optical system is integrated. The different optical systems will be described below.

Depending on the rotor, speeds up to 60 000 rpm can be obtained. This corresponds to gravitational fields of over 290 000 times the Earth's gravitational field g (to give an idea of the huge forces the AUC deals with: a mass weighing one gram in real would weigh 290 kg inside the centrifuge or, in other words, a person of 70 kg would weigh 20 000 t inside such a centrifugal field). The relative gravitational field or, as it is often referred to, the relative centrifugal field (RCF), can easily be calculated according to the following simple equation (2.1):

$$\mathrm{RCF} = \frac{r \cdot \omega^2}{g} \tag{2.1}$$

with the angular velocity ω (in rad/s), the radius r (that is, the distance from the axis of rotation, 5.7 – 7.2 cm in the XL-A/I), and the Earth's gravitational field g ($1g \approx 9.81\,\mathrm{m/s^2}$).

Additionally, on the top of the Optima XL-A/I, Fig. 2.2 shows an analytical eight-cell rotor and some measuring cells.

Temperature Control of the Optima XL-A/I

Exact control of the temperature is essential to any thermodynamic measurement, and control within 0.5 K, or better, has to be guaranteed. Svedberg's first AUC contained a safety chamber that was continuously flooded with gaseous hydrogen to create small drag, and to allow control of the temperature within the chamber. In later developments, the rotor-containing safety chamber was evacuated to guarantee the smallest possible air-friction resistance to the rotor's movement, thus eliminating rotor heating by air friction.

Fig. 2.2. Modern analytical ultracentrifuge Optima XL-A/I produced by Beckman-Coulter

In the former Beckman Model E, cooling of the chamber was achieved by water cooling of a cladding surrounding the rotor inside the chamber. The heating elements were different from these, and were located at the chamber bottom. This led to temperature gradients within the chamber, and therefore within the rotor itself. The temperature was measured in a very specific way: a needle was fixed to the bottom of the hanging analytical rotor, and a temperature-dependent electrical resistance was installed inside the rotor. This needle dipped into a mercury bath, thus closing an electrical circuit. By measuring the heat resistance in this electrical circuit, the rotor temperature was estimated.

In the modern XL-A/I, the vacuum chamber is heated and cooled via Peltier elements located in the aluminum heat sink at the bottom of the chamber, which leads to a homogeneous temperature throughout the vacuum chamber via heat exchange by radiation. The temperature is measured contact-free at the bottom of the rotor via an IR detector, which, of course, requires the rotors to be black-colored. The temperature range covered by the instrument is 0 – 40 °C, according to biochemical needs. Next to the technical problems, physics also complicates the exact control of temperature. Due to adiabatic cooling occurring when the rotor is speeded up to high angular velocities, the temperature in the rotor may vary by 0.1 – 0.3 K. Hence, in high-speed runs, first data collections should not be performed before 10 – 15 min after reaching a constant final speed. This guarantees that heat exchange has reached equilibrium.

Drive and Velocity Control of the Optima XL-A/I (Compared with the Model E)
As well as the exact knowledge of temperature, the exact angular speed $\omega = 2\pi N$ or the running time integral $\int \omega^2 \, dt$, in the case of applying sweeping $\omega(t)$ techniques (see Chap. 3), of the rotor must be determined. It was shown [5] that, in the case of polymeric particles analyzed by applying turbidity optics, the results of the sedimentation analysis were independent of the velocity profile $\omega(t)$. That was due to the exact knowledge and continuous recording of $\int \omega^2 \, dt$. It has to be kept in mind that the angular velocity ω is squared in (1.5) for the determination of the sedimentation coefficients.

In the former Beckman Model E, the electrical drive motor was coupled for safety reasons to the rotor drive shaft via a 1:7 mechanical speed transformation unit. The rotor was linked to the drive shaft by hanging on a steel filament (i.e., the rotor axle) that was only 2 mm in diameter. Sometimes, this 2-mm axle broke, and the rotor fell down within the safety chamber. This Model E configuration allowed the operation of the centrifuge at certain fixed rotor speeds only. The speed was measured indirectly in the drive.

Today, in a safer manner, the rotor of the XL-A/I is put directly onto the drive shaft, which in turn is directly run by an electrical induction motor. The motor operates under computer control, and offers the user an infinitely variable regulation of the speed. The actual velocity ω is measured optically via a stroboscopic ring fixed onto the rotor bottom, which is read out by a photoelectric relay. This ensures that the rotors are not overspeeded, which could lead to rotor explosion, as pointed out above. Another special unit integrated into the drive dampens the rotor oscillations at some critical rotor speeds around 1000 rpm, as well as rotor oscillations created by imbalance. It will stop the rotor if the imbalance is too high. In contrast to the former Model E, the rotor in the XL-A/I is standing on the drive shaft. The driving axle is flexible in such a way that the angle between the rotor axis and motor axis (which is fixed to the vacuum chamber and laboratory floor) can change with rotor velocity. The system is self-stabilizing, like a gyroscope (indeed, it is a gyroscope!), and very safe.

2.1.2 User-Made Centrifuges

Considering that Beckman-Coulter is a more biochemically oriented company, and because most applications of AUC originate from biochemical questions, the design of their last analytical ultracentrifuges, the Optima XL-A/I, is driven by the corresponding requirements. As mentioned above, this book mainly concentrates on the application of the AUC in the research on synthetic polymers and colloids. In this field, we encounter special requirements that are not fulfilled by the XL-A/I. To give only three examples:

(i) Far more than in biochemical applications, the sedimentation velocity experiment is the central experiment type. The colloidal particles or polymers examined are often "large" compared to typical biochemical samples, and therefore sediment much faster. Thus, the need to create fast detectors and variability of rotor speed arises, because a precise and permanent measurement of the running time integral $\int \omega^2 \, dt$ has to be ensured.
(ii) In contrast to biological samples, synthetic polymers and colloids frequently exhibit turbid samples, sometimes in the visible range. Hence, a fast turbidity (= light scattering) detector is important to measure particle size distributions.
(iii) Biological samples are usually measured at the smallest possible concentration in order to obtain physical information on single molecules, or aggregates consisting of a relatively low number of molecules. In contrast hereto, in colloidal systems the intermolecular or, more often, the inter-particle interactions at higher concentrations are of great importance. Very high concentration differences are also created in two-component density gradients, where often one component is not transparent for UV light. Hence, there is a demand for the universal Schlieren optics, which allows measuring high concentrations (up to 100 g/l) and steep refractive index gradients inside the cell.

In order to fulfill these specific and industrially important requirements for the fast detectors (and $\int \omega^2 \, dt$ estimation), turbidity detectors and Schlieren optics, some laboratories have made their own developments of AUC instrumentation. Using the preparative Beckman XL series as a basis for development, extensive modifications have been added (see [5–8] and [12]).

As example, in this section we first describe the modifications (see Fig. 2.3) added to the original Optima XL for the introduction of a new user-made digital Schlieren optical setup [7,8]. Later, in Sect. 2.4.4, we describe two other modifications: Laue's fluorescence detector [12], and a BASF-made turbidity detector [5] to measure precise particle size distributions (see also Sect. 3.5.1).

Figure 2.3 shows the most important modifications for introduction of Schlieren optics into an Optima XL:

- An optical bench containing the light source (1) and Schlieren slit assembly (2) has been mounted to the left side of the XL housing. The flash lamp is adjustable by an x–y–z stage.
- A horizontal hole is drilled into the heat sink (6), which is fixed with the motor (5) and an optical tube containing the collimating lens (3) and the 90°-deflection prism (4) is added.
- For completion of the optical path through the rotor (8), the vacuum chamber (10) has been modified by drilling in two vertical holes, one into the heat sink and the other into the moveable cover plate, that allow the mounting of vacuum-sealed windows (7). For adding a rack that carries the condensing lens (9), existing thread holes in the heat sink are used.

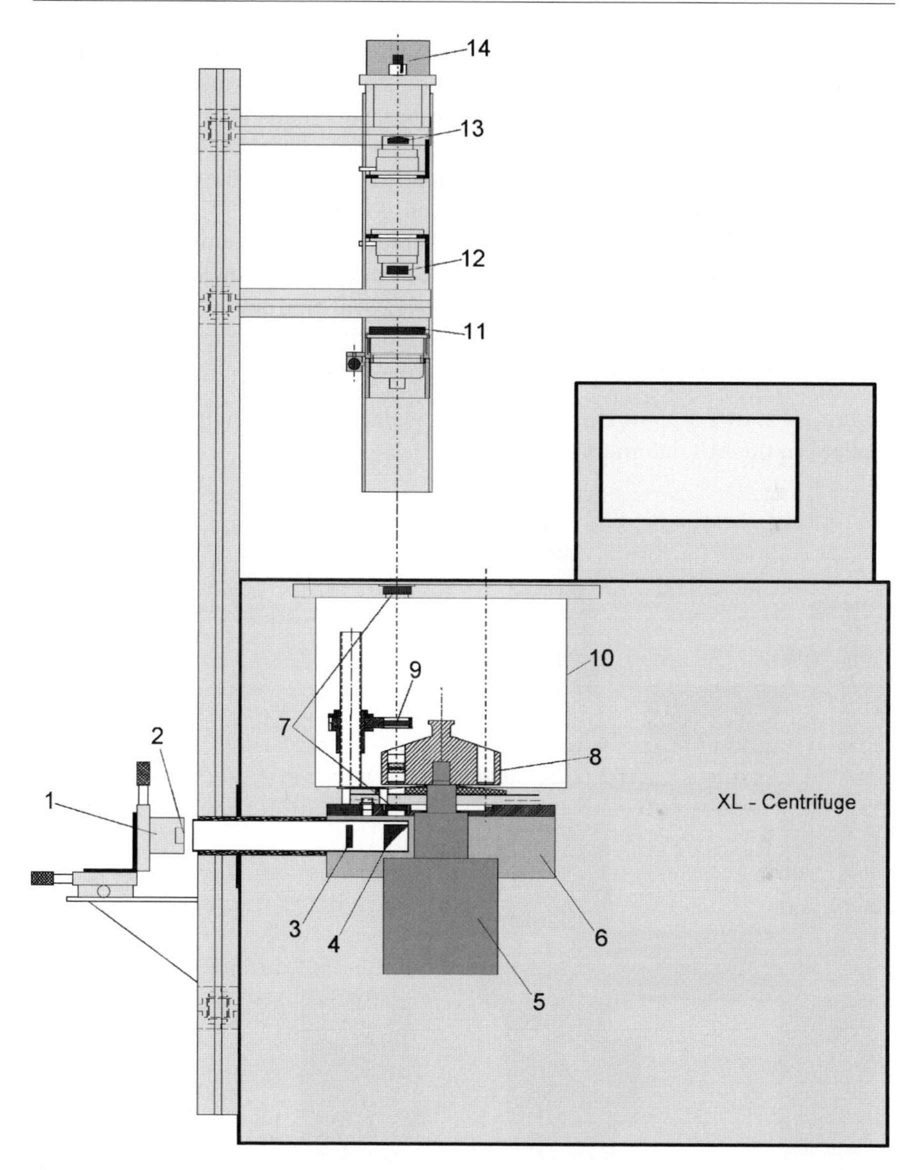

Fig. 2.3. Modifications of an Optima XL with Schlieren optics (and a multiplexer; reprinted with permission from [8])

- A charge-coupled device (CCD) camera (14) was applied as detecting unit. The CCD camera, and the remaining optical elements of the Schlieren optics, i.e., the phase plate (11), camera lens (12), and cylindrical lens (13), are mounted to a scaffold above the vacuum chamber, as indicated in Fig. 2.3.
- Independent from the Optima XL velocity control, a user-made velocity control system to trigger the flash lamp has been implemented (thus, the safety system

> of the original XL, especially over speed and imbalance control, is not changed). By polishing a narrow strip of the rotor bottom, a reflecting mirror was created, and a light barrier has been added to the heat sink under the rotor. This allows us to generate one, single narrow light pulse per revolution that is converted to an electrical signal that, in turn, is applied to measure the exact velocity of the rotor and to trigger the light source of the Schlieren optics. The trigger device is able to trigger not only one measuring cell. In the multiplexer state, it is possible to trigger each single cell of the eight cells measured simultaneously, thus allowing Schlieren optical *superimposition* of one cell with any other cell. The superimposition of the measuring cell with the reference cell is chosen as the standard.

Figure 2.4 shows a picture of the physical Schlieren optics multiplexer setup as realized in the AUC laboratory of BASF.

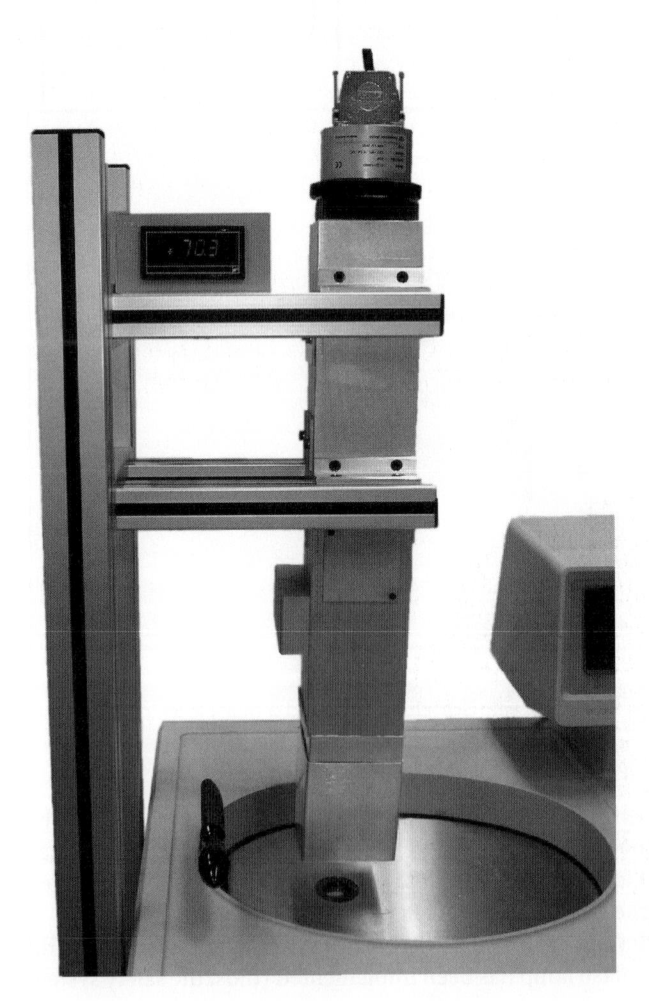

Fig. 2.4. Photograph of the physical Schlieren optics multiplexer setup of a modified Optima XL realized in the AUC laboratory at BASF

2.2 Rotors

In the early days of centrifugation, a lot of work was spent on the design of analytical rotors. Whereas in earlier days steel and later aluminum were the predominately used material, nowadays almost exclusively titanium rotors are applied.

The analytical rotors are built from one piece of titanium. Depending on the rotor type, four or eight holes are drilled into the titanium, leading to four- or eight-hole rotors (see Fig. 2.5). The distances between the midpoints of the holes and the axis of rotation are $r = 6.50\,\text{cm}$. Usually, one of the holes is loaded with a reference cell (also called counterbalance), which contains two radius reference marks at the positions of $r = 5.70$ and $7.20\,\text{cm}$. The reference cell is used to perform the radial calibration at given centrifugal speed. The remaining holes carry up to seven different measurement cells, which in turn can contain up to eight different samples each if multi-channel cells are used. For safety reasons, it is important to know the allowed maximum speed for the different types of rotors. The most frequently used, commercially available analytical rotors have maximum speeds of 50 000 and 60 000 rpm. The lifetime of rotors depends on the time period they have been rotated at maximum speed. If the revolution speed during the experiment is restricted to 95% of the allowed maximum rotor speed, then the lifetime of rotors is nearly unlimited. Compared to the former Model E, the new Beckman Optima XL-A/I is run only with redesigned titanium rotors. The stress levels throughout the rotors are more uniform, and the rotors are lighter. This leads to a more uniform adiabatic temperature change (cooling) while the rotor is accelerated. For every rotor and every run, a continuous running lifetime check protocol has to be written, recording actual running speed and total running time. This is done inside the PC of the XL-A/I automatically for safety reasons.

Fig. 2.5. Eight-hole analytical rotor made of titanium by Beckman-Coulter

2.3 Measuring Cells

The part of an analytical ultracentrifugation system that is in direct contact to the sample is the measuring cell. The cell consists of many pieces. Figure 2.6 shows the cell assembly of a typical analytical ultracentrifugation cell. Characteristically, all these pieces must work under extremely high mechanical stress at maximum rotor speed.

AUC cells must fulfill at least two criteria:

- They should not leak or distort even at high centrifugal fields, which create very high hydrostatic pressures of up to 250 bar [9] at the bottom of the solution column.
- They should allow passage of light through the cell via stable quartz or sapphire windows while the rotor is spinning.

There is a big variety of ultracentrifugation cells available, depending on the kind of experiment they are used for. The differences occur mainly in the choice of the centerpiece, and that of the windows. Because of the high mechanical strength, the most common window materials are optically polished plan-parallel sapphire and quartz glass. The plates exhibit a thickness of 5 mm. It will be pointed out in the corresponding chapters which type of glass is preferably applied in the particular type of experiment. Generally, sapphire is the material of choice, if available. One advantage of quartz glasses is their lower price. Another advantage is the broader spectrum of light that can pass through quartz when UV detectors are applied.

Whereas it is very uncommon to create user-made rotors and other parts of the cell, the centerpieces of AUC cells are frequently built or modified by users in order to adapt them for their special needs. The centerpiece is the heart of an AUC measuring cell (see Fig. 2.7). It is made of aluminum, titanium, Kel-F, Teflon or charcoal/aluminum-filled Epon. In principle, there are four types of analytical ultracentrifugation centerpieces:

(i) The mono-sector centerpiece only contains the sample solution. It is mainly used with Schlieren optics and the turbidity detector.
(ii) The double-sector centerpiece has two separate chambers, one for the sample solution, the other for the solvent. It is used with all detecting systems (interference optics, UV/Vis absorption, and Schlieren optics).
(iii) Multi-channel centerpieces for 3 – 5 different samples can exhibit very different architectures (see Fig. 2.7c,f). In principle, they are double-sector centerpieces. They are applied exclusively in equilibrium runs (see Chap. 5).
(iv) Synthetic boundary centerpieces exist in two different types, the capillary type and the valve type (see Fig. 2.7d,e). They are designed to allow the layering of solvent (or solution) onto a sample solution (or solvent) while the rotor is spinning (see Chap. 3).

In most cases, the channels of the centerpieces are sector-shaped to prevent convection during sedimentation due to the radially sedimenting sample. The standard thickness of centerpieces is a = 12 mm. For special experimental designs, 1-, 2-, 3-

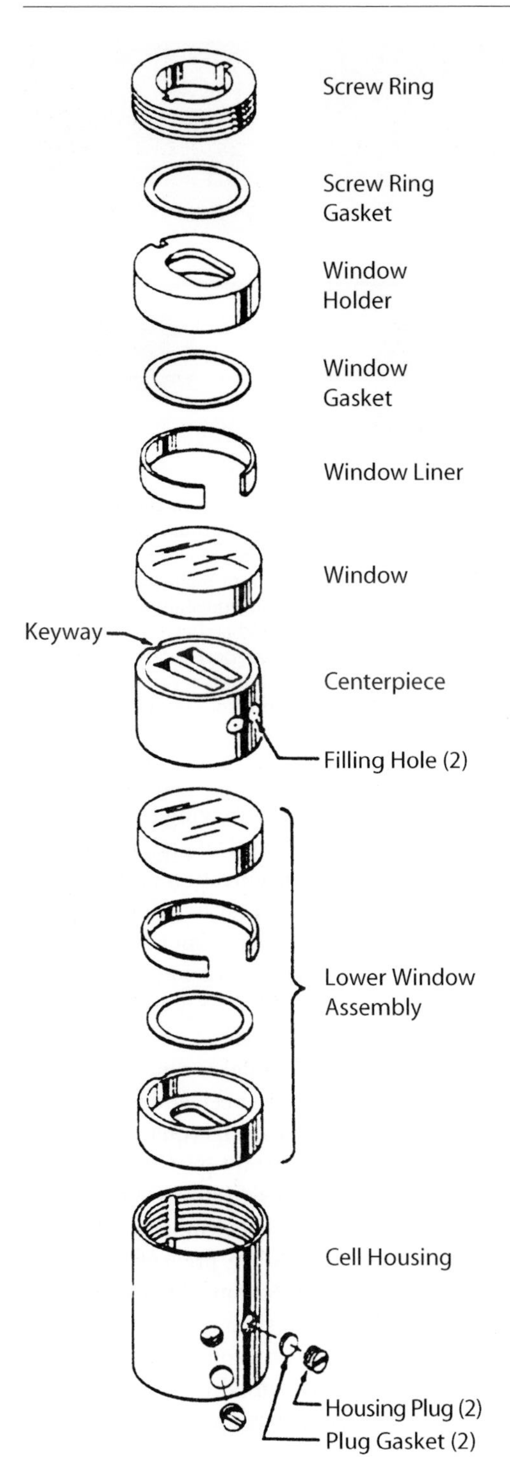

Fig. 2.6. Cell assembly with an (elastic) Kevlar double-sector centerpiece, which needs no gaskets between windows and centerpiece. In the case of (hard) metal centerpieces, two additional gaskets are required (reprinted with permission from [30])

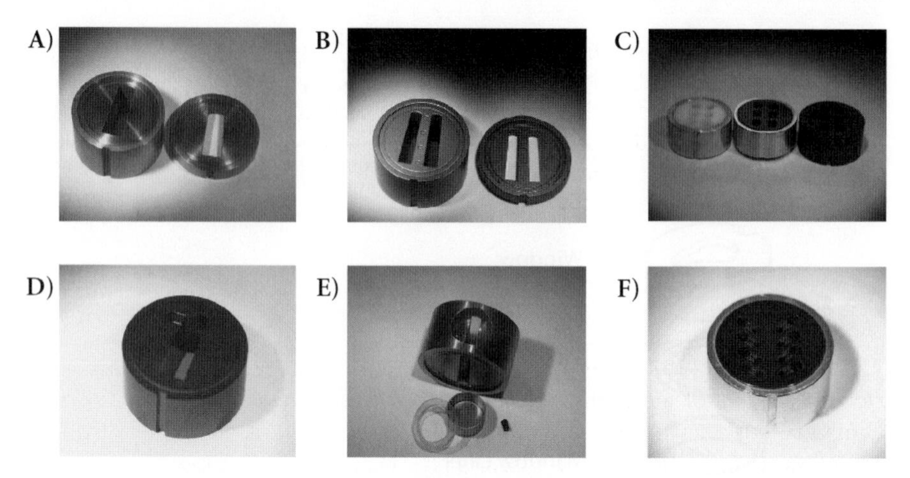

Fig. 2.7A–F. Some examples of different AUC centerpieces: **A** two mono-sector centerpieces (12 and 3 mm) made from aluminum, used for velocity runs applying Schlieren or turbidity optics; **B** two double-sector centerpieces (12 and 3 mm) made from aluminum, used for velocity and equilibrium runs applying interference, absorption or Schlieren optics; **C** three 12-mm multi-channel double-sector centerpieces (made from different materials) for sedimentation equilibrium experiments; in equilibrium experiments, the sector shape of the centerpiece holes/chambers can be omitted; **D** synthetic boundary 12-mm double-sector centerpiece of the capillary type; **E** synthetic boundary 12-mm single-sector centerpiece of the valve type with storage bin, rubber valve, and gasket; **F** multi-channel 12-mm double-sector centerpiece for the study of four solvent–solution pairs in sedimentation equilibriums runs

and 25-mm centerpieces have been developed and used. By varying the thickness of centerpieces, the layer thickness a of the sample in the optical path is varied as well. Depending on the applied detecting system, the concentration of the sample and the sample's optical properties, the optimum thickness may be different. Thus, the proper choice of the centerpiece is important when the experiment is designed. In Fig. 2.7, a selected assortment of different centerpieces for various purposes is given.

The design of the modern Beckman Optima XL-A/I series accommodates the continuous use of the existing cells of the preliminary Model E centrifuge (with exception of the 30-mm cell). Only small modifications have been added to XL-A/I cells, such as smaller window holders, different window material (quartz of higher purity), or anodized screw rings. Some of the most important parts of AUC cells in the case of metal centerpieces, such as aluminum or titanium, are flat elastic gaskets between the centerpieces and the two glass windows (mostly cut out of thin polyethylene foil). They have to tighten the cell, and to prevent leakage of the solutions. Furthermore, they have to diminish the mechanical stress onto the (brittle) windows (see Fig. 2.6). For elastic centerpieces, such as Kel-F, charcoal-filled Epon or Teflon, gaskets are not needed (that is the reason why they are not shown in Fig. 2.6; rather, only the similar gaskets between window and window holder are shown). To close the cell safely, the assembled cell has to be tightened with a torque wrench at defined torque.

The cells have to be inserted into the rotor in a correct manner to avoid convection. Adjusting the horizontal axis of the centerpieces is done by turning the cell inside the rotor hole until this axis is in exact alignment with the rotor axis, indicated by meeting of two marks. Furthermore, to avoid imbalance during the run, cells inserted in opposite holes of the rotor are adjusted to have the same weight (within a maximum difference of 0.5 g).

2.4 Detectors

The essential data collected from an AUC experiment are the radial concentration profile $c(r)$ of the examined sample in the cell at a given time t. The most common detectors make use of three properties of the dissolved or dispersed sample: specific light absorption, light scattering (turbidity), and light refraction. Therefore, different types of detection systems have been integrated into analytical ultracentrifuges. Roughly, they can be divided into three classes:

(i) Detector systems that allow one to obtain specific information on the chemical composition of the sample. The most important example from this class of detectors is the UV/Vis absorption optics integrated in modern XL-A/I analytical ultracentrifuges. Another detector of this type, the fluorescence detector, is recently commercially available. This class of detectors makes use of the sample's specific light absorption/fluorescence.
(ii) The light scattering or turbidity detector is also an absorption detector, but it is not dependent on the specific interaction of light with chemical groups of molecules at characteristic light wavelengths λ. Rather, it is a universal detector dependent on the size of the light scattering particle with respect to Mie's light scattering theory.
(iii) Detector systems that record the overall change of the concentration in the cell during centrifugation. Detectors of this type are interference optics (integrated in XL-A/I machines) and Schlieren optics. Both detectors can be applied whenever the examined dissolved solute particles exhibit different refractive indices compared to the solvent. Thus, this (universal) detector class makes use of the difference in light refraction Δn between the solution and solvent.

Detectors can also be classified as follows (see Figs. 2.8 – 2.13):

- Detectors that scan the cell radially (see Fig. 2.8d). The optical unit is moved stepwise along the radial axis of the cell, i.e., along the cell sectors. The step width is adjustable, and usually lies in the range 10 – 100 μm. At each radial position r, the optical information is recorded. The single measurements are then added to result in a radial concentration profile $c(r)$ of the cell at given time t. The most important example for a scanning type of detector is the UV/Vis absorption detector, as it is realized in the Optima XL-A, and the known fluorescence detectors. The scanning time has to be fast, compared

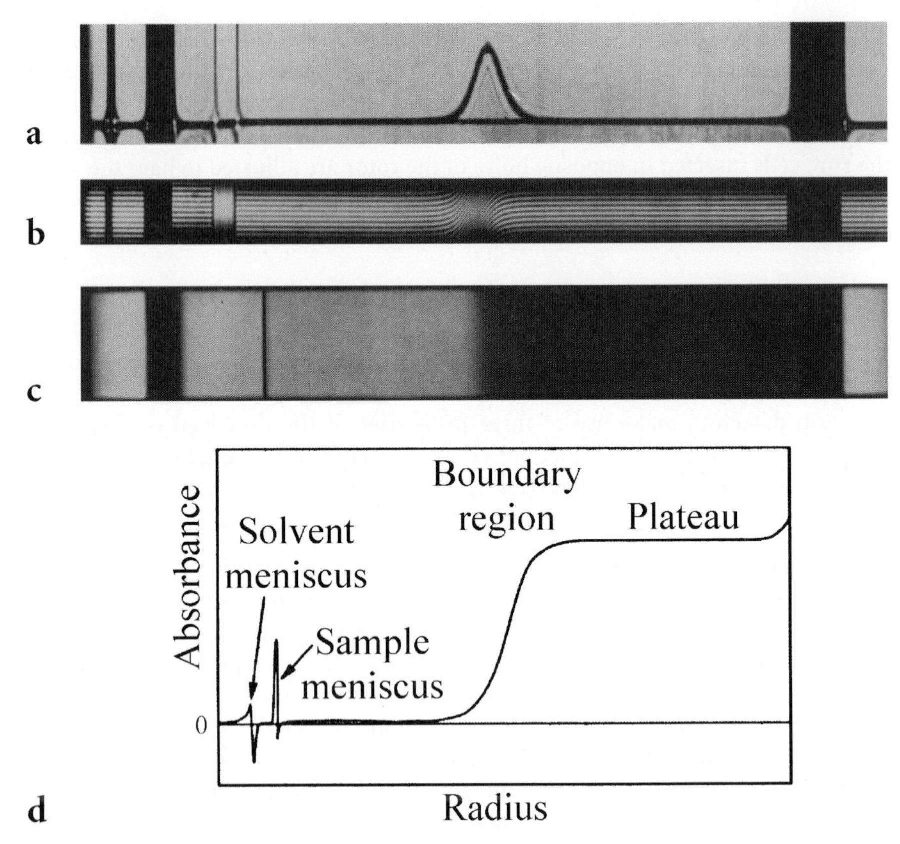

Fig. 2.8a–d. Typical optical patterns as derived by **a** Schlieren optics, **b** interference optics, **c** UV optics via photo-plates, and **d** UV optics via radial scanner (see also Fig. 5.1; reprinted with permission from [30])

to the sedimentation process observed by it. Otherwise, the assumption that the whole concentration profile was measured at once would lead to serious errors. In the case of the UV/Vis absorption optics, especially at low rotor speeds (approximately below 10 000 rpm), this is a real drawback of the XL-A. Scanning detectors are working well in equilibrium runs.

- Another detector type records the optical information at a given fixed radial position r as function of time t. The most important example of this type is the turbidity (= light scattering) detector, where the optical unit is located at the midpoint of the radial axis of the cell (see Fig. 2.13). Here, the change of intensity of the light I passing through the cell is recorded over the whole duration of the experiment, i.e., $I(t)$. The result is a single curve containing the whole experimental information.
- A third class of detectors records the whole concentration profile $c(r)$ at once at a given time t, without applying a scanning unit (see Fig. 2.8a–c). Representatives of this class are the interference and the Schlieren optical system,

where a light flash illuminates the whole cell at once. The result is a picture of the whole cell containing the concentration profile $c(r)$ at a given time t. The picture is recorded on a photo-plate or onto the chip of a CCD camera.

In the following sections (Sects. 2.4.1, 2.4.2 and 2.4.3), the three most important detecting systems – the absorption, the interference, and the Schlieren optical detectors – will be explained in detail.

Typical optical patterns obtained by the three main optical detecting systems are shown in Figs. 2.8 and 2.13 (see also Fig. 5.1). For an extensive description of these optical systems, see [10].

As mentioned above, only the Rayleigh interferometer as an online detector and the absorption optics are still available on modern ultracentrifuges (i.e., the XL-A/I). Few users adapted the Schlieren optics for use in the Beckman Optima XL. As a further detection system, a fluorescence detector was reported. This system is extremely sensitive, and allows the selective investigation of compounds with concentrations as low as 10 ng/ml, even in mixtures with a much larger amount of other components [11]. A prototype fluorescence detector for the Optima XL-A/I ultracentrifuges has recently been constructed, so that now a third detector is commercially available that can be simultaneously used in a modern analytical ultracentrifuge [12].

Amazingly, one of the most important detectors for use especially in industrial R&D is commercially not available: turbidity optics [13, 14]. A turbidity optical system is realized only in very few laboratories throughout the world. It is mainly used to determine particle size distributions. This detector is described at the end of this section. It will also be addressed in Chap. 3 due to its high relevance for particle sizing.

Generally, it is advantageous to combine several optical systems (= multiple detection). Especially the combination of the Rayleigh interference optics and the UV/Vis absorption optics can yield important information about complex systems, for example, when an absorbing component is selectively detected with the absorption optics, whereas the Rayleigh interferometer would detect all components together simultaneously [15, 16]. An example is given in Sect. 7.2 and Fig. 7.4.

2.4.1 Absorption Optics

The basis for applying absorption optics is well known from spectroscopy. Light passing through a solution containing light-absorbing molecules loses intensity in a specific way. The Lambert–Beer law (2.2) describes the interaction quantitatively:

$$A = \lg \frac{I_0}{I} = \varepsilon \cdot c \cdot a \tag{2.2}$$

with absorption A, intensity of light after passing the sample I, intensity of light passing the solvent-filled cell I_0, specific decadic absorption coefficient ε, the concentration of the sample c, and the thickness of the measuring cell along the optical path a.

Figure 2.9 shows the principal setup of the UV absorption optics as realized in the Beckman Optima XL-A/I. A xenon flash lamp with a nearly continuous light spectrum $190 < \lambda < 800$ nm is used as a light source. While the rotor (containing up to eight AUC cells) spins, the lamp is triggered to flash when the cell of interest passes the optical path. Therefore, this XL-A/I UV/Vis scanner is a multiplexer unit, too.

The triggering of the flash lamp is based on the exact measurement of the actual rotor speed. This is realized by a Hall-effect sensing device based on a magnet placed on the bottom of the rotor. Whenever the magnet passes the sensing device placed at the bottom of the vacuum chamber, the magnet induces an electrical signal. The position of the cell, and the sector of the cell of interest with respect to the position of the magnet on the rotor is known, and therefore the time delay from the magnet passing the sensing device until the lamp has to flash can be calculated within the XL-A/I computer, using the known rotor speed.

The duration of a single flash is about 2 – 3 μs, which is sufficient to create data from only one sector of the double-sector cell at once. With a maximum shot frequency of 100 Hz, the lamp can flash every ten revolutions at the maximum centrifugation speed of 60 000 rpm.

Light from the flash lamp passes an aperture (1 mm in diameter) to hit the diffraction grating of the monochromator at the top of the optical arm. Absorbing reflectors that hold back all disturbing light from other, unwanted wavelengths surround the grating. The grating itself is adjustable by high-precision electrical motor gears that allow the adjustment of the desired wavelength. The grating obeys a nominal band pass of 2 nm, and covers the wavelength range 200 – 800 nm. Because the flash intensity is not constant from flash to flash, for standardization the monochromatic light from the grating hits an 8% reflector. That means that 92% of the light passes the reflector while 8% is reflected onto the incident (= standardization) detector. The intensity value measured by this detector is needed to normalize these pulse-to-pulse variations in the light intensity of the xenon flash lamp. This procedure is necessary because one flash illuminates just one sector of a double-sector measuring cell. Below the 8% reflector, a second aperture prevents the illumination of any other, undesired sector of the measuring cell. The monochromatic light then leaves the optical arm, and passes through one of the sectors of the measuring cell containing either the sample or the corresponding solvent. By this optical setup, it is assured that the monochromatic light is parallel with respect to the normal projected onto the window surface of the measuring cell. While passing the measuring cell, the intensity of light is decreased by the interaction with the sample or solvent, according to (2.2), i.e., Lambert–Beer's law. Below the rotor, the remaining light passes to a moveable lens-slit assembly (see insert in Fig. 2.9) that fulfills certain functions:

- The aperture of the lens-slit assembly determines the radial resolution. The width of the slit is 25 μm. The lens-slit assembly moves stepwise as a unit driven by an electrical motor, and allows one to scan the sector of the measuring cell radially. The step width Δr can be adjusted with a smallest step width of 5 μm

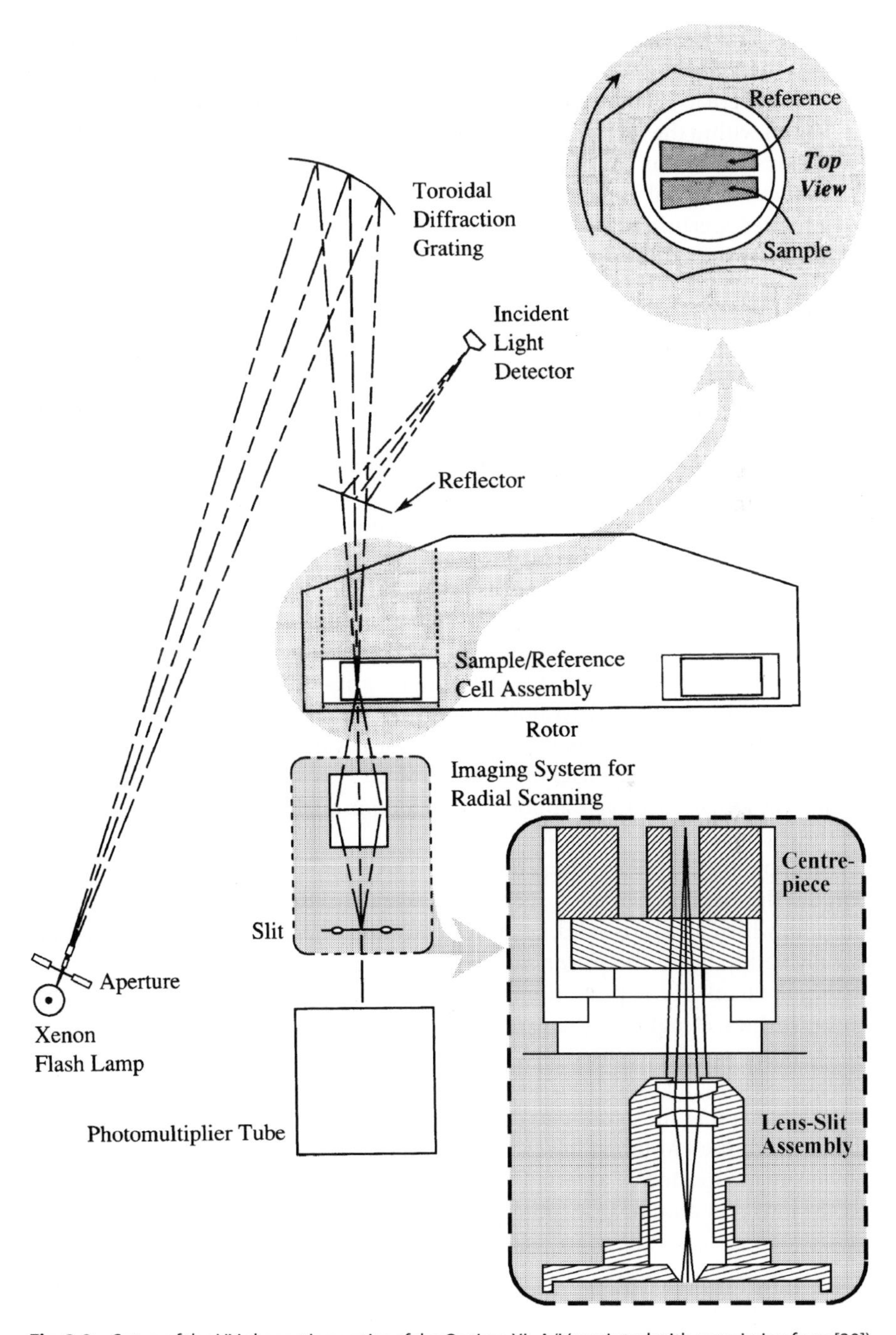

Fig. 2.9. Setup of the UV absorption optics of the Optima XL-A/I (reprinted with permission from [30])

steps. The lens focuses the light onto the photomultiplier (PM) tube, which measures the intensity I of light passing through the measuring cell.

- The camera lens of the lens-slit assembly creates an image of a slice at constant radius within the sector of interest of the measuring cell on the PM tube. A slice of constant radius contains redundant information, and can therefore be used to improve signal quality.
- Another aperture at the bottom of the lens-slit assembly reduces the stray light intensity once more.

The light leaving the lens-slit assembly hits the PM tube, which comprises a large active area that is sufficient to record the complete scanned picture of the cell. Therefore, the PM tube can be kept stationary.

The intensity of the solvent sector I_0, and the intensity of the sample sector I, both recorded on the PM tube, normalized with the intensity from the stationary incident detector, are combined to reveal the desired absorption signal $A(r) = \lg I_0/I$ of the sample, at the radius position r.

The main drawback of the XL-A/I UV/Vis optical system is the long duration of a measurement. Especially at low rotor speeds, such a stepwise radial scan along a single cell at one wavelength, but with, for example, ten replicates at each radial position in order to improve the signal-to-noise ratio, can last up to about 15 min. Nevertheless, for the monitoring of equilibrium runs, where enough time for recording is given, the UV/Vis absorption optics of the XL-A/I is appropriate. For the rapid recording of sedimentation runs, faster detectors like interference or turbidity detectors are mandatory. It is an advantage of the XL-A/I UV/Vis detector that, during an AUC run, complete UV/Vis spectra can be obtained at a constant radial position – although with a poor quality – yielding information on the chemical nature of the solute.

In applying the UV scanning device, it is recommended (i) to choose as wavelength λ the maximum of the absorption peak, and (ii) to select a sample concentration c for which $A < 1.5$ is valid, i.e., a concentration where Lambert–Beer's law (2.2) is fully valid and in the *linear* range. If one uses a λ outside of the maximum, within the flank of the peak, it might occur that small errors in the XL-A monochromator wavelength are increased, and that A is measured incorrectly.

Another problem caused by excessively high concentrations is the so-called Wiener skewing, arising from steep radial refractive index gradients within the cell. Light passing such gradients can be deviated so strongly that it does not reach the target of the PM tube ("black band"). However, if this Wiener skewing is not obviously manifested as a partial black band within the XL-A-UV scan, its effect on the interpretation of absorption gradients may be safely ignored (see [31]).

2.4.2 Interference Optics

The interference optics as realized in the Optima XL-A/I series of Beckman (see Fig. 2.10) is based on the principle of a Rayleigh interferometer [10,17,18]. This, in turn, makes use of the fact that the velocity of light depends on the refractive index of the medium it passes through, i.e., the higher the refractive index of the medium, the lower the velocity of the light. That means that the (small) refractive index difference $\Delta n(r) = (n_{\text{solution}} - n_{\text{solvent}})$, which is proportional to the corresponding concentration $c(r)$ at this radius position, can be monitored very rapidly by interference optics. This is made visible in form of a vertical deviation of the originally parallel interference fringes behind a double (also called Rayleigh) slit (see insert in Fig. 2.10).

The (folded) interference optics as realized in the Optima XL-A/I applies a triggerable laser diode ($\lambda = 675\,\text{nm}$) above the rotor as light source. The monochromatic parallel light from the light source passes through two parallel Rayleigh slits (= double slit) above the measuring cell that allow the *simultaneous* illumination of the reference (= solvent) and the sample sector of the measuring cell. Light from the two Rayleigh slits creates an interference pattern of parallel light and dark fringes (see Fig. 2.8b and insert in Fig. 2.10) behind the cell in the plane of the condensing lens (= CCD camera sensor plane). The "disturbance" of the light by a medium of higher refractive index causes the position of the fringes to shift vertically proportional to the refractive index difference Δn, and thus proportional to c at the radial point of interest. This vertical shift of the interference fringes is counted in numbers $J(r)$ of fringes. Because $J \cdot \lambda = \Delta n \cdot a$ and $\Delta n = (\mathrm{d}n/\mathrm{d}c) \cdot c$, equation (2.3) is valid (with the thickness of the centerpiece a and the known specific refractive index increment $\mathrm{d}n/\mathrm{d}c$).

$$J(r) = \frac{a \cdot (\mathrm{d}n/\mathrm{d}c)}{\lambda} \cdot c(r) \tag{2.3}$$

where $J(r)$ is an absolute measure of the radial concentration distribution $c(r)$ within the AUC cell. This is valid only in the case of sedimentation runs, and synthetic boundary runs where left of the sedimenting boundary the concentration is zero, and so we know the zero fringe. That is not given in the case of equilibrium runs, where no boundary exists and the sample concentration $c(r_\text{m})$ at the meniscus position r_m is finite and unknown, like the absolute fringe number J_m at this position. Hence, only a *relative* fringe shift $\Delta J(r)$ with respect to the meniscus is measurable. Thus, only a relative concentration $\Delta c(r)$ can be determined, too. In Sect. 5.2 (equations (5.7) and (5.8)), it will be shown for the case of equilibrium runs how the mass conservation law can be utilized to calculate $J(r_\text{m})$ and $c(r_\text{m})$.

Again, the detecting system of the older Model E by Beckman served as basis for the development of the new XL-A/I interference optics [18,19]. In comparison, major changes have been made to the light source and the detecting unit. The constantly burning mercury lamp as light source was replaced by a pulsed laser diode. A CCD camera replaced the photographic plate as detecting unit. Nevertheless, the major improvement was the implementation of a computer routine that allows the fast digitalization and a fast Fourier transform analysis of the fringes (first described by T.M. Laue in [11], page 63). These data are collected online. More recently, significant improvements have been realized by changing the CCD camera system to the next camera generation, especially with increased resolution.

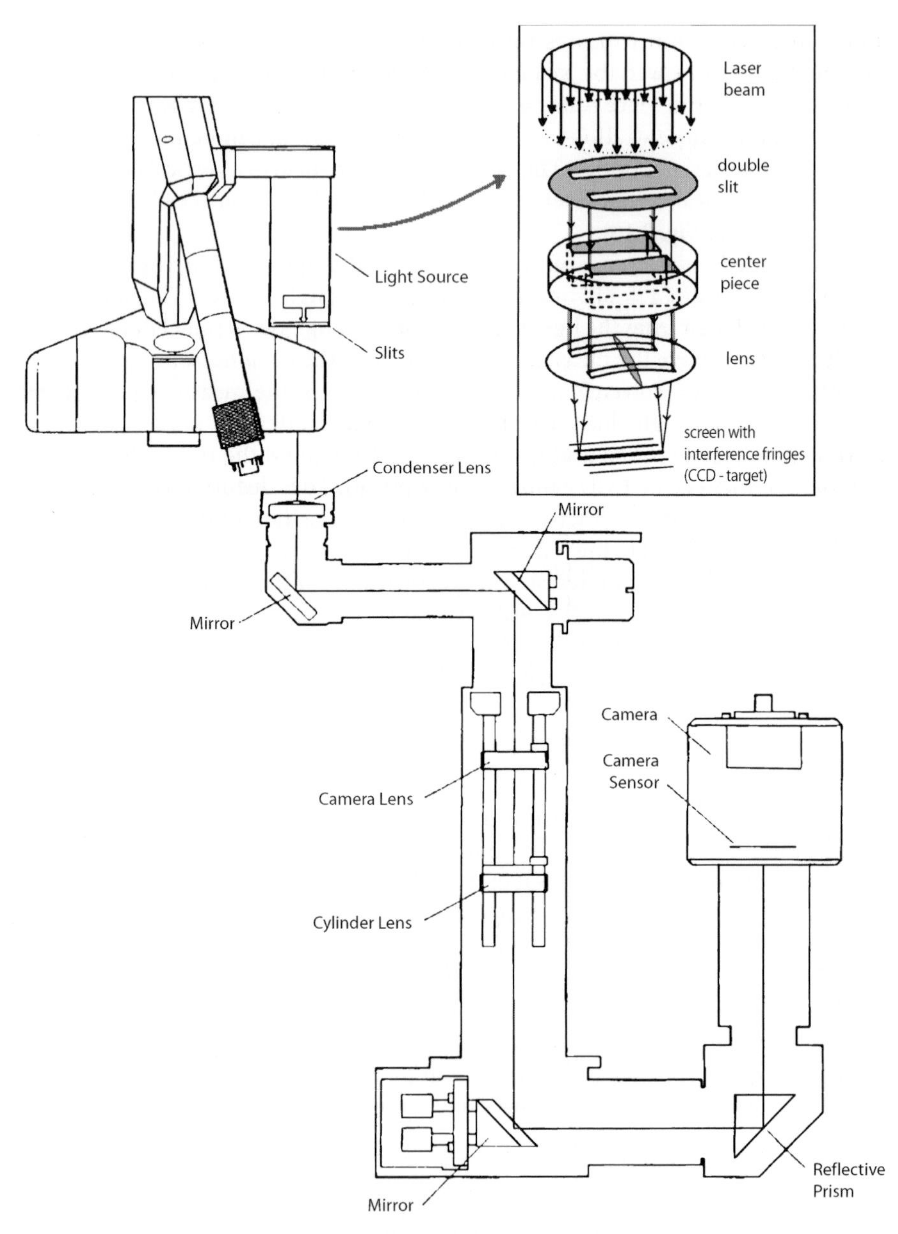

Fig. 2.10. Interference optics detector inside the Beckman Optima XL-A/I. The *insert* shows the principle of a Rayleigh interferometer with the interference fringes behind the double slit (reprinted with permission from [30])

Figures 3.1 and 3.2 show such a series of fast Fourier transform analysis fringes of the XL-A/I. All fringes are standardized on the (left) meniscus side to the same plateau position, $J = 0$. If one starts the fringe scan left of the meniscus, sometimes one or more of the scans are shifted vertically by one or two integer numbers, for example, $\Delta J = \pm 1, \pm 2$, etc. However, this shifting can be corrected easily. Furthermore, if a sample shows too much turbidity, the fringes on the CCD camera monitor appear faintly, and it could be possible that the Fourier transform analysis fails.

2.4.3 Schlieren Optics

The Schlieren optics setup (as described above in Sect. 2.1.2 and Fig. 2.3) is similar to the setup of the Rayleigh interferometer but has a phase plate (or knife-edge) in the focus of the condenser lens as an additional element. As there is no commercially available system anymore, users are obliged to construct their own Schlieren optics, or transfer the Schlieren optics of an older Model E onto the Optima XL platform [7]. As mentioned above, the driving force to do this comes from some unique advantages of Schlieren optics, of which two are given here:

- It allows the use of simple mono-sector cells that are easy to handle, to tighten, and that assure highest reproducibility.
- Even steepest refractive index gradients (resulting often from steep density gradients) can be followed because these gradients can be compensated via wedge windows (as will be shown in Chap. 4, density gradients are among the most important experiments in AUC).

The modification of the preparative Optima XL centrifuge itself, especially the addition of a Schlieren optical system, has been outlined in Sect. 2.1.2. In this paragraph, only the necessary optical components will therefore be described.

The arrangement of the optical elements is basically as follows (see Fig. 2.3): the white light flash lamp (1) illuminates the Schlieren slit (2). Usually, a green filter (wavelength $\lambda = 546\,\text{nm}$) between positions (7) and (11) is applied to create the light of a well-defined wavelength. The light then passes on through the collimating lens (3), a 90° glass prism (4), and then enters the vacuum chamber through a vacuum-sealed window (7). After passing the sample-containing cell inside the rotor (8), the light is focused by the condensing lens (9) through the second vacuum-sealed window (7) onto the plane of the phase plate (11). The new optical arm above the XL housing (see also Fig. 2.4) contains the camera lens (12) and the cylindrical lens (13). The light then falls onto the CCD camera (14) where the Schlieren picture is visible in form of a Schlieren peak or a Schlieren line on a TV monitor (see Fig. 2.8a). The CCD camera is connected to a computer for further image processing. In this arrangement, the light source is triggered mainly because of the high reliability of the stroboscopic light source, and the possibility to control the light source by newly developed multiplexer software. The ease to create superimposed pictures of more than one cell is another reason for triggering the light source. In each centrifugation run, there is one reference cell per rotor (also called counterbalance cell). This cell is used to perform radial calibration during centrifugation by superimposing the Schlieren picture of the reference cell and the Schlieren picture of the cell containing the samples examined. Figure 2.8a shows such a superimposition. Furthermore, superimposition allows the measurement of up to seven different samples in one centrifugation run.

As one possible light source, a Cathodeon C 82007 xenon flash lamp (LOT/oriel, Darmstadt, Germany) can be used (flash frequency 0–100 Hz, flash energy 1 J). A CCD camera (for example, a digital black&white camera CCD 1300 from Vosskühler, Osnabrück, Germany, with a resolution of 1280 × 1024 pixels) serves as a detector unit.

Via a digital RS-644 interface, the camera is controllable by a computer. A Schlieren picture showing a typical Schlieren peak of a sedimentation velocity run is given in Fig. 2.8, in comparison with the results of interference and absorption optics delivered from the same sedimentation experiment (using a double-sector cell). In contrast to the interference optics, where the vertical fringe shift J is proportional to Δn and thus to the concentration c, in Schlieren optics the vertical shift of the Schlieren line is proportional to the radial refractive index gradient ($\mathrm{d}n/\mathrm{d}r$), and thus to the radial concentration gradient ($\mathrm{d}c/\mathrm{d}r$). In other words, integration of the Schlieren line ($\mathrm{d}n/\mathrm{d}r$) reveals redundant information with respect to one interference fringe $n(r)$, and vice versa (see also Fig. 5.1). Thus, the Schlieren peak area A_{schl} is proportional to the sample loading concentration c_0, and can be used to measure c_0 via (3.23). Continuous integration of the Schlieren peak, in combination with the law of conservation of mass, delivers the radial course of concentration $c(r)$ within the cell.

2.4.4 Other Detectors

Fluorescence Detectors

To our knowledge, two types of AUC fluorescence detectors are existing ([11] and [12]). Both are user-made. The latter one [12] is recently commercially available (albeit for a very high price). The manufacturer is Aviv Biochemical, Lakewood, NJ 08701. Thus, we give only a short description here. For more details, the literature or the manufacturers themselves should be consulted.

Especially for biochemical questions, a fluorescence detector is of high interest, mainly because of two inherent advantages:

- A fluorescence detector shows a sensitivity that has exceeded UV optics for decades. For example, Laue [12] performed equilibrium runs (see Fig. 5.12) and sedimentation velocity runs on the green fluorescent protein (GFP) with concentrations as low as $c = 0.012$ and $0.0024\,\mathrm{g/l}$.
- A fluorescence detector can detect single components of a sample selectively, even if the component's concentration is very low compared to those of the other components of the sample.

The principle (see Fig. 2.11) is known from conventional fluorescence detectors: A laser is applied as a light source for the excitation of the fluorescence system. The light from the laser on a scanning stage passes optical elements that illuminate a very small Δr area of the sample sector (spatial resolution $60\,\mu\mathrm{m}$, according to [11]). The response of the fluorescence system is detected either in

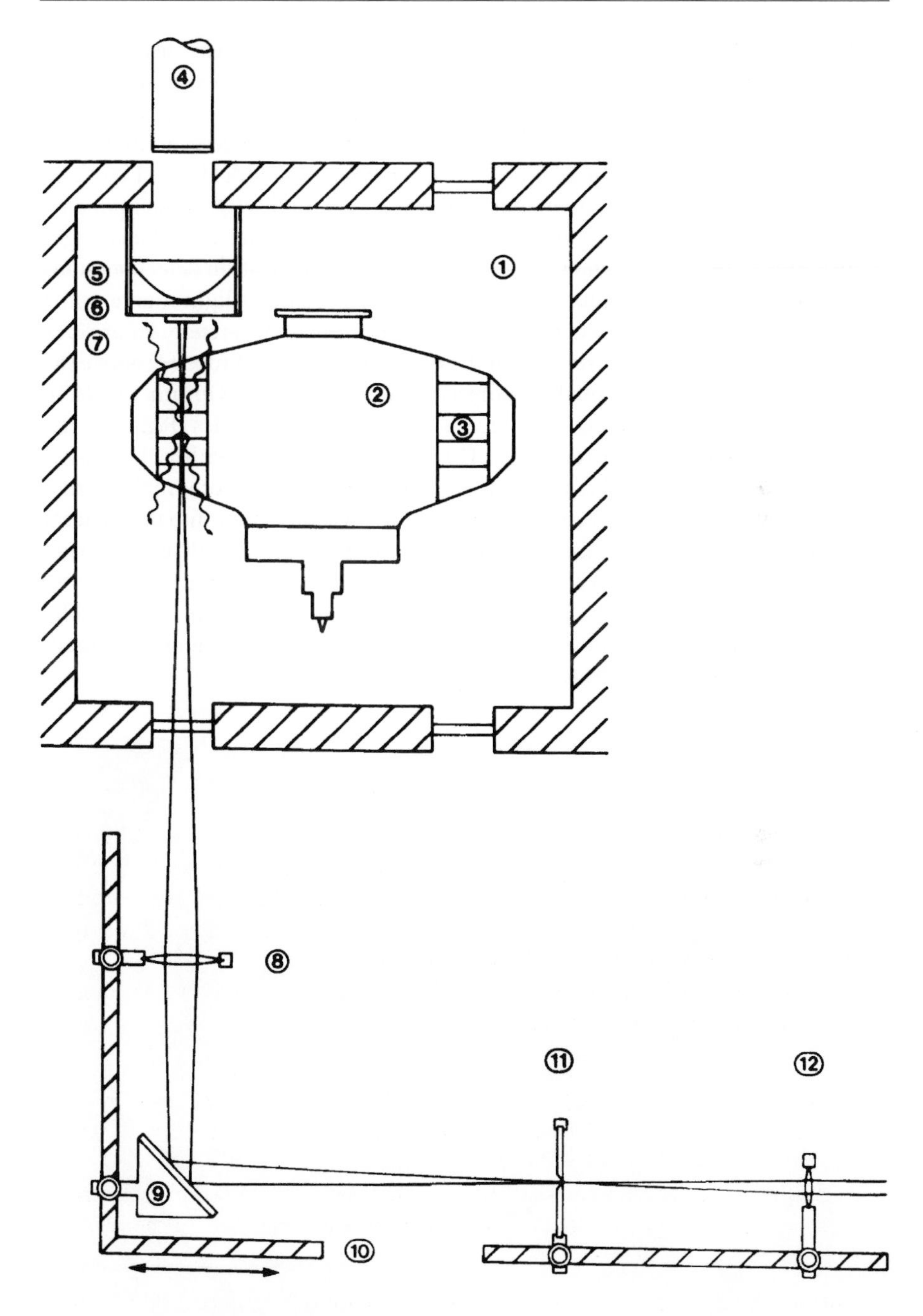

Fig. 2.11. Schematic setup of a user-made fluorescence detector realized by Riesner et al. [11] inside a Model E (reprinted with permission from [11])

reflectance [12] or in transmission mode [11] by a PM tube. A filter is used in front of the photomultiplier to ensure that only fluorescence light passes and excitation light is omitted.

Riesner et al. described the first fluorescence detector adapted to an AUC [11]. They used the geometry of the Schlieren optics channel of a Model E (see Fig. 2.11), and replaced the Model E components of the Schlieren optics by the components of the fluorescence detector. This allowed them to keep the instrumental changes of the Model E minimized, and to run the Model E–UV absorption detector simultaneously. A continuously running argon ion laser served as the light source, while a PM tube was applied to detect the florescence response of the system. Figure 2.11 shows a scheme of Riesner's setup with the filter (6) and the PM tube (4) above the rotor. In contrast, Laue [12] put his fluorescence detector into an Optima XL-A/I in a similar manner as described above for the case of Schlieren optics (Fig. 2.3). The small and compact detector (see Fig. 2.12) is directly installed within the XL-A/I rotor chamber completely above the rotor. There are two new features in this setup: first, the laser excitation light comes from outside via a glass fiber into the vacuum rotor chamber and, second, the light does not penetrate the complete measuring cell. Rather, only the back-scattered fluorescence light is measured by step-wise radial scanning of the whole surface of the cell along a sector. This is reached by a confocal setup of the detector geometry [12]. Figure 5.12 shows a measurement with Laue's fluorescence detector.

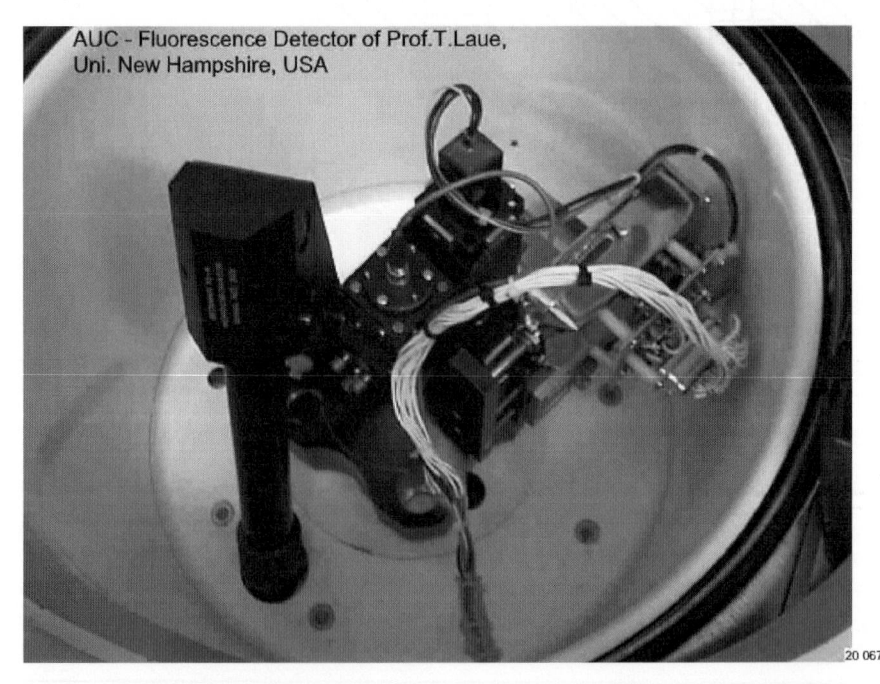

Fig. 2.12. Photograph of a user-made fluorescence detector realized by Laue [12] inside a XL-A/I vacuum chamber (reprinted with permission from [12])

Turbidity Detectors

Similarly to the case of the fluorescence detector, to our knowledge there are just two types of turbidity (= light scattering) detectors [13, 14]. Both are user-made, and therefore not commercially available. However, it is not very difficult to build such a turbidity detector. Again, only a short description is given here.

The principle of a turbidity measurement is comparable to that of an absorption detector, but it is much simpler, because the absorption $A = \lg(I_0/I)$ is measured at only one fixed radial position r_{slit}. Figure 2.13 shows a schematic diagram of the turbidity detector developed at BASF, which is part of a particle sizer.

The turbid dispersion to be analyzed is diluted to about $c = 1$ g/l, and placed into the centerpiece of the 3-mm mono-sector cell. In the first version of the BASF turbidity detector [13], the lower quartz window of the cell is covered by an apertured stop, in the center of which is a 0.2 mm wide slit arranged perpendicularly to the radius of the rotor. The slit picks out a beam from the entering parallel monochromatic light. This is simply created by using a stabilized incandescent lamp as a light source, and a monochromatic light filter ($\lambda = 546$ nm) after the condensing lens.

The intensity I of the beam, which is reduced by light scattering of the latex particles inside the measurement cell, according to Mie's light scattering theory, is registered by the photomultiplier and recorded as a function of the running time t. The concentration c of the dispersions is selected so as to yield an initial light intensity $I_{t=0} = I_0$ of approximately 10% of $I_{t=\infty} = I_{\text{DM}}$, the intensity of the pure dispersant, reached at the end of the run.

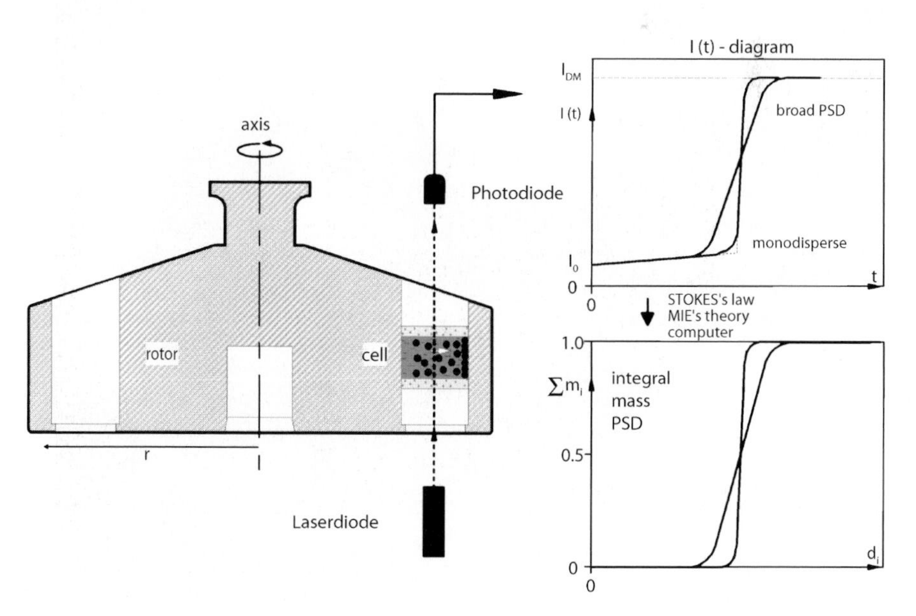

Fig. 2.13. Schematic setup of a user-made particle sizer with a turbidity detector to measure particle size distributions; realized in the AUC laboratory at BASF

A second (newer) version of the turbidity detector in use today [5] applies a narrow continuous light beam of a laser diode (λ = 670 nm) with 0.2 mm diameter, no apertured stop, and a fast photodiode as light detector (see Fig. 2.13).

In the case of a monodisperse latex, all of the particles sediment with exactly the same velocity. This causes a sharp one-step $I(t)$ curve to be obtained (shown in Fig. 2.13), because the intensity jumps from $I_{t=0}$ to I_{DM} at the moment when the sharply defined, sedimenting latex front passes the measuring slit (or the laser beam). The diameter d_{p} of the monodisperse particles can be calculated from the measured jump time t by means of Stokes' law (1.9). In the case of a broadly distributed latex, fractionation by particle size results in a broad $I(t)$ curve (shown in Fig. 2.13) because of large particles running ahead and small particles lagging behind. Details of this particle size distribution measuring procedure, in particular for very broadly distributed particles, are given in Sect. 3.6. Although a mono-sector cell is used, $I(t)$ and I_{DM} can be measured in the same run, and thus the time-dependent absorption $A(t) = \lg(I_{\mathrm{DM}}/I(t))$ at the radius position r_{slit} can be calculated. Figure 2.14 shows a picture of the newest turbidity detector with a thin green laser beam as realized recently at BASF. This green laser light (λ = 546 nm) enters from the outside via a glass fiber into the vacuum rotor chamber of the Optima XL.

Fig. 2.14. Photograph of the turbidity detector system with a thin green laser beam realized in the AUC laboratory at BASF, inside the vacuum chamber of an Optima XL for measuring of particle size distributions; the glass fiber guiding the laser light from outside is visible

2.5 Multiplexer

A pronounced improvement of all kinds of AUC measurements was reached when a multiplexer unit was added to analytical ultracentrifuges. This allowed the measurement of not just one but up to seven samples in an eight-hole rotor in one and the same experimental run. Thus, the effectiveness of the AUC is increased by a factor of 7. In this section, we describe the homemade multiplexer as realized in the Schlieren optical system described in Sect. 2.4.3.

In contrast to earlier multiplexer developments [20, 21], the system is now completely automated. The software-based multiplexer is realized via a Lab-view-based software tool applying a NIDAQ-Board (National Instruments, Austin, USA).

The triggering signal is created by the following homemade system. A photo-electric reflection light gate is fixed to the bottom of the vacuum chamber. This reflection gate consists of a continuously light-emitting LED and a fast photodiode. A "mirror" (= a polished small strip at the rotor bottom) fitted on the base of the analytical rotor reflects the light of the LED. By this alignment, each revolution of the rotor generates one sharp electrical pulse. This signal is imported by the data acquisition software, and used for the calculation of the actual rotor speed and for the calculation of the actual cell positions. The pulse-times of the flash lamp to ensure illumination of the desired superimposed cells are calculated from these data.

2.6 Auxiliary Measurements

As mentioned in preceding sections, a few auxiliary parameters have to be known to interpret sedimentation velocity data (and equilibrium run data, too; see Sect. 5.2). These are the density of the solvent ϱ_s, the viscosity of the solvent η_s, and the particle density ϱ_p or its reciprocal, the partial specific volume of the solute $\bar{v} = 1/\varrho_p$. Furthermore, for detection of the solute/particle concentration c in sedimentation experiments by Schlieren, turbidity, UV absorption or interference optics, the following optical parameters are required: refractive indices of the solvent and the solute/particle, n_s and n_p, respectively, the specific refractive index increment of the solute $(dn/dc)_p$ or the specific decadic absorption coefficient ε according to Lambert–Beer's law. Most of these parameters can be found in table-works [22, 23], or need only simple measurements, such as for η_s with capillary viscometers, or ε with UV spectrometers (the XL-A/I itself is a UV spectrometer!). The measurements of the two most important AUC parameters, $\bar{v} = 1/\varrho_p$ and $(dn/dc)_p$, are presented in the two following sections.

2.6.1 Measurement of the Solvent Density and the Partial Specific Volume

As mentioned above, the partial specific volume $\bar{v}$ is the inverse of the solute/particle density ϱ_p. The partial specific volume $\bar{v}$ is defined as the volume increase obtained if 1 g of solute is added to an infinite amount of solvent. Precise knowledge of this parameter is crucial for the interpretation of sedimentation data. This is because of the small differences in the buoyancy term $(1 - \bar{v} \cdot \varrho_s)$. In fact, it is the most common hindrance for the even more universal use of analytical ultracentrifugation in physicochemical science. Therefore, a lot of work has been spent on this topic, which led to extensive tabulated data of synthetic polymers and biopolymers. For proteins – as mentioned above, they are still the most common materials analyzed in analytical ultracentrifuges today – it is possible to estimate the value of $\overline{v} = 1/\varrho_P$ because this often does not deviate too much from approximately 0.73 cm³/g.

The best way to measure $\varrho_P = 1/\overline{v}$ is by using the well-known Kratky density balance. Figure 2.15 shows a measuring example. For the pure solvent and for ca. three solutions with different concentrations c of the solute, the (absolute) densities ϱ are measured (these solutions can also be used to determine the $(\mathrm{d}n/\mathrm{d}c)_P$ of the solute as well; see following section). Then, the reciprocal of ϱ, the specific volume $v = 1/\varrho$, is plotted as function of c. The slope of a regression line yields the wanted value of $\bar{v} = (\varrho_P)^{-1}$ of the solute. The example presented in Fig. 2.15 is a polybutyl acrylate latex dispersed in water at 25 °C. The results of this plot are $\varrho_P = 1.048\,\mathrm{g/cm^3}$ and $\overline{v} = 0.954\,\mathrm{cm^3/g}$.

Another important possibility to determine the solute density ϱ_P will be subject of Chap. 4: the determination of particle densities in AUC density gradients.

The use of a pycnometer is also possible, but tedious. Furthermore, a classical method to determine solution densities is the use of calibrated sinkers/floaters that make use of the Archimedes principle. Edelstein and Schachman [24] determined for special cases the partial specific volume by variation of solvents. They measured the product $M(1 - \bar{v} \cdot \varrho_s)$, called the effective molar mass, in different solvents with varying known densities, such as H_2O and D_2O, using equilibrium runs (see Chap. 5). The partial specific volume $\overline{v}$ can then be computed from these data. In a similar manner, Lustig et al. [25] and Schubert et al. [26] worked with Nycodenz/water solutions of different compositions, respectively, with two solvents

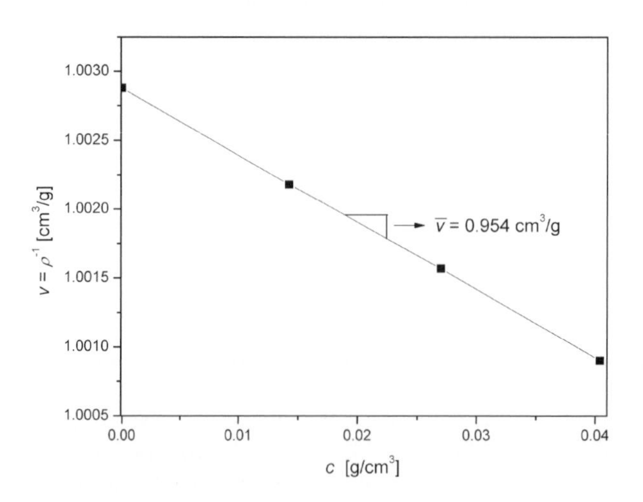

Fig. 2.15. $1/\varrho - c$ plot to determine the partial specific volume $\bar{v} = 1/\varrho_P$ from measurements with a Kratky density balance (polybutyl acrylate latex, water, 25 °C)

of very different densities and some mixtures thereof (see Sect. 5.4.2). A known table-work for $\overline{v}$ values of biopolymers is [27].

2.6.2 Measurement of the Refractive Index and the Specific Refractive Index Increment

In order to measure n_s, n_p and the specific refractive index increment $(dn/dc)_p$ of dissolved macromolecules and dispersed particles, often a Bellingham-refractometer is applied. It allows the measurement at just one wavelength λ. Usually, solutions of the solute of ca. three different concentrations are prepared. It is favorable to use the same solutions that have been prepared to determine the solute density (see Sect. 2.6.1). The (absolute) values of the measured refractive indices n of the pure solvent and the solutions are then plotted as function of c (see Fig. 2.16). The intercept yield n_s and the slope of the resulting line gives the specific refractive index increment $(dn/dc)_p$, that is, the change of the refractive index of the solution per unit mass solute that has been added. The result of this plot is $(dn/dc)_p = 0.130\,\mathrm{cm^3/g}$ (for the same polybutyl acrylate latex as in Fig. 2.15, 25 °C, 589 nm). Well-known $(dn/dc)_p$ table-works are [23] for synthetic polymers, and [28] for biopolymers.

If more detailed information at different wavelengths λ is required, a differential refractometer should be applied. It should be mentioned that the refractive index n, and thus $(dn/dc)_p$, varies with the wavelength λ and with the temperature T.

For *PSD* measurements, using a turbidity detector and Mie's theory, instead of $(dn/dc)_p$, the (absolute) refractive index n_p of the dispersed particles is required. Equation (2.4) gives a formula that allows the roughly approximated calculation of n_p using the measured $(dn/dc)_p$, and vice versa.

$$n_p = (dn/dc)_p \cdot \varrho_p + n_s \tag{2.4}$$

This equation was derived making use of the additivity of molar refractions [29].

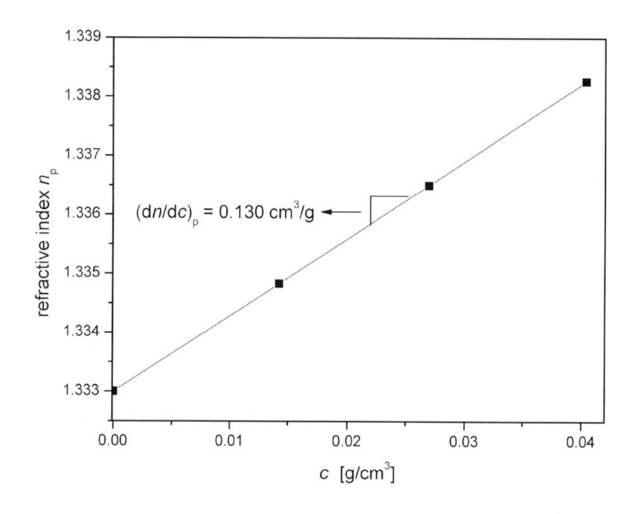

Fig. 2.16. $n-c$ plot to determine the specific refractive index increment $(dn/dc)_p$ from measurements with a Bellingham refractometer (polybutyl acrylate latex, water, 25 °C, 589 nm)

References

1. Svedberg T, Pedersen KO (1940) Die Ultrazentrifuge. Steinkopff, Dresden
2. http://www.ehs.cornell.edu
3. Bowen TJ, Rowe AJ (1970) An introduction to ultracentrifugation. Wiley, London
4. Lewis MS, Weiss GH (1976) Proc Conf Fifty Years of the Ultracentrifuge, 24–26 February 1975, Bethesda, ML, Biophys Chem 5:1–286
5. Mächtle W (1999) Biophys J 76:1080
6. Müller HG (1999) Bayer AG, Leverkusen, Germany, personal communication
7. Mächtle W (1999) Prog Colloid Polym Sci 113:1
8. Börger L, Lechner MD, Stadler M (2004) Prog Colloid Polym Sci 127:19
9. Schachman HK (1959) Ultracentrifugation in biochemistry. Academic Press, New York
10. Lloyd PH (1974) Optical methods in ultracentrifugation, electrophoresis and diffusion. University Press, Oxford
11. Schmidt B, Riesner D (1992) In: Harding SE, Rowe AJ, Horton JC (eds) Analytical ultracentrifugation in biochemistry and polymer science. The Royal Society of Chemistry, Cambridge, p. 176
12. MacGregor IK, Anderson AL, Laue TM (2004) Biophys Chem 108:165
13. Mächtle W (1984) Makromol Chem 185:1025
14. Scholtan W, Lange H (1972) Kolloid Z Z Polym 250:782
15. Mächtle W (1991) Prog Colloid Polym Sci 86:111
16. Böhm A, Kielhorn-Bayer S, Rossmanith P (1999) Prog Colloid Polym Sci 113:121
17. Billick IH, Bowen RJ (1965) J Phys Chem 69:4024
18. Furst A (1997) Eur Biophys J 25:307
19. Rossmanith P, Mächtle W (1997) Prog Colloid Polym Sci 107:159
20. Mächtle W, Klodwig U (1976) Makromol Chem 177:1607
21. Mächtle W, Klodwig U (1979) Makromol Chem 180:2507
22. Lide DR (ed) (2002) CRC handbook of chemistry and physics, 83rd edn. CRC Press, Boca Raton
23. Brandrup J, Immergut EH (eds) (1989) Polymer handbook, 3rd edn. Wiley, New York
24. Edelstein SJ, Schachman HK (1967) J Biol Chem 242:306
25. Lustig A, Engel A, Tsiotis G, Landau EM, Baschong W (2000) Biochim Biophys Acta 1464:199
26. Tziatzios C, Precup AA, Lohmeijer BGG, Börger L, Schubert US, Schubert D (2004) Prog Colloid Polym Sci 127:54
27. Durchschlag H (1986) In: Hinz HJ (ed) Thermodynamic data for biochemistry and biotechnology. Springer, Berlin, Heidelberg, New York, pp. 45–128
28. Theisen A, Johann C, Deacon MP, Harding SE (2000) Refractive increment data book for polymer and biomolecular scientists. Nottingham University Press, Nottingham
29. Mächtle W, Fischer H (1969) Angew Makromol Chem 7:147
30. Ralston G (1993) Introduction to analytical ultracentrifugation. Beckman Instruments, Fullerton, CA
31. Gonzales JM, Rivas G, Minton AP (2003) Anal Biochem 313:133

3 Sedimentation Velocity

3.1 Introduction

The basic information obtained by performing an AUC sedimentation velocity experiment is the sedimentation coefficient s (see (1.5)) and the s distribution $G(s)$ or $g(s) = \mathrm{d}G(s)/\mathrm{d}s$ if the sample is heterogeneous. The gravitational field a_{centr} increases linearly along the radius ($a_{\mathrm{centr}} = \omega^2 r$). This means that also the sedimentation velocity u of the particle increases linearly toward the bottom of the cell.

In general, the sedimentation coefficient s of a sample does not change within the experiment. The exceptions will be considered below in this chapter. In the absence of effects such as concentration dependence of the sedimentation velocity, the sedimentation coefficient can be determined by measuring the radial concentration distribution $c(r)$ of the sample at the given time t within the cell. Both the s measurement (and the corresponding measurement of the s distribution, $G(s)$) and the measurement of the radial concentration distribution $c(r)$ are the basis for the very important measurement of the molar mass distribution (*MMD*), and the determination of particle size distributions (*PSD*).

In this Chap. 3, we will start (in Sect. 3.2) with a simple example of a sedimentation velocity experiment. This example will be our guide through the theory of sedimentation velocity experiments. Subsequently (in Sect. 3.3), we will discuss some more complicated examples exhibiting some new phenomena not represented in the first example. We will then turn your interest to the two most relevant types of sedimentation velocity runs, as far as our industrial work is concerned: (i) in Sect. 3.4, to the s run on dissolved macromolecules, leading to the molar mass distribution, and (ii) in Sect. 3.5, to the s run on dispersed particles, leading to the particle size distribution. Finally, in Sect. 3.6, the technique of synthetic boundary experiments, a special sedimentation velocity run, will be discussed.

3.2 Basic Example of Sedimentation Velocity

Figure 3.1 shows a single interference optics CCD picture and the resulting typical radial XL-A/I interference scan of a sedimentation velocity experiment, $J = f(r)$, performed on a solution of nearly monodisperse calibration polystyrene having a molar mass of $M_{\mathrm{w}} = 106\,000\,\mathrm{g/mol}$ ($M_{\mathrm{w}}/M_{\mathrm{n}} \sim 1.10$). The solvent used was methyl

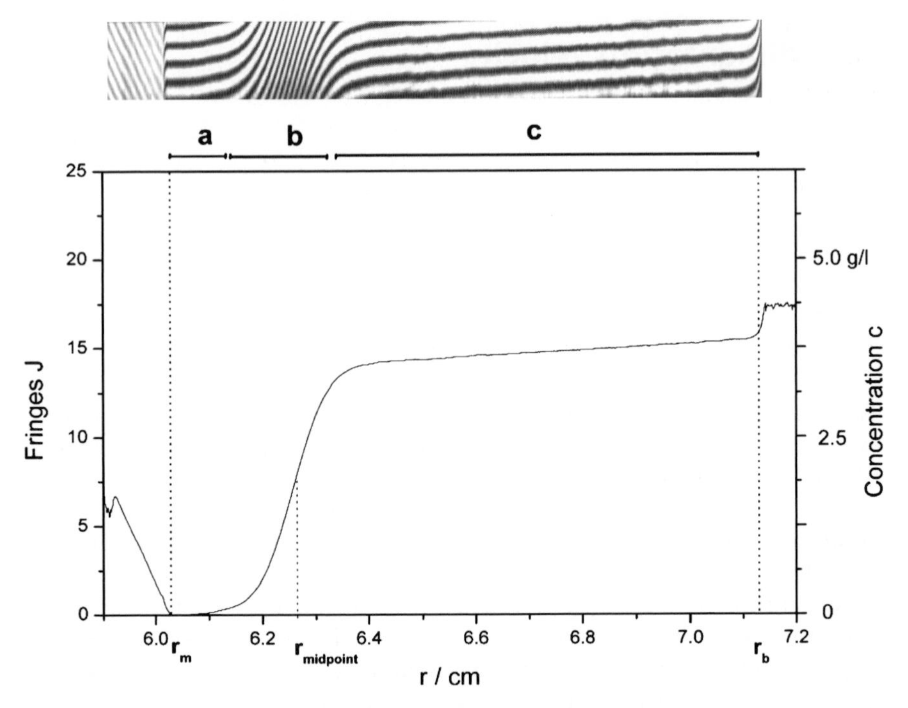

Fig. 3.1. Single XL-A/I CCD interference "photograph" and corresponding radial scan of a sedimentation velocity experiment performed on a nearly monodisperse polystyrene polymer (M_w = 106 000 g/mol) dissolved in MEK (c_0 = 4 g/l). Scan recorded after 46 min at 40 000 rpm. T = 25 °C, 12-mm double-sector cell, sapphire windows

ethyl ketone (MEK). The initial concentration (or loading concentration) of the styrene polymer was $c_0 = 4\,\mathrm{g/l}$, $\omega = 40\,000\,\mathrm{rpm}$, and the running time was 46 min at 25 °C. As the density of the solute is higher ($\varrho_{\mathrm{PS,MEK}} = 1/\bar{v} = 1.1016\,\mathrm{g/cm^3}$) than the solvent density ($\varrho_{\mathrm{MEK}} = 0.7995\,\mathrm{g/cm^3}$), sedimentation occurs. In the opposite case, flotation would be detected. The $J(r)$ scan is the result of a Fourier analysis of the above interference "photograph". It presents the radial course of a single interference fringe. Its vertical deviation J is counted in fringe numbers. $\Delta J = 1$ is identical to the vertical spacing between two neighboring parallel fringes in the interference photograph. The radial course of $J(r)$ is proportional to the radial concentration distribution $c(r)$ of the sedimenting macromolecules within the AUC cell.

Basically, three regions can be distinguished in such a typical, experimental $J(r)$ XL-A/I raw curve (see Fig. 3.1, from left to right):

(i) the solvent region **a**,
(ii) the boundary region **b**, and
(iii) the plateau region **c**.

Furthermore, two special radial positions, the solvent–air meniscus r_m and the bottom of the cell r_b, can be identified. In the solvent region, the concentration of the solute equals zero ($J = 0$ and $c = 0$), whereas in the boundary region the concentration of the solute rises from zero to the value in the plateau region c_p. In the plateau region, c_p is constant (the corresponding J value is J_p). An important point for the determination of the sedimentation coefficient is the inflexion point of the $J(r)$ curve in the middle of the boundary region. The corresponding radius is named $r_{midpoint}$ or $r_{boundary}$. The mathematical derivation of the interference curve according to r exhibits a maximum at this point. If the data are collected via Schlieren optics, then the maximum of the peak, and thus $r_{mid} = r_{bnd}$, is obtained directly (compare, for example, Fig. 2.8a and b). Because the specific refractive index increment of polystyrene in MEK (at $\lambda = 675$ nm and 25 °C) is known, $dn/dc = 0.214\,cm^3/g$, the counted fringe numbers $J(r)$ can be transferred directly with (2.3) into the (absolute) radial concentration distribution $c(r)$ within the measuring cell (see c axis on the right side in Fig. 3.1).

Figure 3.2 additionally contains other scans, collected at other times t from the same experiment. It shows representative raw data curves of $J(r)$ or $c(r)$ collected every 10 min by the digital XL-A/I interference optics. The total experiment duration was about 176 min.

At time zero, the concentration c_0 of the (polystyrene) solute is uniform throughout the measuring cell. With increasing experimental time, the solute sediments toward the cell bottom due to the gravitational field created by the spinning of the rotor inside the ultracentrifuge. The direction of the gravitational force is away from the center of rotation, and therefore leads to a directed movement

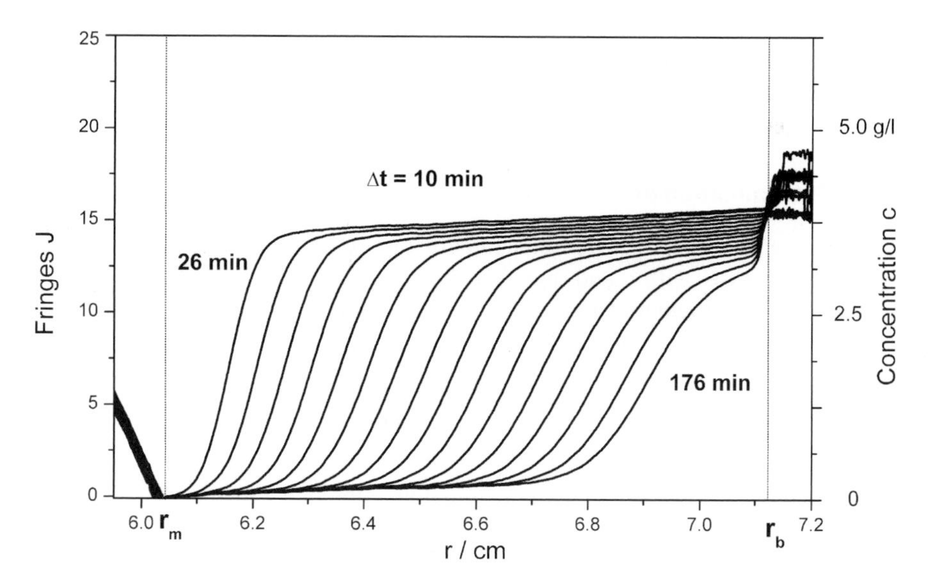

Fig. 3.2. Raw data curves of 15 representative XL-A/I interference optical radial scans collected in a sedimentation velocity experiment performed on a polystyrene polymer ($M_w = 106\,000$ g/mol) dissolved in MEK ($c_0 = 4$ g/l) at 40 000 rpm. Scans were collected every 2 min. The total experiment duration was about 180 min. Shown is every fifth scan starting with the curve after 26 min until the curve after 176 min. $T = 25$ °C, 12-mm double-sector cell

of the solute. The first experimental curve in Fig. 3.2 at 26 min already shows the formation of a sedimentation boundary between the left plateau of the pure solvent near the meniscus region and the right plateau of uniform solute concentration c_p near the bottom region.

The analysis of the radial movement of the midpoint of the boundary, r_{bnd}, across the cell allows us to determine the sedimentation velocity u, and thus the sedimentation coefficient s, too (see (1.5)). This will be shown below. The sedimentation coefficient s is connected to the effective size of the solute, and to the density difference between solute and solvent ($\varrho_p - \varrho_s$), allowing us to determine the molar masses M and particle sizes d_p (see (1.8) and (1.9)). This will be subject of two following sections of this chapter.

A closer look at Fig. 3.2 reveals a spreading/broadening of the boundary itself during the sedimentation experiment, although the sedimenting polystyrene is (nearly) monodisperse. This boundary spreading is due to the undirected movement of the solute molecules caused by diffusion. Therefore, the diffusion coefficient D (of monodisperse samples only!) can be calculated from the measurement of boundary spreading over time.

The ratio of the sedimentation coefficient s and diffusion coefficient D, i.e., the quotient s/D, is a measure of the molar mass M of the solute (see (1.8)). It has to be kept in mind that an additional boundary spreading arises from the sample's heterogeneity if the polymer exhibits a broad molar mass or particle size distribution. Thus, assuming a "monodisperse" sample is analyzed, the "D value" of a broadly distributed sample is falsified: D is too high, and so "M" too small. Again, we emphasize: in a sedimentation velocity run (or short s run), beside the sedimentation coefficient s (and its distribution $G(s)$), the (absolute) concentration distribution $c(r)$ within the measuring cell (see right-hand c axis in Fig. 3.2) is measured.

3.2.1 Determination of s

As mentioned above, the gravitational force $\omega^2 r$ is not constant within the AUC cell. The force increases with r, and therefore the velocity u of the boundary increases as well. Thus, mathematically the sedimentation velocity u has to be expressed as a differential, and the *definition* of the sedimentation coefficient s, (1.5), leads to (3.1):

$$s \equiv \frac{u}{\omega^2 r} = \frac{\mathrm{d}r_{bnd}/\mathrm{d}t}{\omega^2 r} \tag{3.1}$$

Integration of (3.1) gives (3.2) with the radial position of the meniscus r_m:

$$\ln(r_{bnd}/r_m) = s\omega^2 t = s \int_0^t \omega^2 \, \mathrm{d}t \tag{3.2}$$

If ω is not constant during a run, then one replaces $\omega^2 t$ by the so-called running time integral $\int \omega^2 \, \mathrm{d}t$.

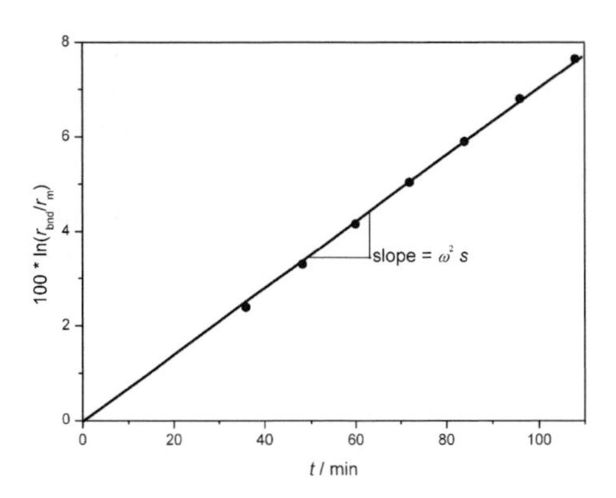

Fig. 3.3. Plot of ln (r_{bnd}/r_m) versus time t to determine the sedimentation coefficient s from a sedimentation velocity run. The slope of the *straight line* reveals a value of s = 6.9 S (Svedberg) for the polystyrene examined (M_w = 106 000 g/mol), dissolved in MEK at c_0 = 4 g/l

Equation (3.2) leads the way to the simplest determination of s by plotting $\ln(r_{bnd}/r_m)$ versus time t. The slope of the resulting straight line is $s\omega^2$. Figure 3.3 shows the plot corresponding to the experimental data given in Fig. 3.2. The result in this case was $s = 6.9 \times 10^{-13}\,\text{s} = 6.9\,\text{S}$. The unit 10^{-13} s, abbreviated as 1 S (= 1 Svedberg or 1 Sved or 1 sved), is the standard unit of sedimentation coefficients, which are laid down in tables. It is named after The(odor) Svedberg, the Nobel prize winner and inventor of the AUC. The error of measurement of s is about 1–3%.

In many cases, samples show either multiple components or even a continuous, broad distribution of particles sizes or molar masses. In these cases, where only r_{bnd} values were used, the s value determined above is a weight *average* value $\bar{s}_w$. Thus, in cases of broad distributions, it is of interest to determine the complete differential sedimentation coefficient *distribution* $g(s)$ or the integral form thereof $G(s) = c(s)/c_0$. This sedimentation coefficient distribution is defined as follows [1]:

$$g(s) = \frac{\mathrm{d}\left(c(s)/c_0\right)}{\mathrm{d}s}\left(\frac{r}{r_m}\right)^2 = \frac{\partial G(s)}{\partial s} \tag{3.3}$$

The function $g(s)$ gives the mass fraction (i.e., the concentration) $\mathrm{d}m = \mathrm{d}c/c_0$ of the sample sedimenting with a sedimentation coefficient between s and $s + \mathrm{d}s$.

The determination of $g(s)$ then allows us to determine (i) molar mass distributions [2], (ii) particle size distributions, and (iii) interaction constants in the case of interacting systems [3].

3.2.2 Standard Conditions for *s* Estimation

If the sedimentation velocity experiment is not run at standard conditions (for biopolymers, these are *aqueous* solutions and at 20 °C), then the collected data can easily be converted to these standard conditions by means of (3.4):

$$s_{20,w} = s_{obs}\left(\frac{\eta_{T,w}}{\eta_{20,w}}\right)\cdot\left(\frac{\eta_s}{\eta_w}\right)\cdot\left(\frac{1-\bar{v}\varrho_{20,w}}{1-\bar{v}\varrho_{T,w}}\right) \tag{3.4}$$

Thus, the standard sedimentation coefficient at 20 °C in water $s_{20,w}$ (usually used in biopolymer tables) can be calculated from the observed sedimentation coefficient s_{obs}, provided that the following data are known: the viscosity $\eta_{T,w}$ of water at the experimental temperature T, the viscosity $\eta_{20,w}$ of water at 20 °C, the viscosity η_s of the solvent at any given temperature, the viscosity η_w of water at the same chosen temperature, the density $\varrho_{20,w}$ of water at 20 °C, and the density $\varrho_{T,s}$ of the solvent at experimental temperature.

3.2.3 Radial Dilution and Thickening

A careful look at Fig. 3.2 shows that the level of the $J(r)$ plateau near the bottom region in the interference optical data decreases with increasing experimental time. This means that the concentration of the solute in the plateau region c_p also decreases continuously with time t. As mentioned in Sect. 2.3 for sedimentation velocity runs, *sector*-shaped measuring cells are applied to prevent convection. This causes an increasing volume per radius unit toward the cell bottom of the measuring compartment (see Fig. 3.4). In other words, the sedimenting sample becomes increasingly *diluted* as it moves downward. This phenomenon is called radial dilution. In the case of a floating sample, the phenomenon exists vice versa, and is called radial thickening.

This phenomenon can be taken into account by application of (3.5):

$$c_p = c_0 \cdot \left(\frac{r_m}{r_{bnd}} \right)^2 \tag{3.5}$$

with the concentration of the solute in the plateau region c_p, the initial concentration c_0, the radius of the meniscus r_m (corresponding to the bottom r_b in the case of flotation), and the radius of the sedimenting (or floating) boundary r_{bnd}.

Equation (3.5) represents the very basic correction for the important radial dilution, which generally has to be performed in every evaluation. This correction for the radial dilution is valid as long as the sample does not show a pronounced concentration dependence. In the latter case, a deviation from ideal behavior occurs when s is determined. According to [4], special care has to be taken if a significant pressure dependence of s occurs (see Sect. 3.3.3).

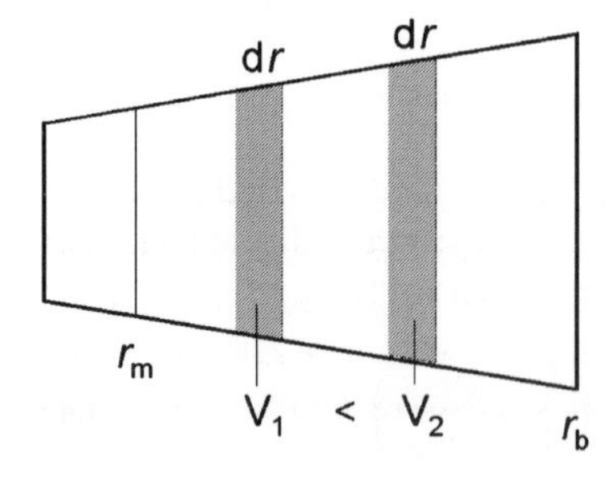

Fig. 3.4. Top view onto a sector-shaped compartment of an analytical ultracentrifugation measuring cell. This scheme illustrates the radial dilution and the corresponding thickening effect. Due to the sector shape, the sample undergoes a dilution while moving toward the bottom, and a thickening while moving toward the meniscus of the cell

3.2.4 Concentration Dependence

In general, sedimentation coefficients are concentration dependent valid to a smaller extent for compact, spherically shaped particles, but distinctly so for longer, coil-like macromolecules. Standard systems of non-associating solutes display a decrease in the observed sedimentation coefficient with increasing concentration. The sedimentation coefficient $s(c)$ at a given concentration c is linked to the limiting (ideal) sedimentation coefficient s_0 (at zero concentration, or more correctly at infinite dilution) via (3.6):

$$s(c) = \frac{s_0}{1 + k_s \cdot c} \quad \text{or} \quad \frac{1}{s(c)} = \frac{1}{s_0}(1 - k_s \cdot c) \tag{3.6}$$

where k_s is the concentration dependence coefficient. The values of s_0 and k_s are measured on a series of three–five different concentrations. Slope and intercept of a straight line through the data points of a $1/s(c)$ versus c plot yield s_0 and k_s (see, for example, Fig. 3.8).

Equation (3.6) is valid only over a limited, moderate range of concentrations [5]. There are mainly three physical reasons for the concentration dependence of the sedimentation coefficient, but only the first listed below is considered to be of significance:

(i) Most importantly, the viscosity of the solution η_s increases with increasing concentration c. Thus, according to (1.5) and (1.7), the sedimentation coefficient decreases.
(ii) The density of the solution is a function of the concentration. A clear proof can be found in the literature (see [6–8]). Gofman showed that lipoprotein molecules, which *sediment* in one region of a centrifuge cell that is free of other proteins, will actually *float* in the same run in another part of the cell that contains a high concentration of smaller protein molecules, and so a higher solution density.
(iii) As the particles are sedimenting, the space they leave must be filled with solvent. This introduces a backward flow of liquid, decreasing the sedimentation velocity.

It is obvious that the concentration dependence of the sedimentation coefficient $s(c)$ depends on the shape of the sedimenting species. In general, it can be noted: the higher the aspect ratio of the particle, the more pronounced the concentration dependence, i.e., the less spherical the shape of the solute, the higher the value of k_s. For a deeper study, see [9] and [43].

These considerations are valid for neutral macromolecules or particles. However, the picture gets much more complicated for polyelectrolytes, i.e., if charge effects have to be taken into account (see Sect. 3.3.5). It has to be noted that charge does not influence only the sedimentation coefficients s itself, but also the concentration dependence coefficient k_s. For a more detailed discussion on this topic, the reader should consult [5].

3.3 Advanced Theory of Sedimentation Velocity Runs

Section 3.2 contained the basic theory of sedimentation velocity experiments as well as the most important and common influences and effects that have to be taken into account. Beside these effects, there are some other phenomena, which – to the authors' knowledge – do not occur frequently. Therefore, readers could leave out Sect. 3.3 if they are not interested in details of sedimentation theory. Nevertheless, the authors believe that some readers may want to have a closer look to theoretical details in order to make sure not to misinterpret experimental data. Some of the readers may draw the line between common and uncommon phenomena differently. In fact, this distinction is rather arbitrary.

The effects described in the following sections influence either the value of the sedimentation coefficient itself or the shape of the sedimentation boundary. Special care has to be taken for the broadening of boundaries. Basically, there are *two* mechanisms resulting in boundary spreading: *diffusion*, and *heterogeneity* of the sample. The separation of these two effects is subject of Sect. 3.3.6.

3.3.1 Johnston–Ogston Effect

A special concentration effect is observed particularly in coil-like solutes (= expanded linear macromolecules) containing mixtures of at least two components, one sedimenting fast, the other slowly (mostly characterized by very different molar masses M_A and M_B). The effect was named after Johnston and Ogston who first explained this phenomenon in 1946 [10]. When a mixture of two solutes A and B, sedimenting at different velocities, is exposed to an s run, two regions of interest in the radial concentration profile can be distinguished (see Fig. 3.5).

In region I, only the slowly sedimenting component A is present. The sedimentation coefficient, which can be measured, is $s_A(c_I)$. The total solute concentration $c_A + c_B$ increases in the boundary region II. Therefore, due to the concentration dependence of $s_A(c)$, the sedimentation velocity of component A in region II drops to a lower value $s_A(c_{II})$. The faster sedimenting component B always moves through a solution of solvent and the slow component A, and therefore exhibits the same sedimentation coefficient $s_B(c_{II})$ throughout the cell. Thus, three sedimentation

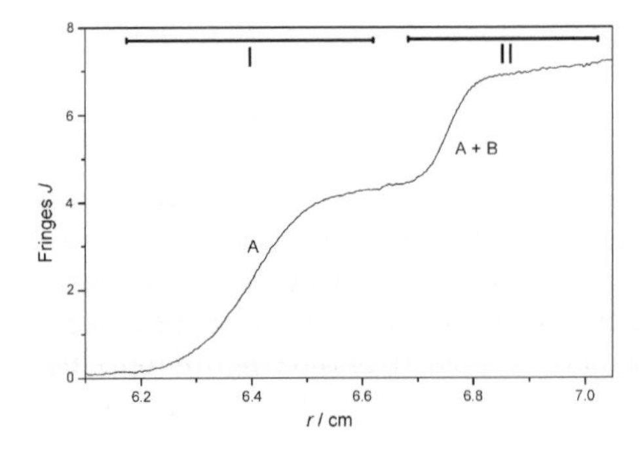

Fig. 3.5. Schematic example of an interference optical radial scan from a sedimentation velocity run. The Johnston–Ogston effect may occur in multi-component systems (a bimodal one, A+B, in this figure) when the components' sedimentation coefficients exhibit pronounced concentration dependence

coefficients have to be taken into account in the system described. Effectively, the slow sedimenting component accumulates at the boundary I. The fringe displacement ΔJ_A (i.e., the concentration c_A), and the corresponding peak area in Schlieren optics $A_{schl,A}$ of the slower component, is larger than the signal created by the faster component (if both initially had a 1:1 concentration relation). The relative concentrations of both components c_A and c_B, determined from the fringe plateaus (or the Schlieren peak area), are incorrect, in that the detected relative concentration of the slower component is too high. The Johnston–Ogston effect vanishes with smaller total concentration c_0, smaller ω, and smaller concentration dependence of the concerned sedimentation coefficients ($k_s \to 0$). To recognize whether the Johnston–Ogston effect is present in a system, the total solute concentration c_0 and/or the rotor speed ω should be varied. To obtain highest accuracy, extrapolation to infinite dilution $c_0 \to 0$ is necessary.

3.3.2 Self-Sharpening of Boundaries

When the concentration dependence of sedimentation coefficients is pronounced (i.e., when k_s is very high), the effect of *self-sharpening* of the boundary may be observed. As example, the sedimentation boundary of a pure monodisperse solute (see Fig. 3.1) will be considered. The concentration of the solute varies through the region of the boundary from the lowest value on the left, normally $c = 0$, to the highest value on the right, that is, $c_p \sim c_0$. Therefore, not only the concentration of the solute varies inside the boundary region but also the sedimentation coefficient itself, if the solute shows a (pronounced) concentration dependence $s(c)$ of the sedimentation coefficient. The sedimentation coefficient inside the boundary is smaller near the bottom region than near the meniscus region. Thus, the boundary region tends to stay narrower. This effect is called self-sharpening. It is often found in polyelectrolyte solutions.

One example of a multi-boundary sedimentation experiment that is believed to show self-sharpening is given in Sect. 3.5.4 and Fig. 3.22. The most prominent example of a self-sharpening phenomenon is the sedimentation of the rod-like tobacco mosaic virus (see [11]). It is not easy to judge whether a self-sharpening effect is present in the experimental data, or not. The reader should simply keep this paragraph in mind in the case of data difficult to interpret. One indication of self-sharpening is the absence of diffusion broadening of the sedimenting boundary.

3.3.3 Pressure Dependence

Inside the measuring cell of an analytical ultracentrifuge, not only is a radial gravitational field $\omega^2 r$ created, but also a radial pressure $p(r)$ gradient due to the fact that the liquid column exhibits different weights at different radial positions. Centrifuges operating at highest rates of angular rotation ω may induce radial pressure gradients $p(r, \omega)$ within the solution column, varying from 1 bar (0.1 MPa) at the meniscus up to 250 bar (25 MPa) at the cell bottom. Depending on the applied solvent, this may have strong influence, and can lead to deviations of up to 30% in values for sedimentation coefficients in some systems (see [5]). The pressure can influence the solvent's density ϱ_s and viscosity η_s (especially of highly compressible organic solvents) as well as the partial specific volume $\overline{v} = 1/\varrho_p$ of the solute. The latter does not need to be taken into account usually, because the common increase in density and viscosity of the solvent with increasing pressure often leads to slower sedimentation. In general, these effects can be neglected in aqueous solutions because the compressibility of water is very low. In highly compressible solvents, the radial pressure gradient $p(r, \omega)$ and its influence on s (see [44]) is usually described by (3.7),

$$s(p) = s(r, \omega) = s_0[1 - \mu p(r, \omega)] \tag{3.7a}$$

with

$$p(r, \omega) = (1/2)\omega^2 r_m^2 \varrho_s^0 \left[(r/r_m)^2 - 1\right] \tag{3.7b}$$

where s_0 is the sedimentation coefficient at zero pressure, μ is a constant dependent on the polymer/solvent system, and ϱ_s^0 is the density of the solvent at $p = 1\,\mathrm{bar}$ (0.1 MPa).

Indirectly, radial pressure gradients may influence sedimentation rates in two main ways:

(i) Associating systems may vary their relative composition following the Le Chatelier principle according to the pressure they are exposed to (see, for example, [12, 13]). The pressure dependence of associating equilibrium systems may therefore be one reason for speed dependence $s(\omega)$ of sedimentation coefficients (see Sect. 3.3.4).

(ii) Some synthetic or biopolymers may undergo structural changes depending on pressure. Thus, their geometry and frictional coefficient may vary with pressure, i.e. during their sedimentation within the radially increasing pressure gradient $p(r)$ in the AUC cell.

For a more detailed discussion of pressure effects, see [2, 5, 14] and [44].

3.3.4 Speed Dependence

Occasionally, it is found that the measured sedimentation coefficient depends on the angular velocity ω of the experiment, for example, as a result of the pressure effect (see Sect. 3.3.3). Sometimes, also in the absence of pressure effects, the observed sedimentation coefficient is found to increase with increasing rotor speed. This is believed to occur through aggregation of the (macromolecular) solute. This aggregation occurs because sedimenting solute takes buffer ions with it. The solvent left behind may then be enriched in macromolecular solute, but may contain less buffer ions.

It is also reported (see [11]) that highly asymmetric molecules, such as the tobacco mosaic virus, are oriented by centrifugation (see also concentration dependence, Sect. 3.2.4), resulting in a speed dependence of $s(\omega)$. This can be overcome by working at the lowest possible angular velocity ω.

3.3.5 Charge Effects

Most biological macromolecules are electrostatically charged, and to maintain electrical neutrality of the solution, each macromolecule (= macro-ion) is associated with a number of counter-ions (of lower molar mass). These counter-ions often have sedimentation coefficients s that are orders of magnitude smaller than those of macromolecules. Thus, when macromolecules sediment, the counter-ions lag behind, creating an electrical field. This field induces an electrical force acting opposite to the sedimentation direction. Therefore, the sedimentation of the macro-ions is slowed down, resulting in a smaller sedimentation coefficient.

Charge effects can be overcome by the addition of (low molar mass) salts, such as NaCl or KCl, to the (mostly) aqueous polyelectrolyte solutions (see Fig. 3.6). A second charge effect has to be mentioned. When polyelectrolytes sediment, their concentration near the cell bottom is increased in relation to the concentration at the meniscus. As they take their counter-ions with them (cf. above), the salt concentration at the bottom is increased as well. Thereby, an osmotic pressure is built up, resulting in a Donnan equilibrium.

A good example of the influence of electrical charge is given in Fig. 3.6, by the investigation of the sedimentation behavior of aqueous solutions of the polyelectrolyte sodium polystyrene sulfonate (NaPSS = negatively charged macro-ions) in the presence of different amounts of salt. The dialyzed sample examined exhibited a molar mass of about M_w = 200 000 g/mol ($M_w/M_n \approx 1.10$). The concentration of the low molar mass salt added (NaCl) was varied from c_{NaCl} = 0 up to c_{NaCl} = 1 mol/l. The concentration of NaPSS itself was held constant at c_{NaPSS} = 2.5 g/l. Figure 3.6 shows the measured dependence of the sedimentation coefficient s_c as a function of the NaCl concentration c_{NaCl}.

As a result of the abovementioned charge effects, at zero salt concentration even highest centrifugal forces do not induce significant sedimentation of NaPSS.

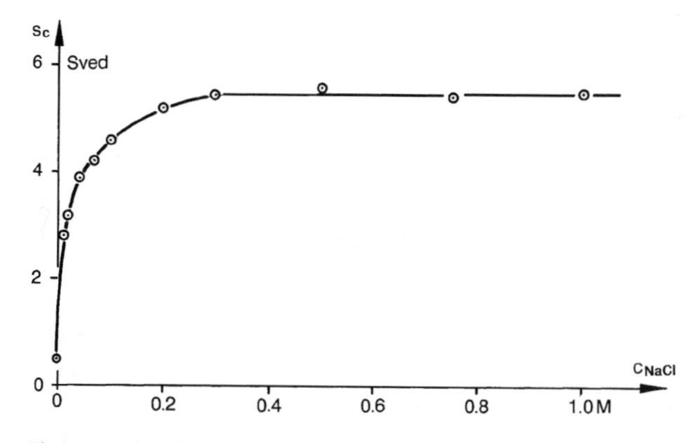

Fig. 3.6. Plot of the sedimentation coefficient s_c of the polyelectrolyte NaPSS (M_w = 200 000 g/mol, c_{NaPSS} = 2.5 g/l = constant) versus aqueous NaCl salt concentration c_{NaCl}. At $c_{NaCl} < 0.3$ mol/l, the measured apparent sedimentation coefficient of the polymer depends strongly on the salt concentration

At higher salt concentration, sedimentation of NaPSS occurs. The minimal salt concentration necessary to allow undisturbed sedimentation of the polyelectrolyte NaPSS can be obtained from Fig. 3.6. Approximately 0.3 mol/l sodium chloride is sufficient to eliminate all influences of charge effects. Further addition of NaCl will not change s_c, which is now constant at $s_c = 5.5$ S.

For further investigations on the sedimentation behavior of NaPSS, described in Sect. 3.4.1, $c_{NaCl} = 0.5$ mol/l was chosen as standard salt concentration. In this aqueous salt solution, the sedimentation behavior of the polyelectrolyte NaPSS is similar to that of uncharged neutral macromolecules.

Some aqueous particle dispersions are electrostatically stabilized, i.e., they contain slightly charged polyelectrolyte particles (see also Sect. 6.1.3). Therefore, if low molar mass salts are added to such dispersions, the stabilization will be destroyed, and the particles will agglomerate partly or completely. This special "salt effect" of aqueous dispersions is the reason why it is not possible to use the versatile water/CsCl density gradient for AUC analysis of such systems. In this case, the electrically neutral water/metrizamide or water/Nycodenz density gradients (see Sect. 4.2.1) have to be used instead.

3.3.6 Separation of Sedimentation and Diffusion

If nanoparticles smaller than 30 nm in diameter or macromolecules exhibiting molar masses well below approximately 100 000 g/mol are investigated by AUC, diffusion broadening of the sedimenting boundaries will occur at a considerable extent due to the high diffusion coefficient D of the smaller species. The smaller the species, the higher is D, and thus the diffusion spreading of the boundary. As a relation between D and size (d_p or M), the following two equations are valid. First, for compact *spherical* particles with diameter d_p, the famous Stokes–Einstein equation (3.8) is valid:

$$D = \frac{kT}{3\pi\eta_s d_p} \tag{3.8}$$

with $k = R/N$ being the Boltzmann constant, and $M = mN = (\pi/6)\varrho_p d_p^3 N$, i.e., $D \sim M^{-(1/3)}$.

Second, for non-spherical particles, especially for macromolecules of all shapes (coils, rods, spheres), the following empirical *scaling law* (similar to that for s_0, see (3.11)) is valid:

$$D_0 = K_D \cdot M^{-b} \tag{3.9}$$

with the two (constant) parameters K_D and b, where b is about $0.5 < b < 0.6$ (in the case of spherical particles, $b = 1/3$ is valid).

There are different approaches to remove or to separate the effects of boundary spreading by diffusion, from spreading by a broad *PSD* or *MMD*: (i) the van Holde-Weischet method, (ii) the time derivative method, (iii) the Lamm equation modeling, and others (see [2] and [15]). Some approaches make use of the finding that broad distributions lead to a boundary spreading proportional to time, whereas diffusion spreading correlates only with the square root of time. In the following, we discuss the first three approaches.

Van Holde-Weischet Method

The procedure of van Holde-Weischet [16], using $D \approx \sqrt{t}$, is as follows: a fixed number of data points from one experimental scan at a defined running time t (such as in Fig. 3.1), which are evenly spaced between the baseline and the plateau, are selected (for example, in 5% steps: $0, 5, 10, ...95, 100\%$ in the c axis direction) and plotted in a so-called van Holde-Weischet plot, such as in Fig. 3.7 (a vertical line in this plot corresponds to one scan at time t). This is repeated for all scans in this series (such as that in Fig. 3.2) at different running times t. Then, an apparent sedimentation coefficient s^* is calculated for each of the data points at the different percentages, and plotted versus the inverse root of the run time t, yielding the typical van Holde-Weischet plot (see Fig. 3.7): straight lines through the different s^* values have the same percentage.

Now, a linear extrapolation of corresponding s^* values, with the same percentage at different experimental times (= one slice in Fig. 3.7) to infinite time t (i.e., $1/t \to 0$), is performed. The *intersection* points of the different extrapolation lines (= slices) with the ordinate axis yield the diffusion-corrected sedimentation coefficients $s_{\%}$ for the different percentages, and thus the complete integral distribution $G(s)$ of the sedimentation coefficients. In the case of a single monodisperse component (such as in Fig. 3.7), the lines intersect in *one* point, and thus only one s value is obtained. If multiple components or a broad *MMD* are present in the sample, a corresponding number or a continuous sequence of different intersection points are obtained, i.e., different values of s (for example, $s_{0\%}, s_{5\%}, ...s_{100\%}$). In the case of non-ideality (i.e., $A_2 > 0$), the intersection point of a monodisperse

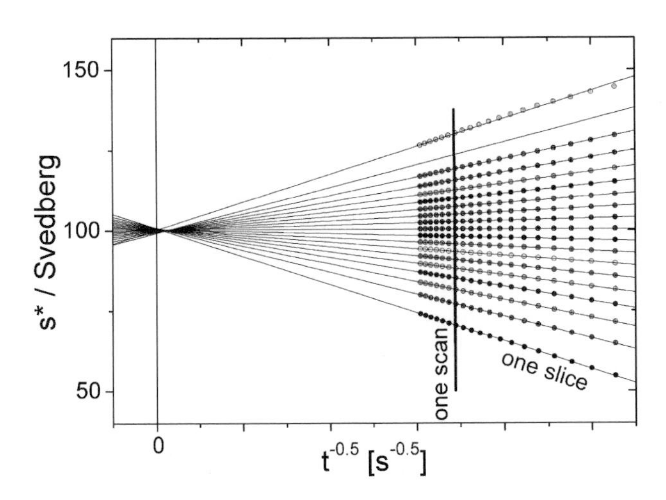

Fig. 3.7. Typical van Holde-Weischet plot for the elimination of diffusion broadening in a sedimentation velocity experiment with dissolved monodisperse macromolecules at one concentration (reprinted with permission from [45])

component is shifted to times less than infinity, i.e., to negative values of $t^{-0.5}$. Therefore, the van Holde-Weischet analysis is also a test for sample homogeneity and/or non-ideality.

The interpretation of van Holde-Weischet diagrams becomes complicated in the case of a broad continuous distribution of components, for non-ideality, and for very small particles. In Sect. 3.4.2 and Fig. 3.11, a modified van Holde-Weischet procedure [2] is described for the separation of diffusion and polydispersity broadening.

Time Derivative Method

In many cases, samples show either multiple components or even a continuous distribution of particle sizes or molar masses. In such cases, it is of interest to determine the diffusion-corrected sedimentation coefficient distribution as mentioned above in the form of $G(s)$, or in the differential form thereof, $g(s)$. Although this is in principle possible by the van Holde-Weischet method, a more suitable method for the determination of $g(s)$ is the time derivative method ([17]; see also [1]). It determines the time derivative of neighboring radial scans – $J(r,t)$, $A(r,t)$ or $c(r,t)$ – acquired at different times t according to (3.10), which is a combination of (3.2) and (3.3):

$$g^*(s)_\mathrm{t} = \left(\frac{\partial\left\{c(r,t)/c_0\right\}}{\partial t}\right)\left(\frac{\omega^2 t^2}{\ln(r_\mathrm{m}/r)}\right)\left(\frac{r}{r_\mathrm{m}}\right)^2 = \frac{\partial G^*(s)_\mathrm{t}}{\partial s} \tag{3.10}$$

with $g^*(s)$ being the *apparent* s distribution. The star indicates that this distribution is not diffusion-corrected.

The apparent $g^*(s)$ distribution equals the true distribution $g(s)$ in cases where diffusion can be ignored. To eliminate the influence of a significant diffusion broadening, an *extrapolation* of the different $g(s^*, t)$ curves, calculated for different times t to infinite time $t = \infty$, yields the true distribution $g(s)$.

An important advantage of the time derivative procedure is a significant improvement of the quality of the calculated distribution $g^*(s)_\mathrm{t}$, because two neighboring scans are subtracted from each other, so that systematic errors in the optical patterns (baseline, window distortion, etc.) are removed and the random noise decreases.

As a drawback, it has to be noted that only scans from a relatively narrow time interval can be used for a single $g^*(s)$ evaluation, because the time derivative method fails for very broadly distributed samples. It works, however, very well for moderately distributed samples exhibiting two plateaus in every scan, one on the left and one on the right of the sedimentation boundary. Figure 7.6 in Sect. 7.3 is an illustration of the above. It shows diffusion-broadened UV optics scans $A(r,t)$ of a sedimentation velocity run of small, moderately distributed gold (colloid) particles, which are evaluated with three different methods: (i) with the simple procedure of (3.2), (ii) with the time derivative methods of (3.10), yielding the non-diffusion-corrected $g^*(s)$ distribution, and (iii) with a diffusion correction to transform $g^*(s)$ into $g(s)$, using the method of Schuck [18] described in the following section.

Lamm Equation Modeling

Very recently, Schuck [18] published a new approach for the analysis of AUC sedimentation data. By modeling of the Lamm equation (1.10), now even sedimentation coefficients and molar mass distributions of *polydisperse* samples can be investigated. The principle of this approach is based on an educated initial guess of the sedimentation coefficients, the frictional coefficients, and the partial specific volume of the sample. This guess is utilized to calculate finite element solutions for a large number of discrete virtual species with different s values. The *calculated* data are adjusted to the *experimental* data by means of maximum entropy regularization, yielding a continuous sedimentation coefficient distribution $g(s)$. This approach offers an easy access to diffusion-corrected sedimentation coefficient distributions $g(s)$ [19]. A drawback of Schuck's method are the essentially non-existing "ghost peaks" in very broad distributions (for practical examples, see Fig. 7.6 and [20]).

3.3.7 Test of Homogeneity

Often, the question arises whether an examined sample is really homogenous. Attention has to be paid if data from AUC sedimentation velocity runs are applied to answer this question. There are several criteria allowing statements on homogeneity of a sample preparation. It must be kept in mind that homogeneity can only be presumed through the absence of detectable heterogeneity. Two important criteria are the following.

Criteria of boundary shape and broadening

There must be a single symmetrical boundary throughout the duration of the sedimentation velocity experiment, such as, for example, in Figs. 3.1 and 3.2. As mentioned in Sect. 3.3.6, it is possible to separate the influence of diffusion on the broadening of a boundary from the influence of heterogeneity.

Criteria of Mass Conservation

The measurable boundary, a step-like $A(r)$ or $J(r)$ scan or a Schlieren peak area, must account for all the material put into the cell, after corrections for radial dilution, throughout the duration of the sedimentation velocity experiment. Depending on the applied detecting system, either the specific decadic extinction coefficient ε or the specific refractive index increment $(\mathrm{d}n/\mathrm{d}c)_\mathrm{p}$ has to be known to allow the exact evaluation of mass conversation. If the plateau concentration c_p near the cell bottom does not reflect the initial concentration c_0 (after correction for the abovementioned effects), heavy and/or large particles may have already settled very fast (in a non-detectable manner) at the bottom of the cell. This is definitely an indication of heterogeneity.

3.4 Sedimentation Velocity Runs of Macromolecules to Measure Average *M* and *MMD*

In the first sections of this Chap. 3, we introduced the technique of sedimentation velocity runs, the basic ideas, but also some limitations and pitfalls. Now, we will use this technique to measure average molar masses M and molar mass distributions (*MMD*) of dissolved macromolecules. Besides measurements of *PSD* (see Sect. 3.5), measurements of *MMD* are the second most important industrial task of the AUC. To carry out *MMD* measurements, we need a relation between the desired M and the measured sedimentation coefficient s, in analogy to that of (1.9) for the determination of particle diameters d_p via s measurements.

Equation (1.8) is a relation between M and s, but we cannot use it to determine M via s runs because it requires additionally the knowledge of the (normally) unknown diffusion coefficient D of the solute. Thus, we have to look for another $M(s)$ relation. This is an empirical *scaling law* in the form:

$$s_0 = K \cdot M^{a} \tag{3.11}$$

where s_0 is the sedimentation coefficient at infinite dilution $c \to 0$, and K and a are the (constant) scaling parameters, valid for the respective homologs of a (special) polymer, in a (special) solvent at a defined temperature T. If one changes the polymer, the solvent or the temperature, also the scaling parameters K and a will be changed. In analogy to the parameter b in (3.9), the parameter a depends on the shape of the dissolved macromolecules and the solvent/solute interactions, too. Typical numerical values of a are $0.2 < a \leq 0.5$ (according to Svedbergs' Eq. (1.8), $a + b = 1$ should be valid).

In Sect. 3.4.1, we will show how it is possible to derive such a scaling law, using a series of (nearly) monodisperse samples (or fractions) of a polymer. In Sect. 3.4.2, we will then use such a scaling law to measure a complete *MMD* of a broadly distributed polymer via s runs, and practice at the same time the radial dilution correction discussed above in Sect. 3.2.3, the concentration dependence correction in Sect. 3.2.4, and the diffusion correction in Sect. 3.3.6.

3.4.1 Evaluation of the Average Molar Mass *M* by Sedimentation Velocity Runs via Scaling Laws

Sedimentation velocity runs can be used to measure the molar mass M of (synthetic) polymers directly via s measurements, if a scaling law such as (3.11) exists. Thus, before such an M determination via s runs is possible, we have to create the corresponding scaling law. This is tedious work, because we need a lot of (homologous) fractions of the corresponding polymer with different, but known M.

In the following, we present the procedure leading to such a scaling law. It will be demonstrated by using the NaPSS example of Sect. 3.3.5, i.e., by using polyelectrolyte samples dissolved in aqueous 0.5 M NaCl, where they behave like neutral uncharged macromolecules (see Fig. 3.6). Figure 3.8 shows the summary of several

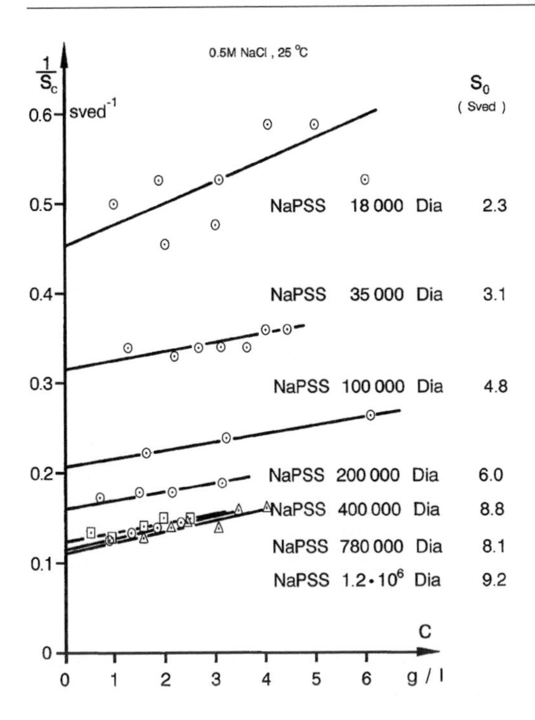

Fig. 3.8. Plots of reciprocal sedimentation coefficients s_c versus polymer concentration c for a series of narrowly distributed sodium polystyrene sulfonate (NaPSS) polyelectrolytes of different molar masses M dissolved in aqueous 0.5 M NaCl

sedimentation velocity runs performed on seven different, nearly monodisperse (dialized) calibration NaPSS samples exhibiting different known molar masses (M = 18, 35, 100, 200, 400, 780, and 1200 kg/mol, and $M_w/M_n \approx 1.10$ for all samples).

On each NaPSS sample in Fig. 3.8, a series of sedimentation velocity runs with varying polymer concentrations was performed ($c = 0.5-6$ g/l). The resulting sedimentation coefficients s_c, measured at the different polymer concentrations c, are extrapolated to zero polymer concentration $c \rightarrow 0$ according to (3.6) to obtain the s_0 value, also indicated in Fig. 3.8. A logarithmic plot of these s_0 values versus the corresponding molar masses M (see Fig. 3.9) reveals a straight (regression) line, allowing to determine the scaling parameter a from the slope of the line, and the scaling parameter K from its intercept. The dimensionless scaling parameters obtained from this linear regression are $a = 0.46$ and $K = 0.024$ [S] for NaPSS in an aqueous 0.5 mol/l NaCl solution at 25 °C. Thus, the resulting scaling law of this system is:

$$s_0 = K \cdot M^{a} = 0.024\,[\mathrm{S}] \cdot M^{0.46} \tag{3.12}$$

For the NaPSS 200 000 sample, discussed in Fig. 3.6, having an average s_0 value of 6.0 S, an average molar mass of $M_s = (s_0/0.024\,\mathrm{S})^{1/0.46} = 163\,000$ g/mol can be calculated from (3.12). This M_s value is 18% lower than the nominal M value of 200 000 g/mol. (This relatively high deviation is due to a slightly falsified nominal M value of NaPSS, which is not a truly measured one. Rather, it was calculated by using the light scattering M_w value of the *precursor* polystyrene of NaPSS.)

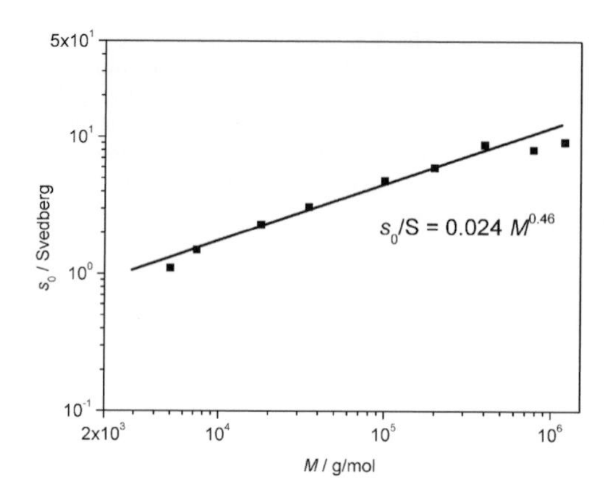

Fig. 3.9. Logarithmic plot of the s_0 values versus the molar mass M of the different NaPSS samples of Fig. 3.8, to yield the scaling parameters K and a of (3.11)

The advantage of such scaling laws is obvious. They allow fast and relatively precise measurements of (average) molar masses M_s of common polymers via s_0 measurements, once a scaling law is set up. For simple routine M_s determinations, s_0 measurements are not always required. It is also possible to work with only one finite fixed concentration c, and create for this c a corresponding scaling law $s_c = K_c \cdot M^{a_c}$. The main advantage of the combination of scaling laws and s runs is not the determination of average M_s values; but rather the measurement of complete molar mass *distributions*. This will be demonstrated in the following section.

3.4.2 Evaluation of Molar Mass Distributions (*MMD*) by Sedimentation Velocity Runs via Scaling Laws

Instead of the NaPSS/0.5 M NaCl example described in the preceding section, for the determination of an *MMD* via s runs and a scaling law, we now will use the well-known system polystyrene/cyclohexane at 25 °C, where also a scaling law exists:

$$s_0 = 0.01343\,[\mathrm{S}] \cdot M^{0.50} \tag{3.13}$$

For details of this measurement and the corresponding computer program, the reader is referred to [2]. The same polystyrene (as in Figs. 3.1 and 3.2), exhibiting a molar mass of $M_w = 106\,000$ g/mol, is used, but here the solvent MEK is replaced by cyclohexane, because the measurements are performed with the digital XL-A/I UV optics. In contrast to MEK, cyclohexane is UV-transparent at the chosen polystyrene absorption wavelength of $\lambda = 257$ nm. We use the four-hole Ti rotor at 40 000 rpm, and three very low concentrations, $c_0 = 0.2$, 0.4, and 0.6 g/l, as the UV optics is very sensitive. All steps of this *MMD*/*s* run measurement are presented in Figs. 3.10 and 3.11.

Figure 3.10a shows the XL-A/I absorption scans $A(r)$ at 257 nm as a function of the radial distance r at one concentration, $c_0 = 0.2$ g/l, and at four different, selected scanning times $t = 133$, 153, 173, and 193 min (for the sake of clarity, not all scans

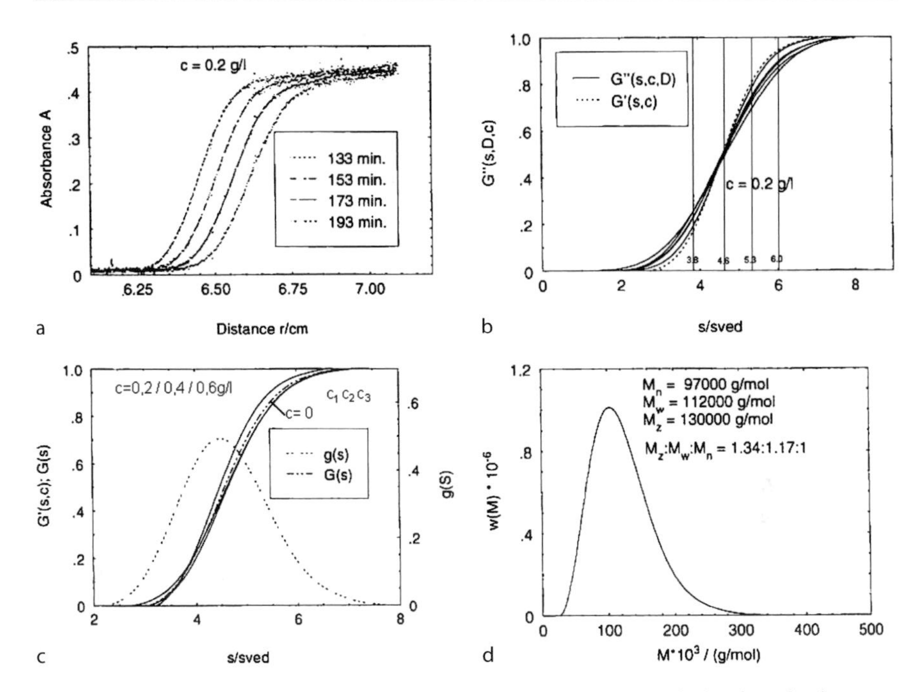

Fig. 3.10a–d. *MMD* determination via *s* runs and a scaling law of a narrowly distributed polystyrene (M_w = 106 000 g/mol) in cyclohexane. c_0 = 0.2, 0.4, and 0.6 g/l, ω = 40 000 rpm, UV optics (257 nm). Plots of **a** *A*–*r*, **b** $G''(s,c,D)$ and $G'(s,c)$, **c** $G'(s,c)$, $G(s)$ and $g(s)$, and **d** $w(M)$ = f(M) (reprinted with permission from [2])

at all times *t* are shown). The small points are the experimental values, and the dashed lines represent the smoothest curves obtained by a special spline-fitting procedure. Now, to obtain Fig. 3.10b, all *r* values in Fig. 3.10a are transformed into *s* values with (3.2), and all *A* values are initially transformed into *c* values via Lambert-Beer's law (2.2). These *c* values are then transformed by means of (3.14) into radial dilution-corrected (see Sect. 3.2.3) apparent integral sedimentation distribution G'' values:

$$G''(s,c,D) = \frac{c}{c_0} \cdot \left(\frac{r_m}{r}\right)^2 \qquad (3.14)$$

The two prime symbols $''$ indicate that these G'' values are not corrected for diffusion and for concentration. The solid lines of Fig. 3.10b show some plots of G'' as a function of *s* for different scanning times *t*. To accomplish the diffusion correction $t \rightarrow \infty$ according to Sect. 3.3.6 and (3.15),

$$\lim_{t\to\infty} G''(s,c,D) = G'(s,c) \qquad (3.15)$$

(to obtain $G'(s,c)$), we do not use one of the three diffusion correction methods, described in Sect. 3.3.6 (i.e., van Holde-Weischet method, time derivative method, and Lamm equation modeling). Rather, we use (see [2]) a modification of van

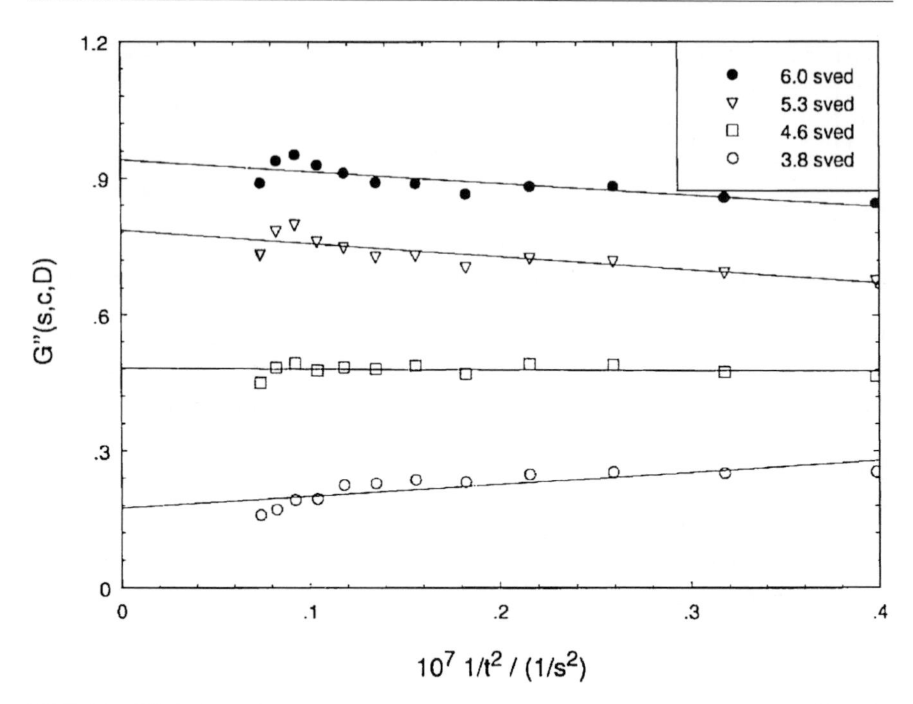

Fig. 3.11. Diffusion correction within Fig. 3.10b to obtain $G'(s,c)$ via an extrapolation of $G''(s,c,D)$ to $t \to \infty$. This is done by plotting $G''(s,c,D)$ values at different constant s values versus $1/t^2$ and extrapolating straight regression lines to $1/t^2 = 0$. Polystyrene ($M_w = 106\,000$ g/mol) in cyclohexane, $c_0 = 0.2$ g/l (reprinted with permission from [2])

Holde-Weischet, which works as follows: we create 100 vertical lines in the G''–s plot of Fig. 3.10b, at constant s values between s_{min} (~ 1.8 S) and s_{max} (~ 7.9 S) in equal Δs sub-steps. These 100 sets of G'' values at the different constant s values (these are the cross points of the vertical lines in Fig. 3.10b with the solid G''–s curves at the different scan times) are plotted in Fig. 3.11 as a function of $1/t^2$. For the sake of clarity, we show in Fig. 3.11 only four of the total 100 G''–$1/t^2$ curves, which all are nearly straight lines. We selected the four curves for s = 3.8, 4.6, 5.3, and 6.0 S (their four representative vertical lines are indicated in Fig. 3.10b). Now, we calculate through all 100G''–$1/t^2$ curves in Fig. 3.11 the corresponding straight regression lines, and extrapolate these to $1/t^2 = 0$. Their 100 intercepts with the ordinate are the $G'(s,c)$ values, i.e., the diffusion-corrected G'' values (this is equivalent to the van Holde-Weischet procedure in Fig. 3.7). All 100 G'/s points are then plotted again into Fig. 3.10b, where they yield the dashed curve. The same diffusion correction procedure is done for the other two concentrations $c_0 = 0.4$ and 0.6 g/l. The resulting $G'(s,c)$ curves of all three concentrations are collected in the plot of Fig. 3.10c as three solid lines.

In the next step, the concentration correction is carried out (see (3.16)):

$$\lim_{c \to 0} G'(s,c) = G(s) \tag{3.16}$$

This means extrapolating the three $G'(s,c)$ curves in Fig. 3.10c to infinite dilution $c_0 \to 0$ in order to obtain the (diffusion- and concentration-corrected) real integral sedimentation coefficient distribution $G(s)$ and its derivative $g(s) = \mathrm{d}G(s)/\mathrm{d}s$ (see (3.3)). Normally, this is done in the same manner as for the above diffusion correction, i.e., creating 100 vertical lines in Fig. 3.10c between $s_{\min}$ and $s_{\max}$, plotting the three $G'(s,c)$ cross points of each line with the three $G'(s,c)$ curves $c_0 = 0.2$, 0.4, and 0.6 g/l as function of c_0, and extrapolating these to $c_0 \to 0$. The 100 intersection points with the ordinate represent the desired $G(s)$ function. This normal extrapolation procedure does not make very much sense for the three $G'(s,c)$ curves in Fig. 3.10c, because they are nearly identical. The reason for that are the very low concentrations chosen because of the sensitive UV detector: they are close to $c_0 = 0$. Thus, for the concentration correction, the following simple procedure was used: the three $G'(s,c)$ curves were averaged to obtain $G(s)$. The resulting $G(s)$ curve and its derivative, $g(s)$, are plotted in Fig. 3.10c (dashed lines).

Finally, this $g(s)$ curve was transferred by the scaling law (3.13) into the differential *MMD* function $w(M)$, which is presented in Fig. 3.10d. The differential molar mass distribution $w(M)$ is connected with $g(s)$ via $w(M)\,\mathrm{d}M = g(s)\,\mathrm{d}s$. Integration leads to

$$W(M) = G(s) \tag{3.17}$$

with $w(M) = \mathrm{d}W(M)/\mathrm{d}M$, $g(s) = \mathrm{d}G(s)/\mathrm{d}s$ and the different *averages* of the molar masses M_β (i.e., M_n, M_w, M_z, ...):

$$M_\beta = \frac{\int_0^\infty w(M) \cdot M^\beta \,\mathrm{d}M}{\int_0^\infty w(M) \cdot M^{\beta-1} \,\mathrm{d}M}\,, \tag{3.18}$$

with the indices $\beta = 0, 1, 2, \ldots$ (n, w, z, ...). M_n, M_w and M_z are the three well-known average molar masses – the number-, the weight-, and the z-average molar masses ("z" is the abbreviation for the German word Zentrifuge = centrifuge. See also Sect. 5.1 and (5.1) for an analogous definition of M_n, M_w and M_z).

The *MMD* in Fig. 3.10d is, as expected, unimodal and relatively narrow. The calculated M averages according to (3.18), $M_\mathrm{n} = 97\,000$ g/mol, $M_\mathrm{w} = 112\,000$ g/mol, $M_\mathrm{z} = 130\,000$ g/mol, and their relation $M_\mathrm{z}{:}M_\mathrm{w}{:}M_\mathrm{n} = 1.34{:}1.17{:}1$, indicate the same narrow molar mass distribution, and are in reasonable agreement with the values given by the supplier of PS 106 000, Polymer Standard Service, Mainz.

The polystyrene sample PS 106 000 discussed above was a relatively narrowly distributed one. Now, in Fig. 3.12, the same *MMD*/scaling law procedure is shown for a *broadly distributed* unimodal sample, namely, the well-known calibration polystyrene NBS 706 of National Bureau of Standards (Washington, DC), also in cyclohexane solution and also measured with the XL-A/I UV optics at 260 nm, 40 000 rpm and 25 °C. The given NBS 706 values are $M_\mathrm{n} = 136\,000$ g/mol,

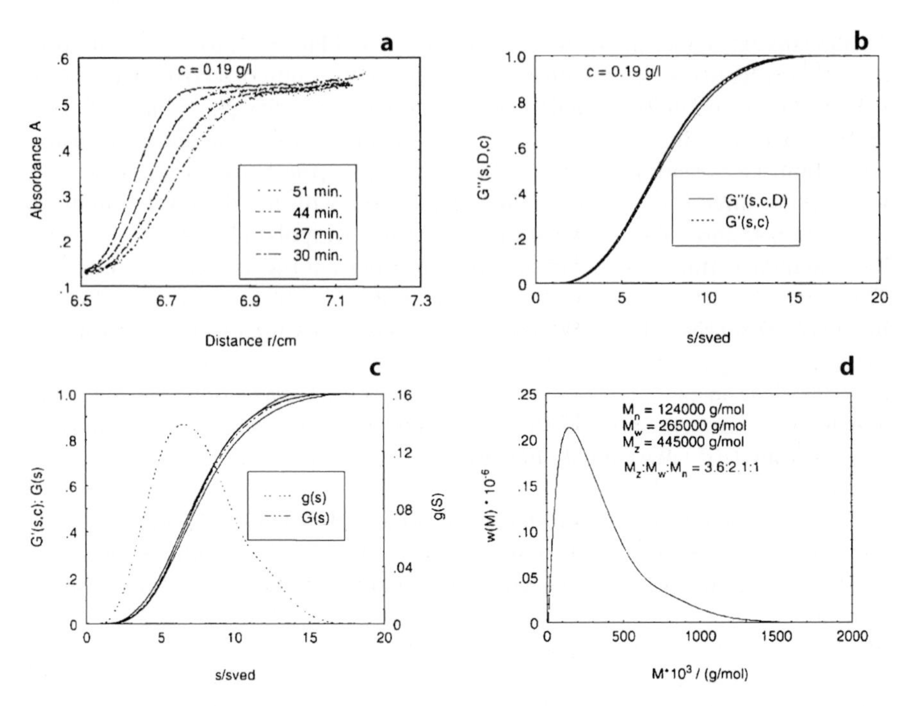

Fig. 3.12a–d. *MMD* determination via *s* runs and a scaling law of the broadly distributed polystyrene NBS 706 in cyclohexane. $c_0 = 0.19$, 0.4, and 0.6 g/l, $\omega = 40\,000$ rpm, UV optics (260 nm). Plots of **a** *A*–*r*, **b** $G''(s,c,D)$ and $G'(s,c)$, **c** $G'(s,c)$, $G(s)$ and $g(s)$, and **d** $w(M) = f(M)$ (reprinted with permission from [2])

$M_w = 257\,800/288\,100$, $M_z = 355\,000/400\,000$ g/mol, and M_z:M_w:M_n = 2.9:2.1:1 (for details, see Sect. 5.3.3). NBS 706 is also intensively characterized in Sect. 5.3.3 via sedimentation equilibrium runs (molar mass distribution and the different M averages).

In spite of measuring with 12-mm double-sector cells, Fig. 3.12a exhibits an excess absorption A_{excess} (i.e., A is not 0 near the meniscus). This is not caused by low molar mass impurities of the solvent cyclohexane. It is rather an artifact of the XL-A/I, which sometimes is present. For further evaluation of the measuring data in Fig. 3.12b–d, this excess absorption ($A_{excess} \sim 0.13$) was subtracted. Figure 3.12b shows that the diffusion correction $G''(s,c,D) \rightarrow G'(s,c)$ is minimal, in contrast to Fig. 3.10b. The reason is the higher molar mass of NBS 706 in comparison to PS 106 000. Furthermore, as in Fig. 3.10c, the difference between the three $G'(s,c)$ curves for the three different concentrations is minimal, too (it demonstrates the high precision of the XL-A data!). Thus, again the three $G'(s,c)$ curves were averaged to obtain $G(s)$ and $g(s)$, and to calculate with the scaling law (3.13) the (broad) differential *MMD* $w(M)$ of NBS 706 (Fig. 3.12d), and also the average values $M_n = 124\,000$ g/mol, $M_w = 265\,000$ g/mol, $M_z = 445\,000$ g/mol, and M_z:M_w:M_n = 3.6:2.1:1. These M averages agree reasonably with the given NBS values.

It is interesting to note that we measured the same average M values (and the *MMD*) by another, completely independent AUC method (see Sect. 5.3.3), namely, by sedimentation equilibrium runs, which resulted in nearly the same average M values and the same *MMD* (within the errors of measurement). That is really a demonstration of the power and the versatility of the AUC! The difference between the two AUC methods is basically as follows: for the s run method, we need a precise knowledge of a scaling law (3.11), whereas for the equilibrium run method we need a precise $\bar{v}$ value. The *MMD*/s run/scaling law procedure described above was demonstrated with UV optics. Of course, this method is applicable also for every other optics system, such as interference, Schlieren or fluorescence optics.

3.5 Sedimentation Velocity Runs on Particles to Measure Average d_p and *PSD*

As mentioned in the introductory Sect. 3.1, the analytical ultracentrifuge is a helpful, but often underestimated instrument for the analysis of nanoparticles and all colloidal systems. Especially spherical colloidal systems can be analyzed in a unique manner by AUC. In this section, we will give some examples to prove this statement. In fact, the topic of this section represent approximately 80% of our daily work in the AUC laboratory of BASF, where several thousand samples, consisting mainly of polymeric latices, are analyzed per year, in most cases with the turbidity detector (plus Mie's light scattering theory) and within the diameter range 30 – 5000 nm. For smaller particles showing nearly no turbidity, interference, Schlieren or UV optics are used, similarly to dissolved macromolecules. In more recent years, the importance of AUC for the examination of such very small *nano*particulate systems is becoming ever more evident [15, 42]. The diameter of such very small nanoparticles may well be below 1 nm, thus consisting of just a few atoms (see, for example, Fig. 3.23).

In principle, the *PSD* determination via s runs is identical to the *MMD* determination of macromolecules via s runs, as described in Sect. 3.4. It is in both cases the determination of $G(s)$ or $g(s)$. However, the *PSD* determination is much simpler, because compact (spherical) particles almost do not show any concentration and diffusion dependence, so that we can neglect the corresponding corrections and measure usually only at one, very low concentration ($0.3 < c < 3$ g/l). Only for particles with $d_p < 20$ nm are these corrections required (see [15] and Fig. 7.6). There are three differences between a *PSD* determination and an *MMD* determination (of which reasons nos. (ii) and (iii) are valid only for the turbidity detector): (i) for the *MMD*, we transform the s values into M values by means of the scaling law (3.11), but for *PSD* we transform the s values into d_p values with the Stokes equation (1.9); (ii) the calculation of the concentration c (for the different fractions) from the measured light intensity values $I(t)$ via the Mie theory is much more complex; and (iii) in contrast to *MMD* determinations, which are measured at constant rotor speed ω, for *PSD* determinations we measure always at *variable*, exponentially increasing rotor speed $\omega(t)$ ($0 < \omega < 40\,000$ rpm within 1.5 h; see also Fig. 3.13).

3.5.1 Particle Size Distribution via AUC Turbidity Detector and Mie Theory

The basis of the measurement of particle size distributions (*PSD*) is the exact determination of the sedimentation coefficient distribution $G(s)$ or $g(s)$ of mono- and polydisperse samples. This includes the measurement of the relative mass portion $m_i = c_i/c_0$ of the different components (= fractions i). The obtained sedimentation coefficients s_i can then be converted into sphere-equivalent particle diameters $d_{p,i}$ by the Stokes Eq. (1.9). The determination of mass portions m_i depends on the applied optical system. Despite the fact that particle size distributions can be obtained using any optical system of the AUC, they are most favorably measured with turbidity optics. This is due to the fact that Mie's light scattering theory [21] allows an exact determination of the component's mass portions m_i by analysis of the reduction of the light intensity I, caused by the particle components in the AUC cell. The procedure will be described in the following.

The turbidity detector itself (see Fig. 2.13) has already been subject of Sect. 2.4.4. As mentioned, in a sedimentation velocity run applying the turbidity detection, the intensity I of the incident parallel light is reduced while traveling through the cell. This is due to Mie's light scattering of the sedimenting particles. The light beam and the detector are located at r_{slit} in the middle of the cell between r_m and r_b (see Fig. 2.13), because one does not know before the experiment whether an unknown sample will sediment or float. Thus, in both cases, the sedimentation/flotation distance is nearly the same. The reduced intensity $I(t)$ is recorded at this position as a function of time t (see Fig. 3.13). Here, the $I(t)$ curve from a sedimentation velocity run performed on an acrylic latex ($d_p \approx 100\,\text{nm}$, $c_0 = 0.49\,\text{g/l}$, BASF-made sample) is shown. Also shown in Fig. 3.13 is the rotor speed $N(t) = \omega(t)/2\pi$ as a function of running time t. This $N(t)$ profile, an exponential increase from 0 up to 40 000 rpm within 1.5 h, is always the same in the BASF particle sizer. It allows one in every run to measure very small (down to 20 nm) as well as very big particles (up to 5000 nm).

Such $I(t)$ curves are the basis for the calculation of the particle size distribution, i.e., for the relation between the mass portion m_i and the particle diameter $d_{p,i}$ of the corresponding different fractions i. In Fig. 3.13 (and Fig. 3.14), the $I(t)$ curve starts at the low initial intensity I_0 given by the initial concentration c_0 of the turbid sample. This initial concentration c_0 (= 0.49 g/l) is chosen to result in initial intensities I_0 of about 10% in relation to the maximum light intensity I_{DM} given by the intensity of the pure solvent (= dispersion medium). Usually, c_0 is in the range of $0.3 < c_0 < 3\,\text{g/l}$. The intensity I_{DM} is reached at the end of the experiment when all particles of the sample have passed the radial position of the slit, or of the detector. In the following, the single steps of the evaluation of an $I(t)$ curve like the one shown in Fig. 3.13 will be described. This procedure to obtain the desired *PSD* (see [22]) will be done by means of the scheme in Fig. 3.14.

As shown in Fig. 3.14, a measured broad $I(t)$ curve is formally considered as a superposition of z theoretical $I(t)$ one-step curves ($i = 1, 2, 3, ..., z$ with $I_z = I_{DM} = I_s$ and $30 < z < 800$). Each of these is assigned to an ideally monodisperse particle fraction having a diameter $d_{p,i}$ and a concentration $c_{0,i}$ (or a relative

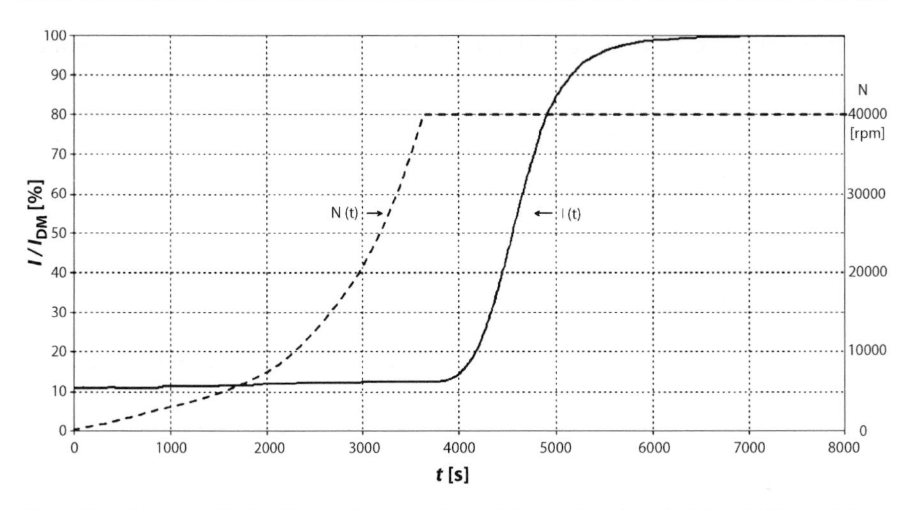

Fig. 3.13. $I(t)$ curve obtained with a homemade particle sizer (equipped with turbidity optics) on a BASF-made sample ($d_p \approx 100$ nm) of an acrylic polymer dispersion. Measured in water, 3-mm single-sector cell, $c_0 = 0.49$ g/l. Also shown is the variation of the rotor speed $N = \omega/2\pi$ as a function of running time t

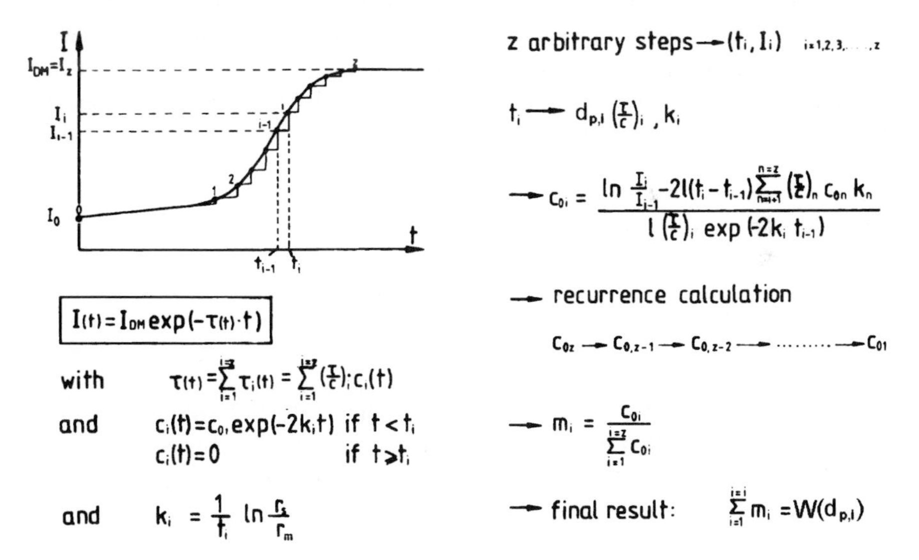

Fig. 3.14. Scheme illustrating the evaluation procedure to convert an $I(t)$ curve into a particle size distribution $W(d_{p,i})$ with a recurrence calculation (reprinted with permission from [22])

mass percentage $m_i = c_{0,i}/c_0$ within the sample). In order to describe the type of superposition mathematically, the following two physical assumptions are made:

(i) the individual particle fractions sediment completely, *independently* of each other, without any mutual interference (for small total concentrations $c < 5$ g/l, this first assumption is virtually always met); and

(ii) the measured $I(t)$ curve can be transformed into a time-dependent turbidity curve $\tau(t)$ using Lambert-Beer's law, (2.2), in the form

$$I(t) = I_s \cdot e^{(-\tau(t) \cdot a)} \tag{3.19}$$

and we assume second that this total turbidity is made up *additively* by the time-dependent turbidities $\tau_i(t)$ of the individual fractions according to

$$\tau(t) = \sum_{i=1}^{i=z} \tau_i(t) \tag{3.20}$$

(for low concentrations $c < 5$ g/l, this second assumption virtually holds always, too).

The turbidity $\tau_i(t)$ of any monodisperse fraction now may be split up, according to

$$\tau(t) = (\tau/c)_i \cdot c_i(t) \tag{3.21}$$

into a product of (i) the time-*in*dependent criterion of matter $(\tau/c)_i$ to be calculated by means of $d_{p,i}$, λ, n_p and ϱ_p in accordance with Mie, and (ii) the time-dependent concentration $c_i(t)$ of the ith fraction in situ of the slit. On the basis of assumption (i) above, the time dependence of $c_i(t)$ can be calculated by means of Stokes' law, (1.9), taking into account the thinning effect (Sect. 3.2.3, (3.5)) in the sector-shaped measuring cells. As a result, we find that $c_i(t)$ is a *step* function of the following kind:

$$\begin{aligned} c_i(t) &= c_{0i} \cdot \exp(-2k_i t) \quad \text{for} \quad t < t_i \\ c_i(t) &= 0 \quad \text{for} \quad t \geq t_i \end{aligned} \tag{3.22}$$

with

$$k_i = \frac{1}{t_i} \ln \frac{r_{slit}}{r_m} \tag{3.23}$$

where k_i is the constant of the thinning effect of the ith fraction, and t_i is its "step" time, i.e., the travel time for the distance $(r_{slit} - r_m)$ between the meniscus and slit. By means of a modified Stokes' equation, t_i yields (the rotor speed ω being constant) the requested particle diameter $d_{p,i}$ as

$$d_{p,i} = \sqrt{\frac{18 \cdot \eta_s \cdot \ln \frac{r_{slit}}{r_m}}{(\varrho_p - \varrho_s)\omega^2 t_i}} = \sqrt{\frac{18\eta_s s_i}{(\varrho_p - \varrho_s)}} \tag{3.24}$$

η_s being the viscosity of the dispersing medium, and $s_i = \left[\omega^2 t_i\right]^{-1} \cdot \ln(r_{slit}/r_m)$ being the sedimentation coefficient of the ith fraction. In the case of a non-constant but time-dependent rotor speed $\omega(t)$, in (3.24) the expression $\omega^2 \cdot t_i$ has to be replaced by the running time integral $\int_0^{t_i} \omega^2 \, dt$. This integral is calculated from the $N(t) = \omega(t)/2\pi$ curve (see Fig. 3.13) digitally registered by the computer. In the case of flotation ($\varrho_p < \varrho_s$), r_m has to be replaced by r_b in (3.23) and (3.24), and instead of a thinning effect, we now have a thickening effect (see, for example, Fig. 3.20).

These theoretical considerations result in a procedure of analyzing a measured broadly distributed $I(t)$ curve as outlined on the right-hand side of Fig. 3.14. This continuous $I(t)$ curve is "digitized" by splitting up into z small steps ($30 < z < 800$). Each step is characterized by a time t_i, its corresponding running time integral, and by a light intensity step $\Delta I_i = I_i - I_{i-1}$. At t_z, the longest travel time of the smallest particles, $I(t)$ becomes time independent, and thus is $I_z = I_s = I_{DM}$. This division into steps means that we consider the turbidity front of the *i*th monodisperse fraction to pass by the slit at the time t_i, and then to disappear from our view ($c_i(t) = 0$ for $t \geq t_i$). From the different travel times t_i of all fractions, for a start their diameters $d_{p,i}$ and their thinning (thickening) effect constants k_i are calculated by means of (3.24) and (3.23). From these $d_{p,i}$ values and the known refractive index n_p and $(dn/dc)_p$ of the dispersed particles (see (2.4)), the computer calculates all the assigned specific turbidities $(\tau/c)_i$, using Mie's light scattering theory as described in [22], and illustrated in Fig. 3.15. From the totality of all values t_i, I_i, k_i, and $(\tau/c)_i$, the requested $c_{0,i}$ values can then be calculated by recurrence according to

$$c_{0,i} = \frac{\ln \dfrac{I_i}{I_{i-1}} - 2a(t_i - t_{i-1}) \cdot \sum\limits_{n=i+1}^{n=z} (\tau/c)_n \cdot c_{0,n} \cdot k_n}{a \cdot (\tau/c)_i \cdot \exp(-2 \cdot k_i \cdot t_{i-1})} \tag{3.25}$$

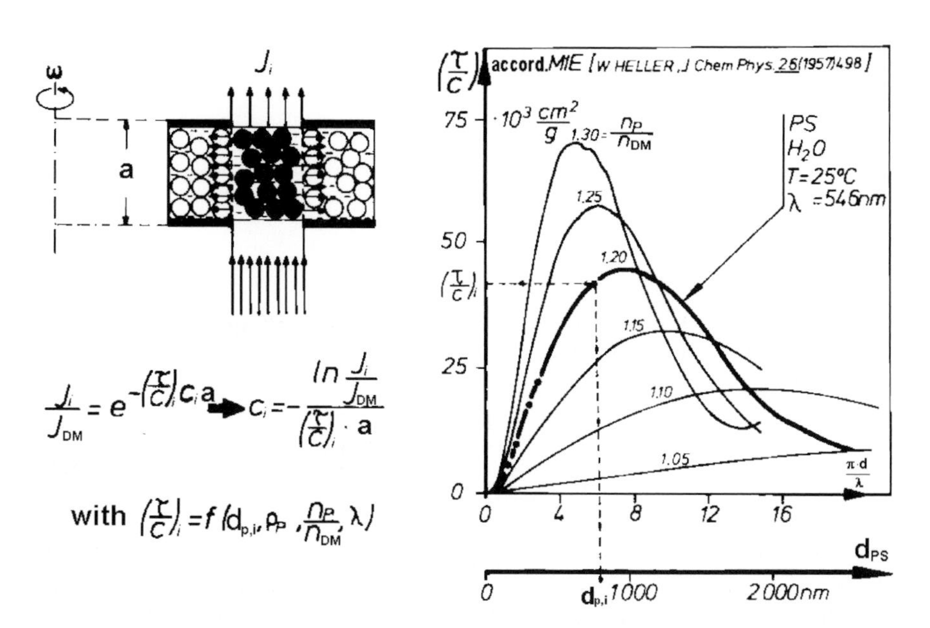

Fig. 3.15. Scheme illustrating the evaluation of the specific turbidity $(\tau/c)_i$ using the known diameter $d_{p,i}$ and the refractive index n_p of the particles according to Mie's light scattering theory [21] and Lambert-Beer's law (reprinted with permission from [26])

The recurrence formula, (3.25), is obtained by inserting (3.20), (3.21) and (3.22) into (3.19), and solving these with respect to $c_{0,i}$. It follows from the indexing of (3.25) that one has initially to calculate $c_{0,z}$, from which $c_{0,z-1}$ is then obtained. Both $c_{0,z}$ and $c_{0,z-1}$ in turn render $c_{0,z-2}$, and so on. In this manner, all $c_{0,i}$ values down to $c_{0,1}$ are calculated successively. The $c_{0,i}$ values then yield the relative mass portions m_i, according to (3.26).

$$m_i = \frac{c_{0,i}}{\sum_{i=1}^{i=z} c_{0,i}} = \frac{c_{0,i}}{c_0} \tag{3.26}$$

Finally, the cumulative distribution curve $\sum_{i=1}^{i=i} m_i = W(d_{p,i})$, the ultimate aim of the analysis, is plotted in the form of the integral (and the differential) *PSD* diagram, as shown in Fig. 3.16 for the starting $I(t)$ curve in Fig. 3.13.

Before we discuss this *PSD* in Fig. 3.16, we will outline by means of Fig. 3.15 how it is possible to determine the $(\tau/c)_i$ values needed in (3.25) to calculate $c_{0,i}$, in a purely theoretical manner, using Mie's light scattering theory (for details, see [22] and [23]).

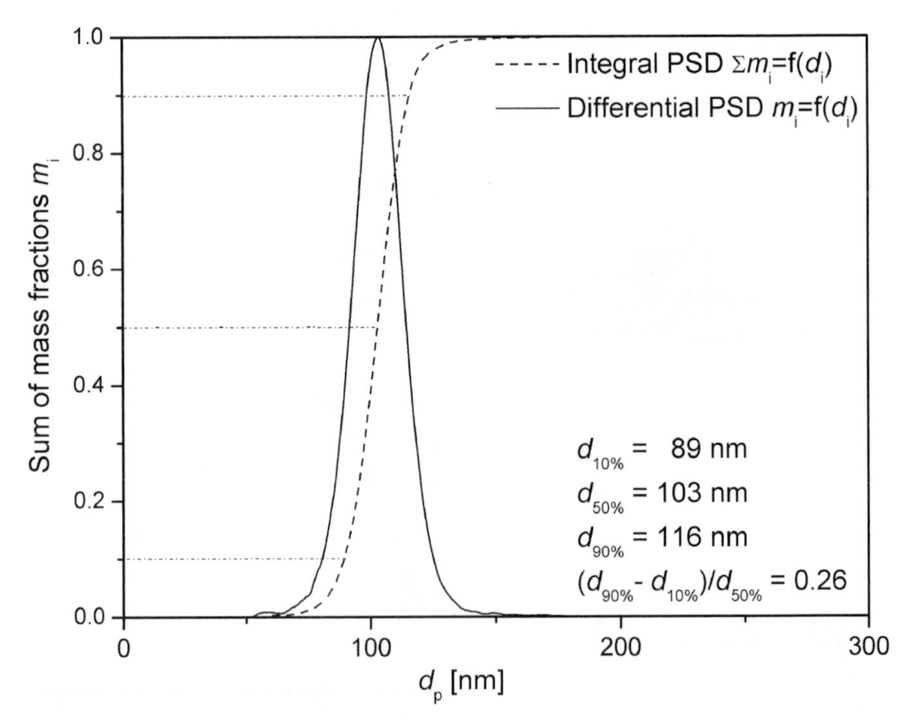

Fig. 3.16. Integral and differential particle size distribution of a BASF-made acrylic polymer dispersion, and the characteristic three diameters, $d_{10\%}/d_{50\%}/d_{90\%}$, defining the polydispersity parameter $(d_{90\%} - d_{10\%})/d_{50\%}$

For this determination, we have to know the diameter d_p, the density $\varrho_p = 1/\bar{v}$, and the refractive index n_p of the particle, the refractive index of the dispersion medium n_s, and the wavelength of light λ. The measurements of the auxiliary parameters ϱ_p and n_p (via $(dn/dc)_p$ measurement) are explained in Sect. 2.6. Commercially available computer programs allow one now to calculate $(\tau/c) = f(d_p, \varrho_p, n_p, n_s, \lambda)$ in the form of well-known Mie diagrams, as presented in Fig. 3.15: (τ/c) as function of the reduced particle diameter $(\pi d_p/\lambda)$, and the refractive index quotient $n_p/n_s = n_p/n_{DM}$.

For every kind of particles with a defined n_p, one of the different curves in Fig. 3.15 (or an interpolated one) is valid. For example, the thick line in Fig. 3.15 is valid for dispersed polystyrene particles in H_2O ($n_{PS}/n_s = 1.59/1.334 = 1.20$ at $\lambda = 546\,\text{nm}$), and as can be seen from the dashed lines in Fig. 3.15: if we enter the Mie diagram with a given $d_{p,i}$, the corresponding $(\tau/c)_i$ value is found in a simple manner for the corresponding PS particles. The Mie theory computer program inside our particle sizer will do this for us. Modern Mie programs work not only for real refractive indices n_p, but also for complex refractive indices, $n_{complex} = n_{real} - i \cdot k$, needed, for example, for metal sols and colored particles (see, for example, Figs. 6.33 and 6.35).

Figure 3.16 shows the particle size distribution (in integral and differential form) obtained from the $I(t)$ curve of Fig. 3.13 in the manner described above. Also indicated in Fig. 3.16 are the usual *three characteristic diameter* values, $d_{10\%} = 89\,\text{nm}$, $d_{50\%} = 103\,\text{nm}$, and $d_{90\%} = 116\,\text{nm}$, which define the polydispersity parameter $(d_{90\%} - d_{10\%})/d_{50\%} = 0.26$, a measure for the broadness of a *PSD*. The percentage index xy% means that within the integral (mass) *PSD*, xy wt% of the particle fractions have diameters below $d_{xy\%}$, and the remainder above $d_{xy\%}$. For a more detailed discussion of this procedure, see [22] and [24].

Beyond the basic theory of particle size analysis in the AUC described above, some experimental details from practice should be addressed here. Firstly, and most obviously, it is possible to measure up to seven different samples in one experiment if an eight-hole rotor is used, as outlined in Figs. 2.14 and 3.17. Secondly, it has proved to be advantageous to vary the angular velocity ω of the ultracentrifugation rotor with time t, rather than running the experiment at a constant speed, outlined in Fig. 3.17 as well. In this case, in all equations given above, the term $\omega^2 t$ has to be replaced by the running time integral $\int \omega^2(t)\,dt$. It has been shown (see [25] and [26]) that the variation of the rotor speed effectively does not influence the resulting particle size distributions. The benefit of this procedure is that the centrifugal field range present in the single experiment is broader, and thus a wider range of particle diameters is covered in a sedimentation run. Thirdly, very small particles, with diameters $d_p < 30\,\text{nm}$, show according to Mie very weak turbidity signals $(\tau/c)_p$, which are near the lower limits of the turbidity detector. This can be compensated by a higher concentration c_0, which, on the other hand, can be too high for a reliable *PSD* determination because of particle interaction. Thus, the d_p values of very small particles measured via turbidity detectors are not very precise. For this reason, it is recommended, if possible, to use interference,

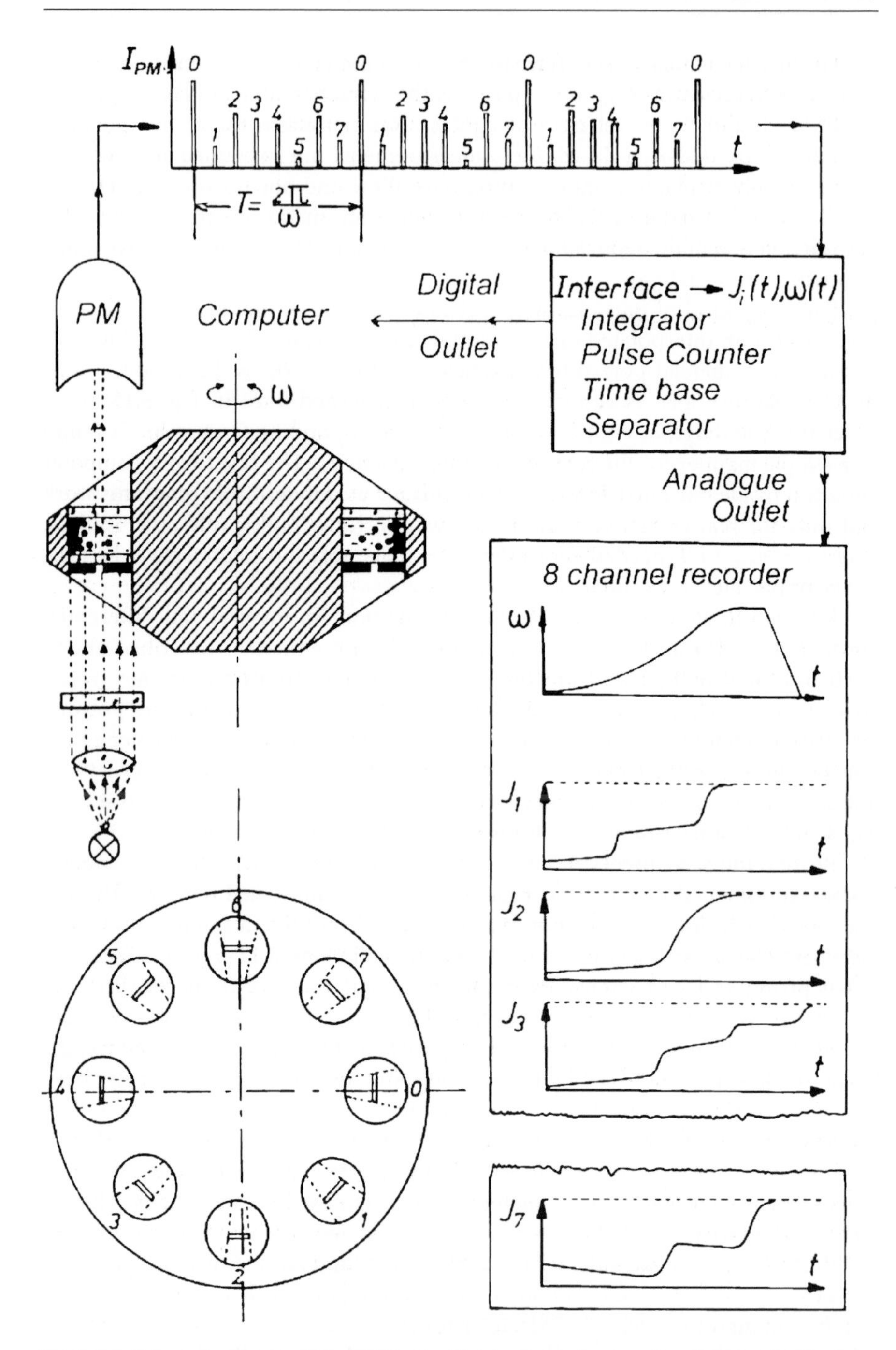

Fig. 3.17. Schematic diagram of the AUC method for the determination of particle size distributions, with an eight-cell rotor, a turbidity detector, and a multiplexer setup (reprinted with permission from [25])

Schlieren or UV optics for that diameter range. These detectors do not require Mie correction to measure correct $m_i = c_{0,i}/c_0$ values, which is often an advantage.

The possibility to measure up to seven samples simultaneously under the same rotor speed conditions $\omega(t)$ with the setup of Fig. 3.17 is very important for the so-called coupling technique (to measure extremely broad *PSD*), and the H_2O/D_2O density variation method (to measure ϱ_p via AUC), which are described in the following two Sects. 3.5.2 and 3.5.3.

3.5.2 Coupling Technique to Measure very Broad *PSD*

There are, of course, many analytical methods capable of yielding information on particle sizes in the range between 1 nm and a few micrometers. Nevertheless, all of these face problems whenever very *broad* particle size distributions are present. The special problem with AUC *turbidity* optics is that, according to Mie (see Fig. 3.15), small particles ($d_p < 30$ nm) show very small, and big particles very high turbidities. For example, 20-nm PS particles exhibit a specific turbidity of $(\tau/c)_{20\,nm} = 17\,cm^2/g$, and 2000-nm PS particles $(\tau/c)_{2000\,nm} = 17\,000\,cm^2/g$, which is not within the linear range of turbidity detectors. To overcome this problem, the so-called *coupling PSD* technique was developed by one of the authors [22], in addition to the standard procedure described above.

In such a coupling *PSD* experiment, *two* different concentrations, $c_I = c_{standard}$ (well suited for big particles), and a higher concentration c_{II} of the same dispersion (used to detect particularly small particles) are measured *simultaneously* in a single ultracentrifugation *s* run with the multiplexer setup of Fig. 3.17. While one concentration, c_I, is chosen as mentioned above to result in an I_0 value of approximately 10% of $I_s = I_{DM}$, the other concentration, c_{II}, is selected 5–30 times higher. The two simultaneously measured $I(t)$ curves resulting from such a sedimentation velocity run are recorded separately (see Figs. 3.18 and 3.19). They are then coupled mathematically, leading to *one* $I(t)$ curve that can be transformed into a particle size distribution in the way described in Fig. 3.14. Figure 3.18 shows schematically two such $I(t)$ curves of the same sample, but with the two different concentrations $c_I = c_{standard}$ and c_{II} ($= c_{high}$). For details of this calculation, see [22].

As coupling point $i = k$ assigned to the travel time $t_c = t_k$, we define that $(I/I_s)_k/t_k$ is the point where, on the $I(t)$ curve for the higher concentration, $(I/I_s) > 0.5$ is valid for the first time. The coupled $I(t)$ curve is now constructed in the following way: from the standard $I(t)$ curve, only those points $(I/I_s)_i/t_i$ holding $0 < i < k$ are taken without any change. To these, those points from the $I(t)$ curve of increased concentration c_{II} holding $k < i \leq z$ are added in a modified form. This is purely a mathematical modification, which transforms the measured light intensities $(I/I_s)_{h,I}$ to the standard concentration c_I according to

$$(I/I_s)_{I,i} = \left(I/I_s\right)_{II,i}^{\frac{c_I}{c_{II}}} \tag{3.27}$$

(Equation (3.27) results from a combination of Lambert-Beer's law, (3.19), and (3.21).) The assigned travel times t_i, however, are taken without modification. The

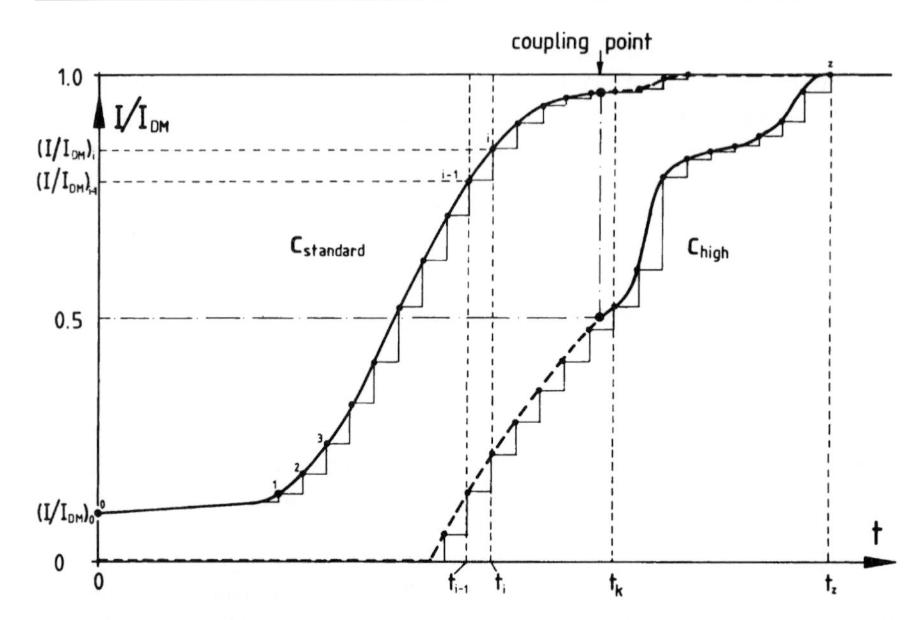

Fig. 3.18. Scheme of two simultaneously measured $I(t)$ curves of an extremely broadly distributed dispersion to illustrate the particle size distribution coupling technique, measured at two different concentrations: $c_I = c_{standard}$ and $c_{II} = c_{high}$ (reprinted with permission from [22])

coupled $I(t)$ curve constructed in this manner is then transformed into a *PSD* curve in the same way as a standard $I(t)$ curve, as described in Fig. 3.14. We call this transformed *PSD* curve the coupling *PSD*. The coupling procedure ensures that both coarse and fine particles are detected appropriately by the turbidity detector. By this procedure, it is possible to measure even 20-nm particles (which show, as mentioned above, extremely low specific turbidity) besides 2000-nm particles by applying turbidity optics. Nevertheless, very small particles should be detected with other detectors, if possible (cf. above).

Figure 3.19 shows a well-known measuring example of a very broad particle size distribution obtained by the coupling *PSD* technique. This is the measurement of a defined mixture of ten different aqueous polystyrene latices exhibiting particle sizes between 67 and 1220 nm. The ten, narrowly distributed standard latices were mixed at 10 wt% each [27]. The upper part of Fig. 3.19 shows the two primary measured $I(t)$ curves at $c_I = 0.35$ and $c_{II} = 3.5$ g/l, the $N(t)$ function, and the coupling point $0.5/t_c$. As expected, at the beginning of the run, $0 < t < 2500$ s, the high concentration signal $I(t)$ is zero. By contrast, at the end of the run in the higher concentration $I(t)$ curve, the $I(t)$ step of the smallest particles, 67 nm, which are not visible in the standard concentration $I(t)$ curve, is now clearly seen.

The lower part of Fig. 3.19, the final coupling *PSD*, shows that both the particle diameters d_p and the mass portions m_i (the original concentration of all ten components was 10 wt%) are reproduced within an error of 5%. The peaks in the differential *PSD* are obtained with baseline resolution for all ten components, also

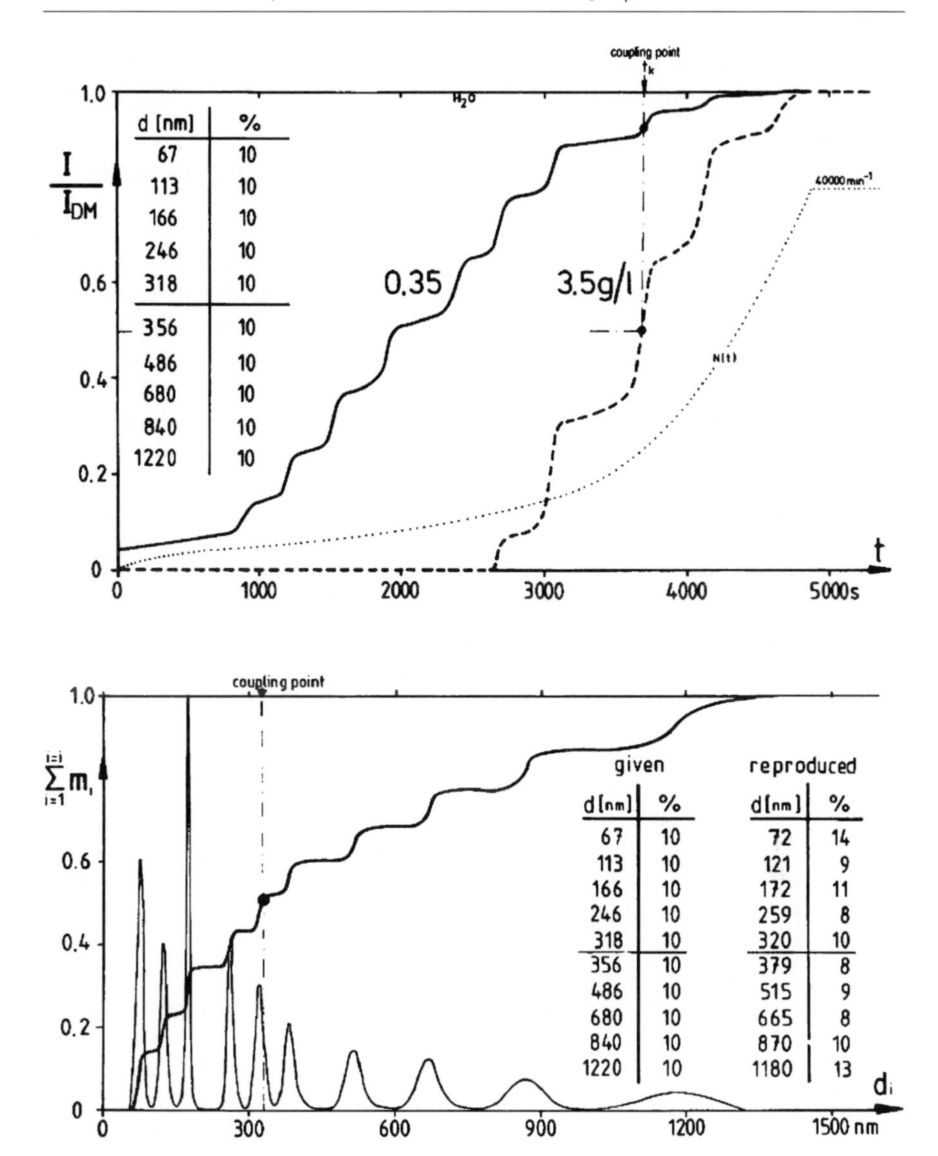

Fig. 3.19. Coupling *PSD* analysis of a mixture of ten narrowly distributed polystyrene dispersions with $67 < d_p < 1220$ nm. *Upper part* two simultaneously measured $I(t)$ curves at different concentrations, $c_I = 0.35$ and $c_{II} = 3.5$ g/l in 3-mm single-sector cells. *Lower part* the resulting integral and differential coupling *PSD*s (reprinted with permission from [27])

for the two neighboring ones at 318 and 356 nm, which are very close to each other. This result was obtained in a standard coupling experiment without further optimization. The example presented clearly shows the high potential of AUC as an apparatus for the measurement of high-resolution particle size distributions, also in the case of very broad distributions (see also the repetition of this *PSD* experiment with sedimentation field flow fractionation described in Sect. 6.2.1).

3.5.3 H_2O–D_2O Density Variation Method to Measure Particle Densities via Sedimentation Velocity Runs

In order to overcome one disadvantage of the AUC determination of particle size distributions via turbidity detector, namely, that the particle density ϱ_p and the refractive index of the polymer n_p have to be known, an offshoot of the AUC *PSD* measurement via a multiplexer described above has been developed by several authors ([26, 28]): the density variation method, or H_2O/D_2O analysis (HDA). The approach (details can be found in [26]) is to perform sedimentation velocity experiments on the same sample simultaneously at least in two, or better, in three (or even more) solvents (= dispersion media) having different densities ϱ_s in a setup such as that in Fig. 3.17. Generally, the solvents used are D_2O (ϱ_s = 1.1004 g/cm^3), a 1:1 mixture of D_2O and H_2O (ϱ_s = 1.050 g/cm^3), and H_2O (ϱ_s = 0.99804 g/cm^3 at 25.0 °C). The basis for this analysis is the assumption that the particle size d_p as well as particle density ϱ_p are identical in all three media. This is surely true for compact particles of hydrophobic polymers such as polymer latices. Figure 3.20 shows three $I(t)$ curves of such an H_2O/D_2O analysis experiment,

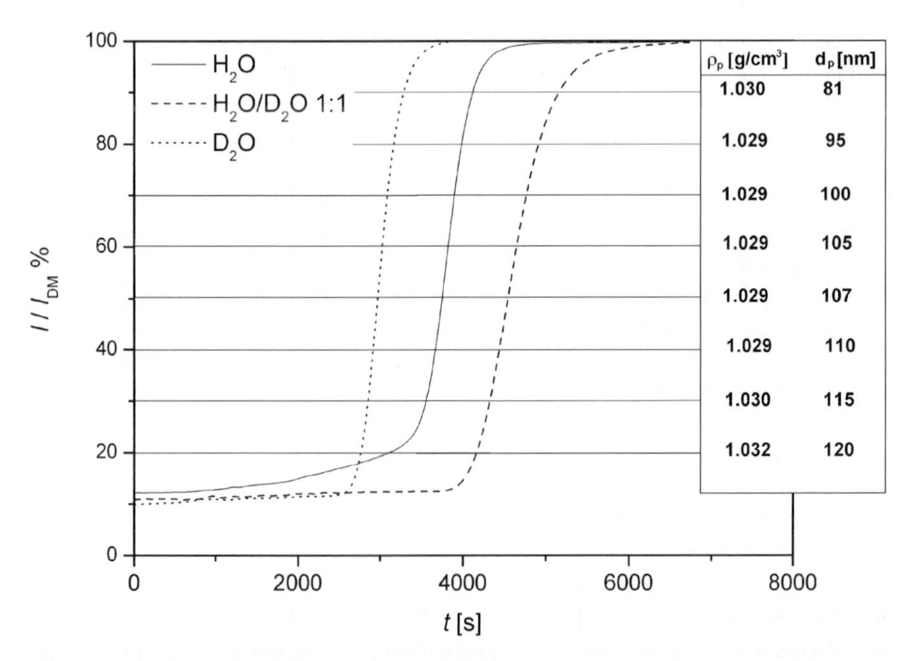

ρ_p [g/cm^3]	d_p [nm]
1.030	81
1.029	95
1.029	100
1.029	105
1.029	107
1.029	110
1.030	115
1.032	120

Fig. 3.20. Three simultaneously measured $I(t)$ curves, in H_2O, in an H_2O/D_2O 1:1 mixture, and in D_2O, of an H_2O/D_2O analysis experiment (HDA) on the BASF-made acrylic polymer dispersion of Fig. 3.13, in 3-mm single-sector cells at c_0 = 0.49 g/l. Analysis of the running times in the three different media reveals the *PSD* as well as the particle density distribution, given in this figure in the form of a table

measured simultaneously and performed on the same sample as that subjected to a standard *PSD* experiment in Sect. 3.5.1 (see Figs. 3.13 and 3.16). Again, the sample concentration in all three media is $c_0 = 0.49\,\text{g/l}$.

The three $I(t)$ curves reveal some useful information at first sight without any calculation, only via a simple logical consideration. The three $I_{50\%}$ running times $t_{50\%}$ of the particles (i.e., the time the particles need to move from either the meniscus (for sedimenting particles) or from the bottom (for floating particles) to the measuring slit in the middle of the cell) differ, depending on the media. In the 1:1 mixture of H_2O/D_2O, in which particles are floating (as can be recognized due to the thickening effect at the beginning of the $I(t)$ curve), the particles move at lowest velocity ($t_{50\%} = 4500\,\text{s}$). This is followed by the particles in H_2O, which sediment with a medium velocity ($t_{50\%} = 3800\,\text{s}$), as can be recognized by the thinning effect at the beginning of the $I(t)$ curve. In D_2O (where the particles float again), the fastest movement is observed ($t_{50\%} = 1900\,\text{s}$). These findings yield information on the particle density ϱ_p upon rearrangement of (1.9) to (3.28):

$$s = \frac{d_p^2 \cdot (\varrho_p - \varrho_s)}{18 \cdot \eta_s} = \left(\ln r_{\text{slit}}/r_{\text{m,b}}\right) \omega^2 t \qquad (3.28)$$

Beside the parameter η_s and the particle diameter d_p (which does not differ in the different media), the sedimentation velocity (i.e., s as well) depends solely on the density difference between particle and solvent ($\varrho_p - \varrho_s$), according to (3.28). The running times $t_{50\%}$ of the particles in the different media, which are inversely proportional to the different s values, are therefore a direct measure of this density difference. Thus, the analysis of the three different running times in Fig. 3.20 yields the following results: (i) the density difference between the particles and the 1:1 H_2O/D_2O mixture is the lowest, because it shows the longest running time (4500 s, which means the lowest velocity); (ii) the density difference between the particles and H_2O is medium; and (iii) the largest density difference is found in the experiment performed in D_2O (1900 s, which means the highest velocity). This shows, without further calculation, that the density ϱ_p of the particles must be found between the density of H_2O, $1.00\,\text{g/cm}^3$, and that of the 1:1 H_2O/D_2O mixture, $1.05\,\text{g/cm}^3$. Moreover, particle density is closer to the value of the H_2O/D_2O mixture. Thus, this simple, rough estimation shows that the HDA particle density value is $1.025 < \varrho_p < 1.035\,\text{g/cm}^3$.

Of course, it is helpful to recognize whether the sample has sedimented or floated in the three media. Usually, this can be judged by the shape of the $I(t)$ curve at the beginning, showing a thinning or a thickening effect.

Additionally, like in all AUC experiments, it is recommended to thoroughly check the samples after the run (cf. are the particles gathered at the cell bottom or at the meniscus?). This helps to identify problems that may have occurred, but also allows one to decide whether sedimentation or flotation (or both, in the case of a heterogeneous sample!) has occurred during the run in general.

Certainly, an HDA experiment is certainly not analyzed in the simple manner described above. This consideration was presented only in order to show how much information even the untreated raw data reveals. For a quantitative mathematical analysis of the three $I(t)$ curves of Fig. 3.20, several horizontal lines are drawn into the $I(t)$ plots (in practice, every 2% with respect to the intensity signal), which correspond to a particle species of defined diameter d_p and defined particle density ϱ_p. In Fig. 3.20, for the sake of clarity, only eight exemplary horizontal lines are shown. By means of these lines, each diameter fraction is correlated to one running time per "solvent". In total, for each horizontal line three running times are collected: t_{H_2O}, t_{D_2O} and $t_{1:1}$.

Equation (3.28) contains two unknown parameters, d_p and ϱ_p. In standard sedimentation experiments, the solvent and its parameters (ϱ_s, η_s and n_s) are well known. The "trick" of the density variation method is to perform the sedimentation velocity experiment in more than one solvent in order to transform the mathematical system of one equation with two unknown parameters into a system of at least two equations with two unknown parameters (see (3.29)). The third dispersion medium and the third equation are needed only to decide automatically, during the computer evaluation, whether the sample is sedimenting or floating in the two other media [26]. The density $\varrho_{p,i}$ of the particle fraction belonging to a horizontal cross section i can be calculated from the viscosities of solvent 1, η_{s1}, and of solvent 2, η_{s2}, the corresponding densities ϱ_{s1} and ϱ_{s2}, and the corresponding runtimes of the particles in both solvents, $t_{1,i}$ and $t_{2,i}$.

$$\varrho_{p,i} = \frac{\eta_{s1} \cdot \varrho_{s1} \cdot t_{1,i} - \eta_{s2} \cdot \varrho_{s2} \cdot t_{2,i}}{\eta_{s1} \cdot t_{1,i} - \eta_{s2} \cdot t_{2,i}} \tag{3.29}$$

The corresponding diameter $d_{p,i}$ is calculated by applying either (3.24) or a modification of this, (3.30), with the measured two sedimentation coefficients of the species i in solvent 1, $s_{1,i}$, and solvent 2, $s_{2,i}$.

$$d_{p,i} = \sqrt{\frac{18 \cdot (\eta_{s2} \cdot s_{1,i} - \eta_{s2} \cdot s_{2,i})}{\varrho_{s1} - \varrho_{s2}}} \tag{3.30}$$

Figure 3.20 shows the densities and diameters obtained by this procedure for the species indicated (= eight representative horizontal lines) in form of a table. These data clearly indicate that the analyzed acrylic polymer dispersion has a constant particle density of $\varrho_p = 1.029 \pm 0.003\,\mathrm{g/cm^3}$ but a moderately broad *PSD* with $70 < d_p < 130\,\mathrm{nm}$. By means of HDA, it is thus possible not only to estimate an *average* ϱ_p but also to measure the complete particle density *distribution*. Additionally, according to the Mie theory (see Fig. 3.15) with the result of a HDA, the four known (average) values, (i.e. diameter d_p, density ϱ_p, wavelength λ, specific turbidity $(\tau/c)_p$) additionally also the refractive index of the HDA-analyzed polymeric particles, n_p, can be derived as well (not presented here; for details, see [26]).

Each of the three HDA $I(t)$ curves in Fig. 3.20 can be transformed into a *PSD*. The resulting three particle size distributions of this HDA experiment on the BASF-made 100-nm acrylic latex particles, using the HDA average value $\varrho_p = 1029\,\mathrm{g/cm^3}$, are plotted in Fig. 3.21. The conformity of the three independent particle size distribution measurements is good. The average values of the three characteristic diameters, $d_{10\%} = 90\,\text{nm}$, $d_{50\%} = 108\,\text{nm}$, $d_{90\%} = 124\,\text{nm}$, and the broadness parameter $(d_{90\%} - d_{10\%})/d_{50\%} = 0.31$ agree well with the corresponding values in Fig. 3.16. This agreement indicates that the HDA particle density used for the evaluation of the *PSD* can be considered to be the correct one – in other words, the HDA is a reliable fast method to measure particle densities ϱ_p with high precision.

In practice, the authors start the analysis of every unknown sample always by an HDA experiment, because this reveals much basic information helping to design the ongoing experiments in a more sophisticated way. The most important basic information is the following: (i) the unknown sample is either homogeneous, or heterogeneous in terms of ϱ_p, and (ii) its *PSD* is either narrow, or broad (perhaps multi-modal).

For the demonstration of the basic features of HDA in Figs. 3.20 and 3.21, a simple, nearly monodisperse latex, also having a homogeneous, uniform density ϱ_p, was examined. However, also inhomogeneous samples showing a particle density distribution can be measured. In these cases, the table contained in Fig. 3.20 represents a particle diameter as well as a particle density distribution. For advanced examples (also for complex samples where the HDA fails), the reader is referred

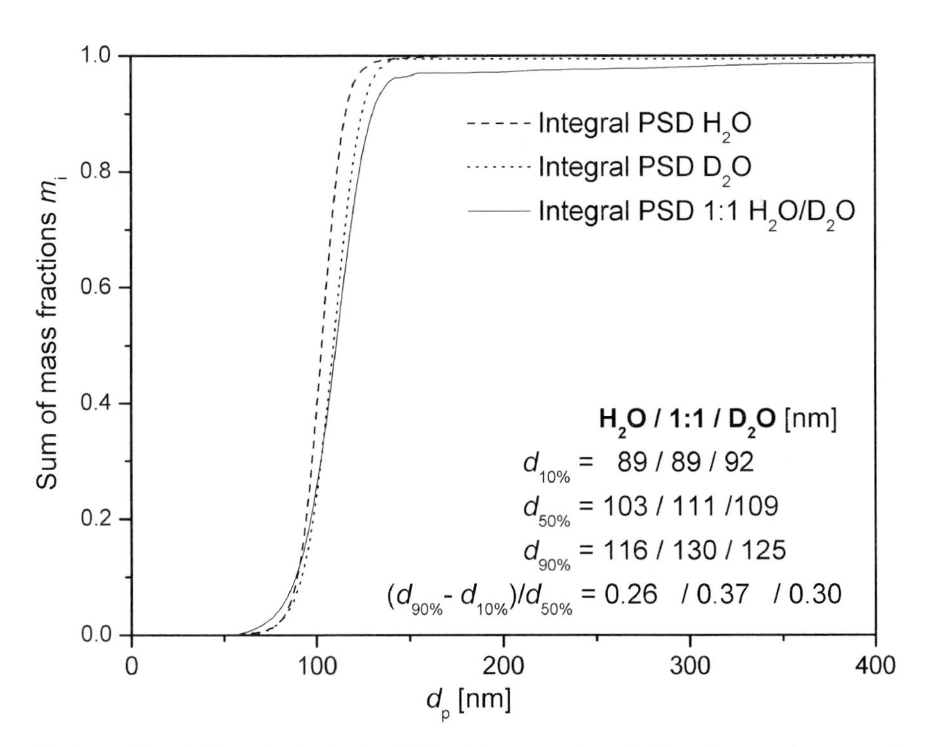

Fig. 3.21. Three independently obtained *PSDs* of the same polymeric dispersion sample measured in the HDA experiment of Fig. 3.20 in three different dispersion media (H_2O, H_2O/D_2O 1:1 mixture, and D_2O)

to [26]. This paper presents HDA samples in the density range $0.85 < \varrho_p < 6\,g/cm^3$, including inorganic particles with a high density and small diameters of only 15 nm. Thus, HDA should be applicable to organic/inorganic hybrid particles with high densities. In principle, according to (3.29), the accessible HDA density range is unlimited, but in practice the errors of measurement of the three run times t_{H_2O}, $t_{1:1}$ and t_{D_2O}, and of the three run distances $(r_{slit} - r_{m,b})$ restrict this range considerably.

3.5.4 *PSD* Measurement of very Small Platinum Clusters Using UV Optics

The measurement of particle size distributions is, of course, not only possible by applying turbidity optics, as described in Sect. 3.5.1. In principle, particle size distributions are available from any sedimentation velocity run, irrespective of which detector has been applied. The diameters are always calculated from Stokes law, (3.24), which requires the knowledge of ϱ_p. Depending on the detector type, the mass portion $m_i = c_i/c_0$ is determined differently, either by the measurement of the refractive index or its gradient (interference or Schlieren optics), or by Lambert-Beer's law with the knowledge of the decadic extinction coefficient ε (UV

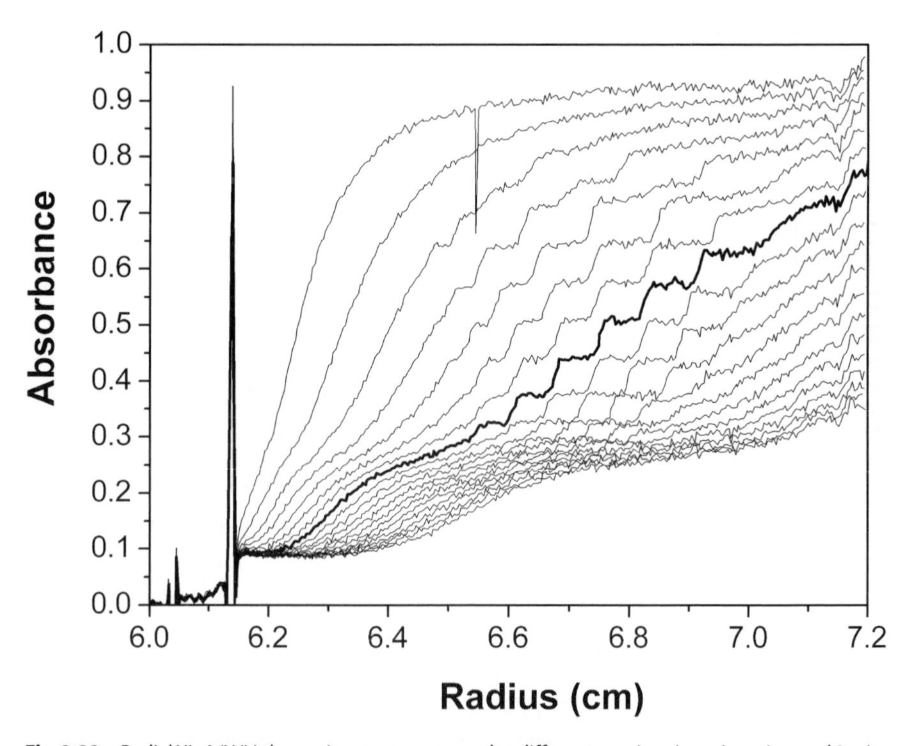

Fig. 3.22. Radial XL-A/I UV absorption scans measured at different running times (scan interval 2 min, ω = 60 000 rpm, λ = 380 nm, 12-mm double-sector cell) of a sedimentation velocity experiment on Pt colloids (reprinted with permission from [29])

absorption optics). In contrast to the turbidity detector, these detector signals do not need any Mie correction. One example for a *PSD* obtained from a sedimentation run applying UV absorption optics is presented in the following [29]. As an example, very small platinum clusters with diameters of only $0.5 < d_p < 2.0$ nm are chosen (see also our introduction sample on gold colloids in Fig. 1.2). This example (presented in Figs. 3.22 and 3.23) demonstrates also the capability of AUC to investigate even smallest nanoparticles.

A platinum cluster system was synthesized by reduction of platinum tetrachloride in water with formic acid in the presence of the stabilizer alkaloid dihydrocin quinoline, as described in [30]. The colloid was dispersed in a mixture of acetic acid and methanol (5:1 by volume), using ultrasonic stirring for 5–10 min. The density of the solvent at 25 °C was determined as $\varrho_s = 1.0182$ g/ml, and its viscosity as $\eta_s = 0.01167$ Poise. The density of the stabilized platinum particles ϱ_p was found to be close to that of bulk platinum, $\varrho_{Pt,25\,°C} = 21.5\,g/cm^3$ [31]. The concentration c_0 was not known precisely, but it was so low that one worked within the linear range of the UV detector ($\lambda = 380$ nm), which means $A < 1.5$ was always valid.

From the different XL-A/I UV scans of Fig. 3.22 at different running times, it becomes clear that one must carefully select the radial scan that is used for the evaluation of the particle size distribution. In the early stages of the experiment, the complete fractionation of the different species in the mixture is not yet achieved, whereas in the late stage, bigger particles may already have reached the cell bottom,

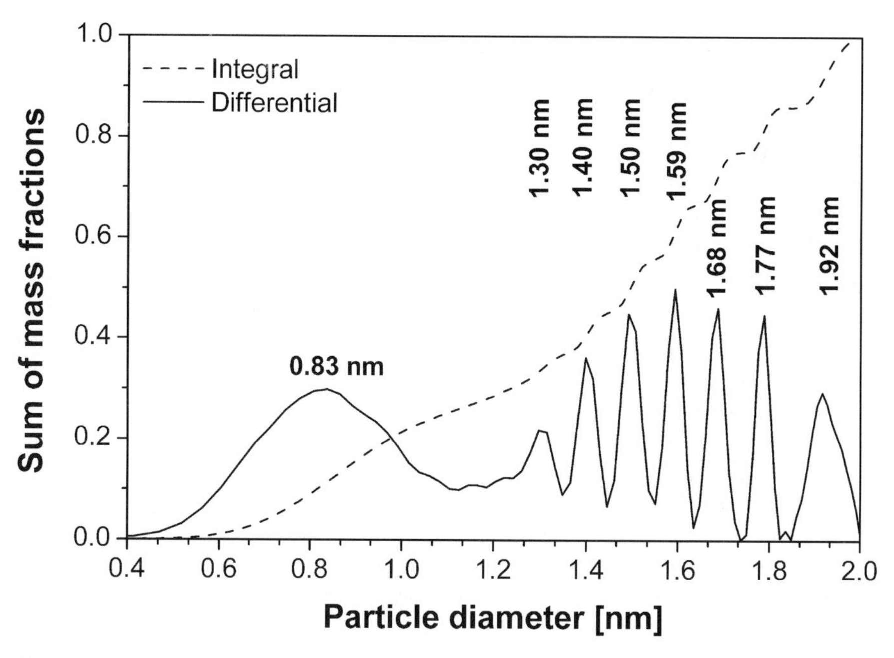

Fig. 3.23. Integral and differential particle size distribution of the Pt colloid in Fig. 3.22. The *thickly marked* scan in Fig. 3.22 (at running time 18 min) was used for the *PSD* evaluation (reprinted with permission from [29])

and would therefore not be detected. As a suitable radial scan, we chose a scan at 18 min from the middle of the run, indicated in Fig. 3.22 by means of a thick line. The corresponding particle size distribution, both in integral and in differential form, is presented in Fig. 3.23.

This multi-modal (!) particle size distribution is characterized by a distinct number of monodisperse species. The evaluation and comparison of the particle size distributions for further radial scans of Fig. 3.22, taken of the same Pt colloid at different running times t, yielded (within the errors of measurement) the same particle size distribution. This means that there is only a weak diffusion broadening of most of the components within this particle size distribution (or perhaps a self-sharpening of the different boundaries; see Sect. 3.3.2).

Upon analysis of the same Pt colloid by electron microscopy (EM), a continuous, only *bi*modal (apparent) particle size distribution is detected, with average particle sizes of 1.0 and 2.2 nm for the two main components. Nevertheless, this rough agreement is a proof that the absolute AUC d_p values in Fig. 3.23 are in the correct order. The striking feature in Fig. 3.23 is the presence of seven different, resolved monodisperse species that differ only by 0.1 nm in their particle diameter. As such Pt clusters are very small, the broad peak around 0.83 nm (component number 8) must be considered with some caution, because it does not reflect the true particle size distribution, due to the expected high diffusion effect for such small clusters.

3.6 Synthetic Boundary Experiments

Synthetic boundary experiments in the AUC have been performed since the 1950s. They are a special form of sedimentation velocity experiments. Earlier work on this topic has been published by Kegeles [32] and Pickels [33].

The measuring cells applied in synthetic boundary cells have already been described in Sect. 2.3 and Fig. 2.7. Whereas most AUC laboratories use synthetic double-sector boundary cells of the *capillary* type (available for the XL-A/I), in the authors' laboratory synthetic boundary cells of the *valve* type are applied predominantly: mono-sector valve-type cells (older Model E cells) are used for Schlieren optics (see Fig. 2.7e), whereas double-sector valve-type cells (homemade at BASF) are preferred for interference and UV optics. The essential part of each kind of synthetic boundary cell is the corresponding centerpiece.

Since this versatile and useful double-sector *valve*-type centerpiece is at present not available from the manufacturer (Beckman-Coulter), our own homemade centerpiece is published as a sketch in Fig. 3.24.

This sketch may be used by those interested to have produced a replica in a good mechanical workshop. Figure 3.24 shows a workshop drawing and a photograph. In principle, our centerpiece is only a modification of the mono-sector valve-type centerpiece shown in Fig. 2.7e, as used in the older Beckman Model E. Nevertheless, there are three differences between the BASF-made and the older Beckman valve-type centerpieces: (i) our centerpieces, in particular the double-sector centerpiece,

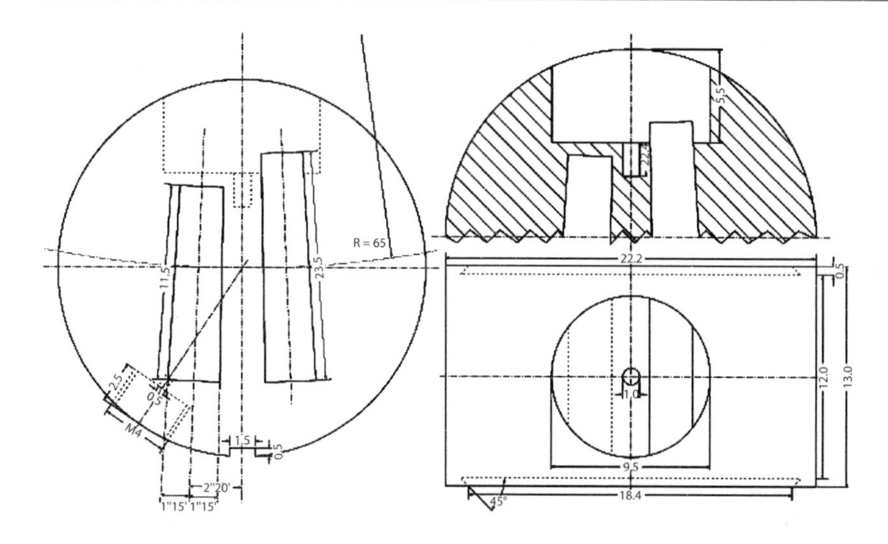

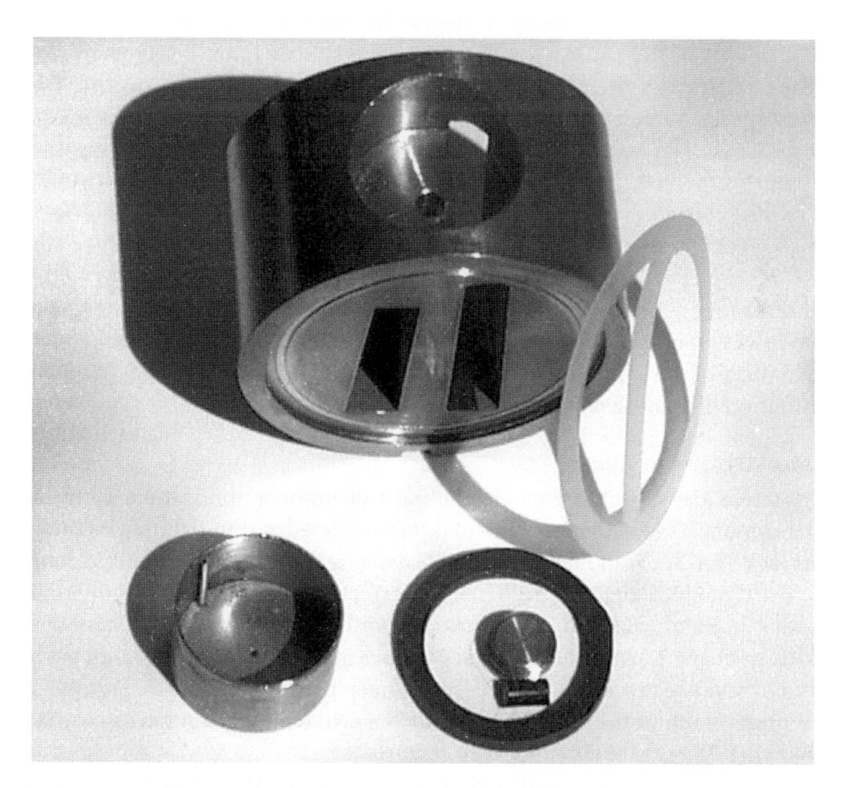

Fig. 3.24. Workshop drawing and photograph of a homemade (BASF) 12-mm double-sector valve-type titanium centerpiece for synthetic boundary runs, required for interference and UV optics. The photograph shows also the storage bin (reservoir) with the thin ventilation pipe, the filling hole screw, the cylinder-shaped rubber valve, and two gaskets

the storage bin with its small ventilation pipe, and the tightening screw, are all made of *titanium*. It is therefore possible to use these for all kinds of solvents, also aggressive ones; (ii) the filling hole, and the tightening screw of the solvent sector are not on the meniscus side, as usual, but rather on the bottom side; and (iii) as rubber valve, we use a cylinder-shaped piece (diameter 1.2 mm), cut out from chemical-resistant Kalrez O-rings for organic solvents, and from Viton for aqueous systems. By varying the length of this rubber cylinder, it is possible to vary the rotor speed ω_{overlay}, at which overlaying starts.

In the following, only experiments performed in cells of the valve type or equivalent cells are described. These valve-type cells (like the capillary-type cells) regulate the overlaying of a liquid from a reservoir (storage bin) onto the liquid column in the sector-shaped chamber of the measuring cell. This results in a sharp boundary between the two liquids, which is additionally stabilized by the centrifugal field. It has been shown that is advantageous if the liquid to be overlaid by the liquid from the reservoir exhibits a slightly higher density (in the magnitude of $\Delta\rho = 10^{-4}\,\text{g/cm}^3$, [34]), in order to avoid turbulences. The sedimentation experiment itself is usually performed in an angular velocity range of $10\,000 < \omega < 40\,000\,\text{rpm}$, depending on the actual analytical problem. The design of the capillaries, as well as the rubber functioning as a valve, allow the user to vary the rotor speed ω_{overlay} at which overlaying occurs. Typical rotor speeds to start overlaying are in the range 5000–10 000 rpm for the valve-type cells. The valve-type cell facilitates a more defined overlaying than the capillary-type cell, by varying the hardness (crosslinking) or the cylinder length of the rubber valve, especially for organic solvents with their low interfacial tensions (these interfacial tensions create serious problems with capillary-type cells).

There are four main applications of synthetic boundary experiments, described in the following four subsections, these being (i) dynamical density gradients, (ii) determination of very small s values, (iii) determination of D values, and (iv) determination of loading concentrations c_0 and $(\mathrm{d}n/\mathrm{d}c)_\mathrm{p}$.

Dynamical Density Gradients

Fast dynamical density gradients are the most prominent application in the authors' laboratory. This is another kind of fast (completed within 10 min!) H_2O/D_2O analysis (see Sect. 3.5.3) to measure particle densities ρ_p, and their possible distribution. The dynamical density gradient is described in this subsection (rather than the density gradient; cf. Chap. 4) because it vividly illustrates how the valve-type cell works to create a synthetic boundary. Figure 3.25 shows that Schlieren optics and simple mono-sector valve-type cells have been used in most cases. Figure 2.7e shows a photograph of the centerpiece of such a cell, with the storage bin, rubber valve and ring-like gasket outside of the centerpiece.

A gravity valve (a compressed rubber cylinder) under the small hole of a pot-like storage bin (reservoir), filled with water (or other media), opens at a rotor speed of about 10 000 rpm, and subsequently all light H_2O inside this storage bin will be overlaid onto the heavier D_2O inside the cell sector. Because of the

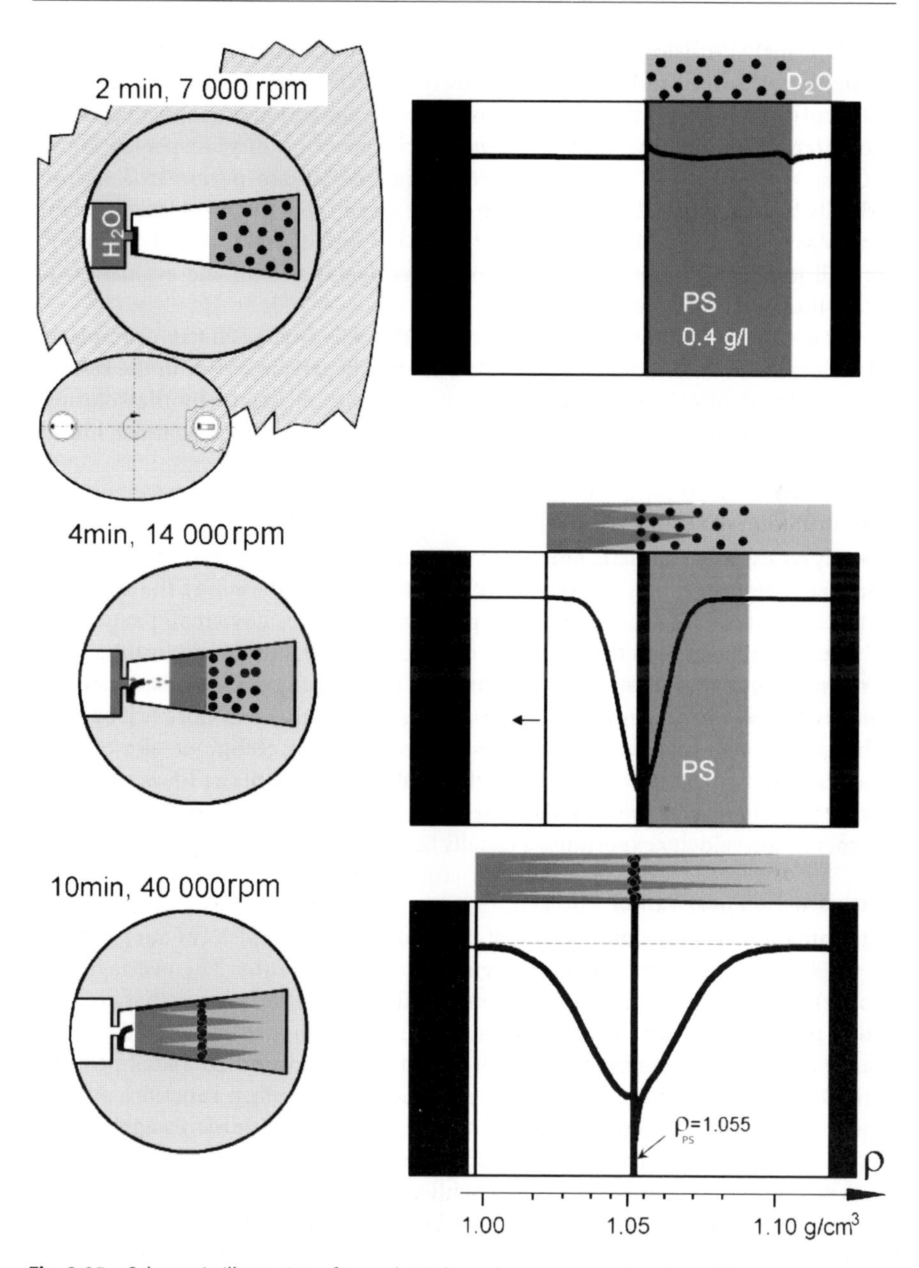

Fig. 3.25. Schematic illustration of a synthetic boundary experiment, using a 12-mm mono-sector valve-type cell/centerpiece (*left*). As an example, real Schlieren photographs of a dynamic H_2O/D_2O density gradient run of polystyrene latex particles, $c = 0.4$ g/l, are shown (*right*)

H_2O/D_2O inter-diffusion (visible as a negative Gaussian Schlieren peak in Fig. 3.25), within these 10 min a radial dynamic density gradient $\varrho_s(r)$, from $\varrho_s = 1.00$ up to $1.10\,g/cm^3$, is built up within the cell (visualized as a ϱ axis in the lowest Schlieren photograph). Polystyrene latex particles (which show turbidity) dispersed in this D_2O ($c_0 = 0.4\,g/l, d_p \approx 100\,nm$) gather within this 10-min period in a narrow turbidity band at that radius position where the densities of the particles ϱ_p and the density gradient are identical. The experiment in Fig. 3.25 yields $\varrho_p = 1.055\,g/cm^3$, the well-known value of polystyrene particles. For details of the evaluation to obtain these ϱ_p values, see Sect. 4.3.

For a standard synthetic boundary run with a valve-type cell (for example, to detect low molar mass solute in a solution or in a dispersion), H_2O in the storage bin is replaced by the pure solvent, and D_2O in the cell sector by the solution. Usually, the solute concentrations are in the range 0.3 – 10 g/l, but also much higher concentrations reaching 100 g/l are possible (see Fig. 3.26).

Determination of very Small *s* Values

In many cases, only synthetic boundary runs allow us to determine *s* values of very small particles/molecules by creation of an artificial boundary over the air/liquid meniscus. This facilitates avoiding the problem that especially small particles or polymers tend to stick to the air/liquid meniscus due to interfacial tension. Thus, in many sedimentation velocity experiments performed on slowly sedimenting samples (mostly because of low molar mass), the meniscus is not cleared even at highest centrifugal rates. In such cases, the overlaying technique allows us to clear the meniscus, and to measure sedimentation coefficients as low as $s = 0.2\,S$ (e.g., for saccharose having a molar mass of $M = 300\,g/mol$; [5, 34]). Conventional sedimentation velocity experiments usually have a lower limit of approximately 1 S.

Determination of *D* Values

Determination of diffusion coefficients D of dissolved samples (M range $100 < M < 100\,000\,g/mol$) can be done by synthetic boundary runs. The overlaying of (slowly) sedimenting molecules/macromolecules in solutions (or in dispersions) with a pure solvent (or a dispersion medium) results in a steep, step-like radial change of the sample concentration $c(r)$ within the cell around the radial position of the synthetic boundary $r_{overlay}$ (see Fig. 3.26). The steep step function, directly visible via interference optics (see Fig. 3.26a), broadens with increasing experimental time due to diffusion of the sample molecules into the pure solvent. Via Schlieren optics, a narrow (Gaussian) Schlieren peak is visible at this boundary (see Fig. 3.26b). Measurement of the broadening of the interference fringes or of the Schlieren peak as a function of running time t can directly be correlated to the diffusion coefficient D of the sample (see, for example, [35]). The results are of high accuracy if the sample is monodisperse. In fact, this type of measurement has been state of the art in diffusion coefficient determination until laser technology and dynamic light scattering (DLS) were introduced. Still today, this is a method that can be applied favorably for complex systems, or if only a very low amount of sample is available.

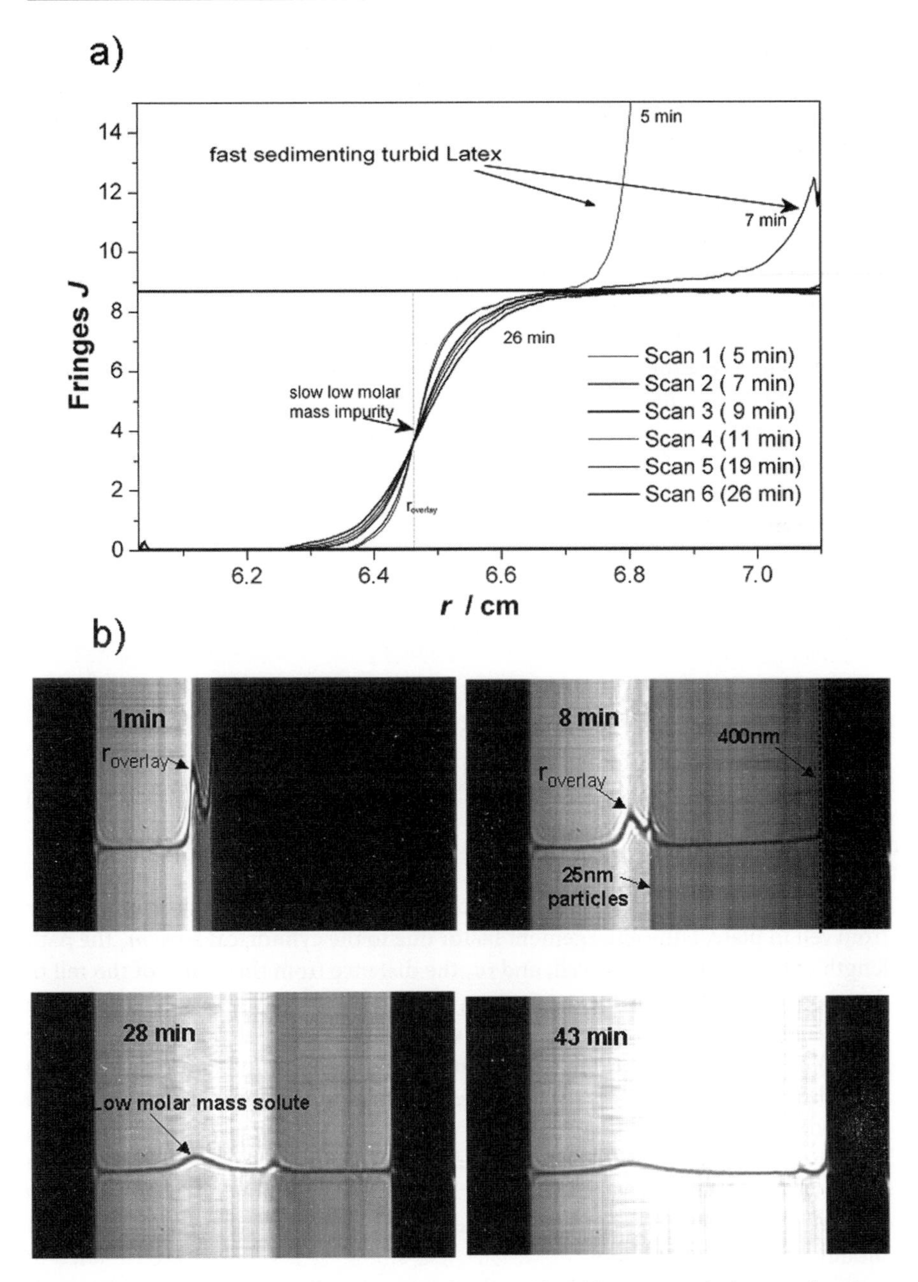

Fig. 3.26. **a** Radial XL-A/I interference optics scans, and **b** Schlieren optical photographs of two synthetic boundary experiments, taken at different experimental running times. In both cases, (turbid) latices (c_0 = 100 g/l) with additional low molar mass solute (about 3 g/l) in the aqueous serum were measured. The diffusion broadening of the boundary due to the (non-sedimenting) low molar mass solute can be well recognized at the radial overlay position $r_{overlay}$

Determination of Loading Concentrations c_0 and $(dn/dc)_p$

Synthetic boundary runs are also used to measure loading concentrations c_0 and $(dn/dc)_p$. At the radial overlaying position $r_{overlay}$ (see Fig. 3.26), the function of the refractive index over the radius in the cell $n(r)$ shows a steep step due to the corresponding increase of the solute concentration $c(r)$. This measurable change of refractive index $\Delta n = (n_{solution} - n_{solvent})$ is detected either as a step ΔJ (interference optics, Fig. 3.26a) or as a peak area A_{schl} (Schlieren optics, Fig. 3.26b). The detected signal Δn can either be utilized for high-accuracy measurement of the solute concentration c_p if the specific refractive index increment $(dn/dc)_p$ is known, or vice versa, the signal can be used to measure the specific refractive index increment if the loading concentration is known. For the case of interference optical detection of the synthetic boundary experiment (see (3.31)), a modification of (2.3) gives the correlation between the fringe displacement ΔJ, the light wavelength λ used, the specific refractive index increment $(dn/dc)_p$, the length of the optical path through the cell a, and the concentration difference of the solute before and after the concentration step Δc_p (usually $\Delta c_p = c_0$):

$$\Delta c_p = \frac{\Delta J \cdot \lambda}{a \cdot \left(\frac{dn}{dc}\right)_p} \tag{3.31}$$

For the case of Schlieren optical detection (see Fig. 3.26b), the corresponding relation between Δc_p and the Schlieren peak area A_{schl} is given as

$$\Delta c_p = \frac{A_{schl} \cdot \tan \Theta}{L \cdot a \cdot m_x \cdot m_y \left(\frac{dn}{dc}\right)_p \cdot E^2} \tag{3.32}$$

In (3.32), Θ is the Philpot angle of the Schlieren optics, E the magnification factor from cell to plate, L the enlargement factor due to the cylindrical lens, m_x the path length of the solution in the cell, and m_y the distance from the center of the cell to the plane of the phase plate.

Such synthetics boundary runs to determine c_p (especially with interference optics) are also important for sedimentation equilibrium runs to measure M (see Sect. 5.2). They allow one to check whether the law of conservation of mass is fulfilled.

Beside the above mentioned four main applications there are other applications of the synthetic boundary technique, such as the measurement of differential sedimentation coefficients [36], and the determination of extinction coefficients [37] should simply be noted here.

Figure 3.26 shows two typical synthetic boundary experiments to detect small amounts of low molar mass solutes in the serum of aqueous particle dispersions, (i) with a double-sector valve-type cell, using the XL-A/I interference optics, and (ii) with a mono-sector valve-type cell, using the XL-SO Schlieren optics. In both examples, much higher total concentrations of $c_0 = 100$ g/l are chosen to detect, beside the major component of the (fast) turbid particles, with diameters of about

$d_p = 250$ and 400 nm, respectively, also any minor components of (slow) low molar mass solutes. In both cases, pure water was overlaid onto the aqueous polymer dispersions.

In the interference optics scans of Fig. 3.26a, we see the large ΔJ step of the very fast, 250-nm particles only in the first two scans. Later, we see only the smaller step ($\Delta J = 8.7$), and the diffusion broadening of the slow ($s = 0-0.2$ S) low molar mass solute (abbreviated "lmms") at the overlay position $r_{overlay}$. If we assume $(dn/dc)_{lmms} = 0.15\,cm^3/g$, a concentration of $c_{lmms} = 3.3$ g/l follows with (3.31). This is about 3.3 wt% of the total solid content of the analyzed dispersion.

In the four Schlieren optical photographs of Fig. 3.26b at different running times, the fast turbidity front of the 400-nm particles is visible only in the first Schlieren photograph (taken after 1 min). Surprisingly, however, we see (for the first time in the second photograph, taken after 8 min) a second, more slowly sedimenting, weak turbidity front (correlated with a small Schlieren peak). This was an unexpected 2-wt% component of small, 25-nm latex particles. Additionally, we found in all Schlieren photographs, permanently at the $r_{overlay}$ position, the small Schlieren peak of the low molar mass solute ($s = 0-0.2$ S). From its Schlieren peak area $A_{schl,lmms}$ follows, with (3.32), a concentration of $c_{lmms} = 3.0$ g/l. This corresponds to about 3 wt% of the total solids content of this dispersion. The diffusion broadening of this low molar mass solute Schlieren peak is clearly seen, too.

In the following Sect. 3.6.1, we would like to present one unusual application of a synthetic boundary experiment, namely, synthetic boundary crystallization. Despite the fact that this experiment is not of common use and pronounced interest, it should inspire creativity and show that new types of AUC experiments are still conceivable.

3.6.1 Synthetic Boundary Crystallization Ultracentrifugation

The basic idea of this method is to make use of synthetic boundary cells to perform chemical surface reactions inside the AUC while the centrifugal field acts. From the field of biological systems, two interesting approaches of (chemical) reactions within a synthetic boundary cell should be mentioned:

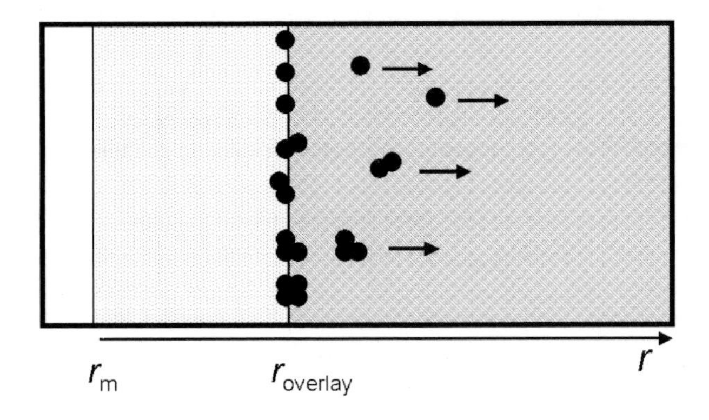

Fig. 3.27. Schematic representation of the synthetic boundary crystallization process inside the cell of an analytical ultracentrifuge

(i) the active enzyme centrifugation, where the chemical reaction between an enzyme and its substrate is investigated [38], and
(ii) the investigation of the formation of a polyelectrolyte membrane [39].

The synthetic boundary crystallization method was developed to investigate the early stages of crystallization. So far, it has been applied only to a system of low solubility, such as cadmium sulfide [40]. In this example, a 10 mM sodium sulfide solution was overlaid onto a 5 mM cadmium chloride solution, which additionally contained 0.5 mM stabilizing thiols. At the sharp (synthetic) boundary between the two solutions, the nucleation of the hardly soluble cadmium sulfide particles takes place (see scheme in Fig. 3.27). For detection, the XL-A/I UV scanner ($\lambda = 370$ nm) was used. The experiments were performed in 12-mm synthetic boundary double-sector cells of the Vinograd type, which is a capillary type with a small reservoir for the liquid used for overlaying (similar to the mono-sector capillary-type cell in Fig. 2.7d).

Upon their formation, the very small CdS particles either continue to grow in the boundary region, or sediment toward the bottom of the cell into the cadmium chloride solution. This movement quenches any further growth, and allows us therefore to investigate the particles formed early in the process, as well as their particle size distribution.

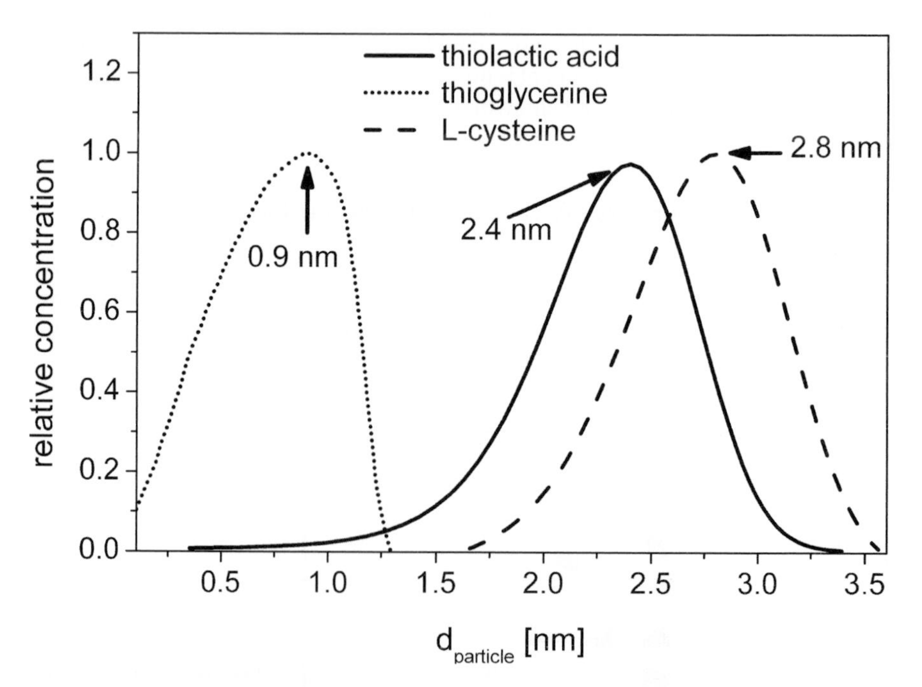

Fig. 3.28. Comparison of (apparent) particle size distributions obtained from synthetic boundary crystallization experiments from three differently stabilized CdS nanoparticle systems via XL-A/I UV scans at $\lambda = 370$ nm (reprinted with permission from [40])

The efficiency of the stabilizers in such experiments has been successfully studied by Börger and Cölfen [41]. The particle sizes obtained in these experiments are in the magnitude of 1 – 3 nm (see Fig. 3.28), using $\varrho_P = 3.2\,\mathrm{g/cm^3}$ for the density of the CdS particles. Figure 3.28 shows also that, of the three investigated stabilizers, thioglycerine is the most effective because it yields the smallest CdS particles, having an (average) diameter of only 0.9 nm.

References

1. Stafford WF (1992) In: Harding SE, Rowe AJ, Horton JC (eds) Analytical ultracentrifugation in biochemistry and polymer science. The Royal Society of Chemistry, Cambridge, p. 358
2. Kehrhahn JH, Lechner MD, Mächtle W (1993) Polymer 34:2447
3. Stafford WF (1994) In: Schuster TM, Laue TM (eds) Modern analytical ultracentrifugation. Birkhäuser, Berlin, p. 119
4. Fujita H (1975) Foundations of ultracentrifugal analysis. Wiley, London
5. Schachman HK (1959) Ultracentrifugation in biochemistry. Academic Press, New York
6. Gofman JW, Lindgren FT, Elliott H (1949) J Biol Chem 179:973
7. Mächtle W (1984) Colloid Polym Sci 262:270
8. Börger L, Kühnle A (2003) Unpublished data
9. Creeth JM, Knight CG (1965) Biochim Biophys Acta 102:549
10. Johnston JP, Ogston AG (1946) Trans Faraday Soc 42:789
11. Schachman HK (1951) J Am Chem Soc 73:4453
12. Gilbert GA (1960) Nature 186:882
13. Bowen TJ, Rowe AJ (1970) An introduction to ultracentrifugation. Wiley, London, p. 107
14. Lechner MD, Mächtle W (1995) Prog Colloid Polym Sci 99:120
15. Lechner MD, Mächtle W (1999) Prog Colloid Polym Sci 113:37
16. van Holde KE, Weischet WO (1978) Biopolymers 17:1387
17. Yphantis DA (1984) Biophys J 45:324
18. Schuck P (2000) Biophys J 78:1606
19. Schuck P, Rossmanith P (2000) Biopolymers 54:328
20. Schuck P, Perugini MA, Gonzales NR, Howlett GJ, Schubert D (2002) Biophys J 82:1096
21. Mie G (1908) Ann Phys 25:377
22. Mächtle W (1988) Angew Makromol Chem 162:35
23. Heller W (1957) J Phys Chem 26:498
24. Scholtan W, Lange H (1972) Kolloid Z Z Polym 250:782
25. Mächtle W (1999) Biophys J 76:1080
26. Mächtle W (1984) Makromol Chem 185:1025
27. Mächtle W (1992) Makromol Chem Macromol Symp 61:131
28. Müller HG, Herrmann F (1995) Prog Colloid Polym Sci 99:114
29. Cölfen H, Pauck T (1997) Colloid Polym Sci 275:175
30. Bönnemann H, Braun GA (1996) Angew Chem Int Ed 35:1992
31. Lide DR (ed) (2002) CRC Handbook of chemistry and physics, 83rd edn. CRC Press, Boca Raton
32. Kegeles G (1952) J Am Chem Soc 74:5532
33. Pickels EG (1952) Methods Med Res 5:107
34. Schachman HK, Harrington WF (1954) J Polym Sci 12:379
35. Aminabhavi TM, Munk P (1979) Macromolecules 12:1194
36. Hersh R, Schachman HK (1955) J Am Chem Soc 77:5228
37. Voelker P (1995) Prog Colloid Polym Sci 99:162
38. Kemper DL, Everse J (1973) In: Hirs CHW, Timasheff SN (eds) Active Enzyme Centrifugation and Methods in Enzymology, vol XXVII. Enzyme structure, part D. Academic Press, New York, p. 67
39. Wandrey C, Bartowiak A (2001) Colloid Surf A Physico Chem Eng Asp 180:141
40. Börger L, Cölfen H, Antonietti M (2000) Colloids Surf A 163:29

41. Börger L, Cölfen H (1999) Prog Colloid Polym Sci 113:23
42. Cölfen H (2004) ACS Symp Series 881:119
43. Harding SE (1995) Carbohydrate Polym 28:227
44. Oth J, Desreux V (1954) Bull Chim Belges 63:133
45. Schilling K (1999) PhD Thesis, Potsdam University, Potsdam

4 Density Gradients

The two major types of basic AUC experiments are the sedimentation velocity experiment (subject of Chap. 3), and the sedimentation equilibrium experiment (subject of Chap. 5). The *density gradient* experiment, topic of this Chap. 4, can be understood as a special kind of the sedimentation equilibrium experiment. Figure 4.1 illustrates the transition from Chap. 3 to Chap. 4, which means from the time-dependent sedimentation velocity runs to the time-*in*dependent *static* (equilibrium) density gradient runs.

In Fig. 4.1a, this transition is experimentally done for dissolved macromolecules by simply adding 25 wt% of a heavy co-solvent, such as di-iodomethane (with a density of $\varrho_{\mathrm{DIM}} = 3.2\,\mathrm{g/cm^3}$), to the light solvent of the s run, tetrahydrofuran (with a density of $\varrho_{\mathrm{THF}} = 0.8811\,\mathrm{g/cm^3}$ at 25 °C). The density gradient experiment then takes considerable time, about 17 h in this example, until a radial exponential density gradient $\varrho(r)$ is built up within the cell. The density range covered in this case reaches from 1.0 to 1.2 g/cm^3. The gradient is built up due to enrichment of the heavy co-solvent near the cell bottom. Simultaneously within this 17 h, the polystyrene molecules ($M = 1\,800\,000\,\mathrm{g/mol}$, $c = 1\,\mathrm{g/l}$) sediment (positive Schlieren peak in Fig. 4.1a) or float (negative Schlieren peak in Fig. 4.1a) to that radius position where the densities of the gradient and the particles/macromolecules are identical. This radial position is called the *isopycnic* position r_{iso}.

The higher the molar mass M of the sample examined, the narrower becomes the detected final (constant) *double* Schlieren peak. The reason is diffusion broadening around this isopycnic position.

In Fig. 4.1b, this transition from sedimentation run to equilibrium run is done for dispersed nanoparticles by simply adding 11 wt% of a heavy iodinated sugar, metrizamide ($\varrho = 2.155\,\mathrm{g/cm^3}$ at 25 °C), to the (light) water. Again, within 16 h a radial exponential density gradient, $1.03 < \varrho(r) < 1.10\,\mathrm{g/cm^3}$, is built up within the cell, and the starting broad turbidity band of the dispersed polystyrene particles ($d_{\mathrm{p}} = 160\,\mathrm{nm}$, $c = 1\,\mathrm{g/l}$) is compressed to a narrow turbidity band at a radius position r_{iso} with $\varrho = 1.053\,\mathrm{g/cm^3}$, the known density value of polystyrene particles.

The evaluation of density gradients such as the one in Fig. 4.1, the calculation of the shape and broadness of the double Schlieren peak and of the relation $\varrho(r)$, as well as the introduction of other types of density gradients (such as *dynamic* density gradients) is the topic of this Chap. 4.

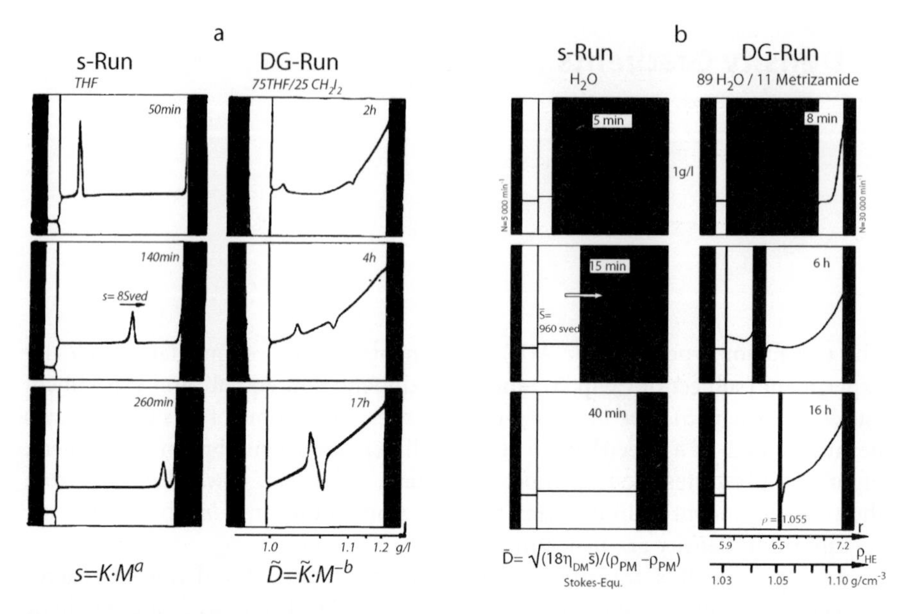

Fig. 4.1. **a** Sedimentation velocity experiment (*s* run) in THF and organic static 75 THF/25 DIM density gradient experiment (DG run) of narrowly distributed dissolved polystyrene molecules (M_w = 1 800 000 g/mol). **b** *s* run in H_2O and aqueous static 89 H_2O/11 metrizamide DG run of polystyrene latex particles (d_p = 160 nm). For all runs: N = 40 000 rpm, 12-mm mono-sector cells, c = 1 g/l, Schlieren optics, Philpot angle = 70° (reprinted with permission from [19])

The chapter is structured as follows: Sect. 4.1 explains the basic principle of density gradients. In Sect. 4.2, the *static* (or sedimentation equilibrium) density gradient, the most frequently applied type of density gradient experiments, will be described, including the theory, data evaluation, experimental procedure, gradient materials, and examples. Section 4.3 deals with the fast *dynamic* density gradient, not commonly used by the AUC community, but of outstanding importance in the AUC laboratory at BASF. The experimental details as well as details of data evaluation are given. In Sect. 4.4 one exotic dynamic, but mandatory type of density gradient experiment, the aqueous dynamic Percoll density gradient, is presented.

4.1 Introduction

In density gradient centrifugation experiments, a radial density gradient $\rho(r)$ is induced in the measuring cell either by a heavy auxiliary additive added to a light solvent or by a mixture of two solvents exhibiting very different densities (see Figs. 4.1 and 4.2). The binary mixture has to consist of agents of different densities that are separated or accumulated differently within the measuring cell. The component with the higher density accumulates more in direction of the cell bottom. The relative composition of the solvent is different at each radial position r of the measuring cell. Thus, the density of the binary mixture varies throughout

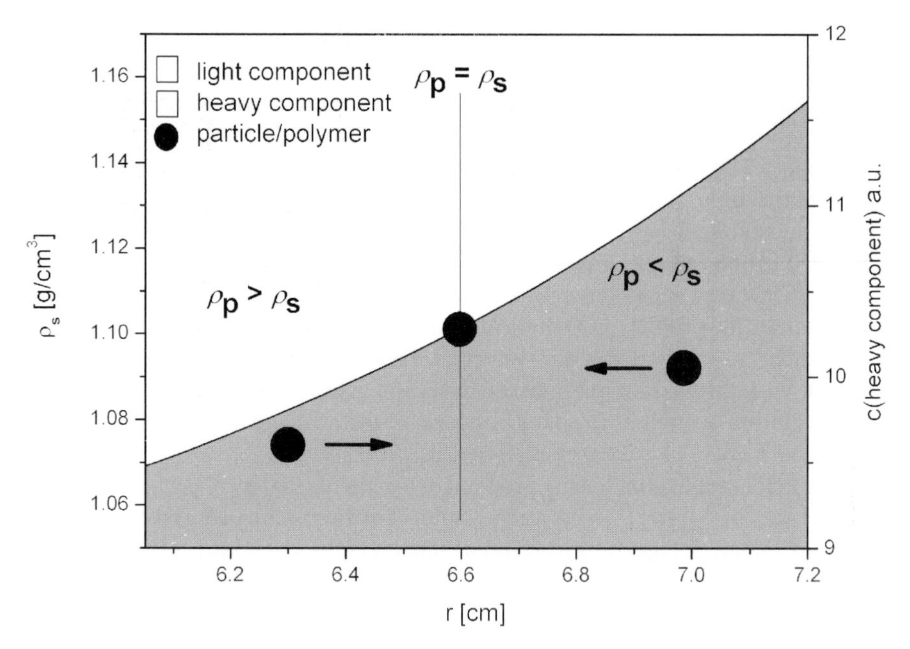

Fig. 4.2. Schematic presentation of a static AUC density gradient. The chemical composition, and therefore the density $\varrho_s(r)$ of the binary solvent DG mixtures varies with the radius r in the measuring cell. Depending on its density in comparison to the surrounding solvent density, the sample sediments or floats. At one radial position in the cell, the solvent density matches the sample's density ϱ_p. The sample accumulates at this position

the cell, exhibiting higher densities at the bottom of the cell and lower densities at the meniscus. In the following section, we will refer to the binary mixture as *solvent*, meaning that we will use the parameter $\varrho_s(r) = \varrho_{DG}(r)$, having in mind that this is arguable.

The sedimentation behavior of a sample (containing macromolecules or nanoparticles) added to such a (binary) density gradient is determined partly by the buoyant term $(1 - \overline{v} \cdot \varrho_s)$, according to (1.5), or by the density difference between particle density and density of the density gradient $(\varrho_p - \varrho_s)$, according to (1.9), where $\varrho_p = 1 / \overline{v}$. If the sample component occupies a radial position in the density gradient exhibiting a solvent density that is lower than its own density, then the particle or polymer will sediment toward the cell bottom. In the opposite case, the component will float toward the meniscus. If particle/macromolecule density ϱ_p and solvent density ϱ_s are equal, no movement will occur (see Figs. 4.1 and 4.2). The sample component has then reached its *isopycnic* position r_{iso} within the cell.

Together with the choice of an appropriate gradient material and proper experimental conditions, the main challenge in isopycnic centrifugation is to build up the relation between the radial position r and the (binary) density gradient mixture density, i.e., $\varrho_s(r)$. Several approaches for this (see [3]) are established. Some of these will be described in Sects. 4.2.1 and 4.3. Appropriate optical detectors locate the radial position r_{iso} of the sample (Schlieren optics is most important, as in

Fig. 4.1, but also all other detectors are possible, if the density gradient medium is transparent).

Cheng and Schachman were the first to perform a kind of density gradient experiment in 1955 [1], where they made use of the high compressibility of fluorotoluene. The high compression at high rotor speed induces a radial density gradient throughout the AUC cell. In a double-sector cell, two droplets were added to this fluorotoluene system: first, one droplet of an aqueous buffer solution, and second, a droplet of the same buffer solution containing additionally a dissolved polymer (having a slightly increased density). As water and the organic solvent are not miscible, the droplets as a whole moved within the gradient to a radial position where their known densities were matched. This allowed an accurate evaluation of the relative density difference between the buffer solution and polymer-containing buffer solution, as well as the determination of the absolute density values. These droplets of known density are called density *markers*.

The first equilibrium density gradient experiments that were performed in the manner sketched above [2] were already subject of discussion in the introduction to this book (Chap. 1, Figs. 1.4 and 1.5). The impact of the H_2O/CsCl density gradient experiment of Meselson and Stahl on double-stranded DNA in scientific history is beyond controversy. Also nowadays, however, density gradients are of great help in analyzing manifold (macromolecular/nanoparticulate) systems. Thus, it is not astonishing that isopycnic centrifugation is well established in biochemical research, with H_2O/CsCl as the standard density gradient. Again, to the authors' surprise, the use of density gradient methods in the field of synthetic polymers and colloids is not common, despite the fact that density gradient experiments are excellent tools for the characterization of dispersed nanoparticles in the diameter range 10 nm–10 μm (see examples in Chap. 6). Also, dissolved synthetic macromolecules can be well subjected to density gradients, taking advantage of the high fractionation resolution of analytical ultracentrifuges. As the densities of particles and macromolecules depend on their chemical composition, the AUC density gradient experiments are beneficial for highly sophisticated analysis of complex polymeric or colloidal systems. Due to this ability, density gradients have also been called "*density spectroscopy*" [3].

This "density spectroscopy" is visualized in Fig. 4.3 (see also Figs. 4.5 and 4.12). The upper part of Fig. 4.3 shows schematically the particle densities of the most frequently used aqueous *homo*polymer dispersions (latices) arranged as turbidity bands in a density gradient along a ϱ axis according to their particle densities ϱ_P. Via *co*polymerization (and reactions on the surfaces of all kinds of particles), thousands of other particles with nearly all densities between them can be created. AUC density gradients allow one to measure these particle densities (and their distribution, too!) very precisely, and with an extremely high resolution of $\Delta\varrho_P = 0.0003\,g/cm^3$ (in special cases). Thus, such reactions can be analyzed via density gradients, and the expression "density spectroscopy" is indeed justified. Chapter 6 presents real application examples.

The lower part of Fig. 4.3 shows the accessible density ranges of some differently composed water/metrizamide density gradients.

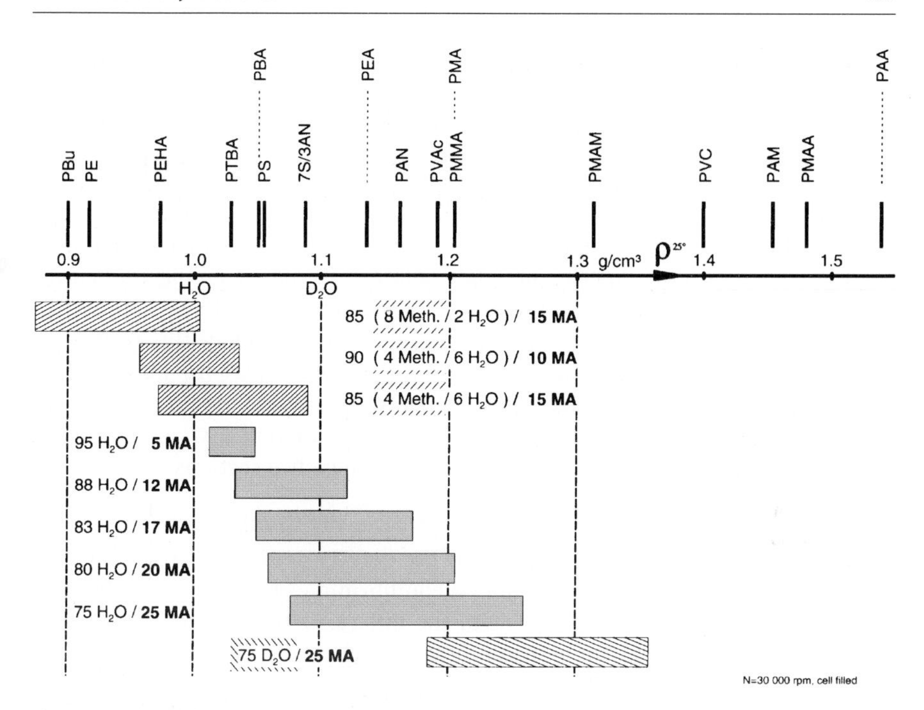

Fig. 4.3. Particle densities of the most common *homo*polymer latices, arranged along a density ϱ axis (*upper part*), and accessible density ranges of nine differently composed static water/metrizamide AUC density gradients (*lower part*). This figure illustrates what is called "AUC density spectroscopy" (reprinted with permission from [3])

Preferential solvation of one of the two density gradient components does not play any role in the analysis of bigger (compact) nanoparticles, i.e., if $d_p > 30\,\text{nm}$ is valid. For this reason, absolute particle densities can be measured with an accuracy of $\Delta\varrho_p = \pm 0.002\,\text{g/cm}^3$. For very small particles, however, and especially for dissolved macromolecules, preferential solvation can play a role and falsify the measured density values. Thus, for example, in Fig. 4.1a, because of preferential solvation of the light THF, the polystyrene density value of the dissolved macromolecules is shifted slightly to an apparently lower value $\varrho = 1.04\,\text{g/cm}^3$, if one compares this with the "true" value $\varrho = 1.055\,\text{g/cm}^3$ of compact polystyrene particles in Fig. 4.1b. Nevertheless, also in the presence of preferential solvation, density gradients are a valuable analytical tool because of their high *fractionation* power according to chemical composition. Again, this is demonstrated in Chap. 6.

4.2 Static Density Gradients

Static equilibrium density gradients are the most frequently used density gradient experiments. In the authors' industrial R&D laboratory, they are tools of daily use as they allow easy and unambiguous analysis of known and unknown complex colloidal and/or polymeric systems.

4.2.1 Theory of Static Density Gradients

The theoretical description of density gradients has to cover the following aspects:

(i) Formation and evaluation of the radial static density gradient $\varrho_s(r)$.
(ii) Radial distribution of the examined sample within the density gradient.

From the literature, several approaches for the determination of particle and/or polymer densities from experimental data obtained in static equilibrium density gradient experiments are known (see [3]). The aim of all of these approaches is to build up a solvent density over radius function $\varrho_s(r)$. Once this relation is known, only the radial (isopycnic) positions r_{iso} of the sample's fractions have to be known to determine their densities. There are two ways to obtain $\varrho_s(r)$: (i) via theoretical calculation, and (ii) via calibration with different markers of known particle density ϱ_p. Both ways are described in the next two sections.

Formation and Calculation of the Radial Static Density Gradient $\varrho_s(r)$
The correct treatment of the formation of the density gradient $\varrho_s(r)$ can be carried out similarly to that of sedimentation equilibrium experiments (see Chap. 5) applying thermodynamic laws, but with the difference that in the case of density gradients the (heavy) equilibrium-forming component is usually present in a very high concentration. Therefore, activities a_c, rather than concentrations c, have to be considered (this is the weak point of all density gradient theories!). For the following derivation, we continue to use concentrations, rather than activities, bearing this restriction in mind. The gradient has formed when equilibrium between sedimentation and diffusion of the binary mixture, forming the density gradient, is reached. Therefore, the net mass flow at each radial position $\mathrm{d}m/\mathrm{d}t$ within the cell is zero. Starting from (1.13), the mass transport equation, this leads to (4.1):

$$\frac{\mathrm{d}m}{\mathrm{d}t} = \phi a r \left[c s \omega^2 r - D \frac{\partial c}{\partial r} \right] = 0 \quad \Rightarrow \quad c s \omega^2 r = D \frac{\partial c}{\partial r} . \tag{4.1}$$

In (4.1), the parameters s, D and c, as well as the parameters $\overline{v}$ and M in the following part of this section, refer to the heavy component (for example, CsCl, di-iodomethane (DIM), metrizamide, or Nycodenz) of the binary density gradient mixture. Besides this, (4.2) holds:

$$\frac{\partial c}{\partial r} = \frac{\partial c}{\partial \varrho_s} \cdot \frac{\partial \varrho_s}{\partial r} \tag{4.2}$$

Replacing the quotient s/D of the sedimentation coefficient s and the diffusion coefficient D in (4.1) by the expression from the Svedberg equation (1.8), and inserting (4.2), the following (4.3) is obtained:

$$\frac{\partial \varrho_s}{\partial r} = \frac{\partial \varrho_s}{\partial \ln c} \left(1 - \overline{v} \cdot \varrho_s \right) \frac{M \omega^2 r}{RT} \tag{4.3}$$

This can be rearranged to give (4.4):

$$\frac{\partial \varrho_s}{\partial r} = \frac{\omega^2 r}{\beta(\varrho_s)} \tag{4.4}$$

where $\beta(\varrho_s)$ is given by (4.5):

$$\beta(\varrho_s) = \frac{\partial \ln c}{\partial \varrho_s} \frac{RT}{(1 - \bar{\upsilon} \cdot \varrho_s)M} \tag{4.5}$$

As the concentration c of the heavy gradient component determines the solvent density, the function $\beta(\varrho_s)$ varies only with the density ϱ_s.

In flat density gradients (i.e., at low c and low rotor speed ω), which exhibit no intense variation of the solvent density, $\beta(\varrho_s)$ can be assumed to be constant over radius. Thus, an empirically evaluated value can replace it (for aqueous CsCl gradients, the value was found to vary between ca. 1.13×10^9 and $1.21 \times 10^9\,\mathrm{cm^5\,g^{-1}\,s^{-2}}$; [4]). The density at any given radial point in the cell can then be calculated by (4.6), which is obtained by integration of (4.4) under the assumption that β is constant:

$$\varrho_s(r) = \varrho_{s,0} + \frac{1}{2\beta}\omega^2(r - r_0)^2 \tag{4.6}$$

For the calculation, a reference point at a radial position r_0 with known density $\varrho_{s,0}$ has to be given. This problem is often overcome by addition of *marker* particles with known densities to the density gradient. The detection of the radial position of these makers delivers r_0, and allows us to calculate the density at any other radial position within the measuring cell.

Ifft et al. [5] calculated β values for five aqueous gradient-forming substances (CsCl, KBr, RbBr, RbCl, and sucrose) from listed values for density and activity over the full density gradient range covered by the materials (roughly 1.0–2.0 g/cm^3). The resulting plots of $\beta(\varrho_s)$ versus density were then fitted by polynomial functions such as (4.7):

$$\beta(\varrho_s) = \beta_0 + \beta_1 \cdot \varrho_s + \beta_2 \cdot \varrho_s^2 + \beta_3 \cdot \varrho_s^3 + \ldots + \beta_k \cdot \varrho_s^k \tag{4.7}$$

The determined constants, β_i, are listed for the five systems in [5], leading to reliable results when used in the evaluation of the five aqueous density gradient systems mentioned above.

In the same paper, the concept of the *isoconcentration point* is introduced. The basic idea is as follows: if a measuring cell, filled with a density gradient solution of concentration c_i of the heavy gradient-forming compound, is centrifuged until equilibrium is reached, a position r_i in the gradient must exist that exhibits exactly the initial concentration c_i of the solution, and thus the initial (constant) average density $\varrho_{s,i}$ (see Fig. 4.4). This point is called the isoconcentration point, r_i.

Again taking β as being constant, (4.6) becomes (4.8) in this special case:

$$\varrho_s(r) = \varrho_{s,i} + \frac{1}{2\beta}\omega^2(r - r_i)^2 \tag{4.8}$$

The advantage of this isoconcentration approach is that r_i can be calculated via geometrical considerations (see Fig. 4.4). As approximation to r_i, the arithmetical mean between r_m, the radius of meniscus and r_b, the radius of bottom, was proposed. A better approach for sector-shaped cells is the geometrical mean given in (4.9) [5]:

$$r_i = \sqrt{\frac{r_m^2 + r_b^2}{2}} \tag{4.9}$$

It has to be kept in mind that this $\varrho_s(r)$ evaluation does not take into account pressure effects (see Sect. 3.3.3), and therefore it is applicable only in restricted cases, or not applicable at all in (often strongly compressible) organic density gradients. Still, it works well in aqueous systems.

Hermans and Ende first addressed *organic* density gradients in 1963 [6]. These authors introduced a new approach for the evaluation of density gradients: the so-called Hermans–Ende theory. This theory describes density gradients under the assumption that the organic binary density gradient system behaves like an *ideal mixture*. The derivation of this theory will be described in the following.

Given is an ideal binary mixture of two components (light/heavy) with indices 0 and 1, and volume fraction φ_0 of components 0 and volume fraction φ_1 of component 1, defined by (4.10):

$$\varphi_k = \frac{n_k V_k^0}{\sum\limits_k n_k V_k^0} = \frac{m_k v_k^0}{\sum\limits_k m_k v_k^0} \qquad k = 0, 1\,, \tag{4.10}$$

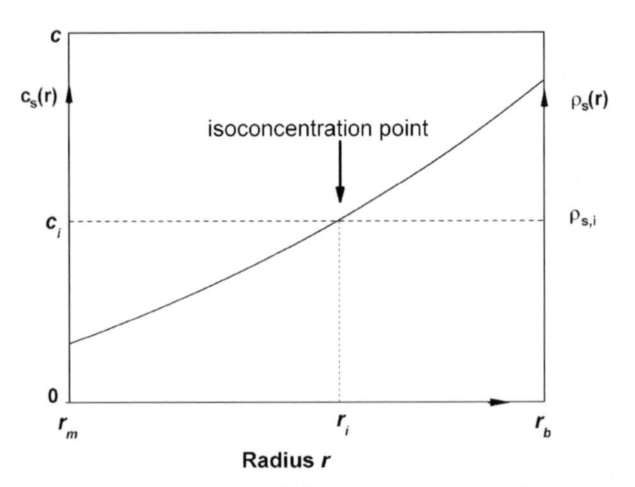

Fig. 4.4. Definition of the isoconcentration point in a static density gradient. At the beginning of the experiment, a uniform concentration c_i of the gradient-forming material is given in the whole measuring cell. After the equilibrium is reached, one point in the cell at radial position r_i exhibits the initial concentration, and therefore the initial known solvent density $\varrho_{s,i}$

with the molar amount of substance n_k, the mass m_k, the molar volume V_k^0, and the specific volume v_k^0 of the component k.

Equation (4.10) leads to (4.11) in the case of binary density gradient mixtures:

$$\varphi_0 + \varphi_1 = 1 \tag{4.11}$$

According to Hermans and Ende [6], the ratio of the volume fractions as a function of the radius is given by (4.12), a kind of barometrical equilibrium equation:

$$\frac{\varphi_0}{\varphi_1} = \alpha \cdot e^{\beta \cdot r^2} \tag{4.12}$$

Equation (4.13) gives the expression for the integration constant α in (4.12) for sector-shaped cells:

$$\alpha = \frac{\exp\left(\beta \cdot \varphi_1^{\mathrm{in}} \cdot \left(r_\mathrm{b}^2 - r_\mathrm{m}^2\right)\right) - 1}{\exp\left(\beta \cdot r_\mathrm{b}^2\right) - \exp\left(\beta \cdot \varphi_1^{\mathrm{in}} r_\mathrm{b}^2 + \beta \cdot \varphi_0^{\mathrm{in}} r_\mathrm{m}^2\right)} \tag{4.13}$$

with the initial volume fraction of component k, φ_k^{in}. The constant β is given by (4.14):

$$\beta = \frac{\omega^2 \cdot \left(\frac{M_1}{\varrho_1}\right)(\varrho_1 - \varrho_0)}{2RT}, \tag{4.14}$$

with the molar mass of the heavy component 1, M_1, and the densities of the components 0 and 1, ϱ_0 and ϱ_1.

The density of an ideal solution is given by (4.15):

$$\varrho_\mathrm{s} = \varphi_0 \cdot \varrho_0 + \varphi_1 \cdot \varrho_1 \tag{4.15}$$

The combination of (4.11), (4.12) and (4.15) results in the so-called Hermans–Ende equation, (4.16):

$$\varrho_\mathrm{s}(r) = \frac{\varrho_0 + \varrho_1 \cdot \alpha \cdot \exp\left(\beta \cdot r^2\right)}{1 + \alpha \cdot \exp\left(\beta \cdot r^2\right)} \tag{4.16}$$

This Hermans–Ende equation is applicable for all kinds of static equilibrium density gradients, for organic ones and for aqueous ones.

Under the assumption that $\alpha \cdot \exp(\beta \cdot r^2) \ll 1$, the simplified Hermans–Ende equation (4.17) holds:

$$\varrho_\mathrm{s}(r) = \varrho_0 + \varrho_1 \cdot \alpha \cdot \exp\left(\beta \cdot r^2\right) \tag{4.17}$$

The assumption that $\alpha \cdot \exp(\beta \cdot r^2) \ll 1$ is very often valid (see [3]). As stated above, the Hermans–Ende theory is valid only for ideal mixtures. In real mixtures, the parameters α, β, ϱ_0 and ϱ_1 have to be corrected. Therefore, (4.17) can be put into the following form:

$$\varrho_\mathrm{s}(r) = a_\mathrm{c} + b_\mathrm{c} \cdot \exp\left(c_\mathrm{c} \cdot r^2\right), \tag{4.18}$$

with the three adjustable parameters a_c, b_c and c_c. Use is made of (4.18) in a recent approach [3] of data evaluation of density gradient experiments, applying a calibration approach with density marker particles.

All numerical values (r_m, r_b, φ^{in}, M_1, ρ_1, ω, etc.) in (4.13) and (4.14) are known (see, for example, the water/metrizamide system in [3]), and so $\rho_s(r)$ can be calculated using the Hermans–Ende equations (4.16) or (4.17). This calculation yields satisfactory results if $\varphi_1^{in} < 0.10$ is valid, not only for organic but for water/sugar density gradients, too, and it is thus the standard procedure for static density gradient evaluation.

For higher φ_1^{in} contents, the calculated $\rho_s(r)$ value becomes increasingly falsified (see [3] and Fig. 4.6), especially near the cell bottom where the φ_1 content reaches a maximum. Nevertheless, also at high φ_1^{in} values, the radial ρ resolution of such density gradients continues to be very high, and additionally the falsifying is systematic. That means that, although the absolute particle density values of two neighboring sample fractions are falsified if one uses Hermans–Ende, their calculated relative density *difference* $\Delta\rho$ is nevertheless very precise within $\pm 0.0005\,\text{g/cm}^3$.

The reason for the failing of the Hermans–Ende theory at high φ_1^{in} values is that the assumption of an ideal mixture is not fulfilled. There were some attempts [21,22] to improve the Hermans–Ende theory, but the results are not very satisfactory and nearly impracticable. Thus, at present, the only practicable solution to this problem of high φ_1^{in} is the marker calibration of density gradients [3], described in the following section. This marker calibration allows further proving of every new density gradient theory very precisely.

Evaluation of the Radial Density Gradient $\rho_s(r)$ by Calibration with Markers

In the AUC laboratory at BASF, mainly synthetic polymers and colloidal particles are investigated. Many of these (mostly aqueous) colloidal systems are synthetic rubber particles that are stabilized by an electrostatic mechanism. Consequently, these systems are in general not of high electrolyte stability, and tend to agglomerate if salts (e.g., CsCl) are added. Therefore, H_2O/CsCl density gradients cannot be applied with these systems. The iodinated electrically *neutral* heavy sugars, metrizamide ($\rho_1 = 2.155\,\text{g/cm}^3$ at 25 °C, $M_1 = 789.1\,\text{g/mol}$) or Nycodenz ($\rho_1 = 2.060\,\text{g/cm}^3$, $M_1 = 821\,\text{g/mol}$), do not create agglomeration in aqueous colloidal systems. Thus, they are proved to be the most convenient density gradient compounds (see Sect. 4.2.2) for synthetic polymers and colloidal systems. In the following, we discuss the marker calibration of such water/sugar density gradients.

As indicated above, the Hermans–Ende theory fails if

(i) the equilibrium is not reached completely,
(ii) a density gradient mixture of *three* components is used to build up the density gradient (metrizamide, water and methanol is the most common case – see middle part of Fig. 4.3), and
(iii) the mass percentage of metrizamide is very high ($\varphi_1^{in} > 10\,\text{wt\%}$).

The latter point is perhaps the most important in practice. To overcome these restrictions, a marker calibration method was established [3]. Vice versa, this method allows proving the Hermans–Ende theory (and all other density gradient theories).

In common emulsion polymerization reactions, 11 marker particles with different chemical compositions resulting from different monomer ratios of ethylhexyl acrylate (EHA) and methyl acrylate (MA) were synthesized by co-polymerization. The sizes of all of the marker particles were in the range of 200 nm in diameter. The chemical EHA/MA compositions of the 11 markers and the particle densities, $\varrho_p = \varrho_{Kr}$, obtained with the Kratky density balance (see Sect. 2.6.1), are given in Table 4.1.

A *mixture* of this 11 marker particles with equal mass percentages was subject to seven different, composed static (water/methanol)/metrizamide density gradients (see Fig. 4.5, and the lower part of Fig. 4.3). The running conditions for all seven gradients in the BASF Optima XL apparatus equipped with Schlieren optics (XL-SO) were identical: 30 000 rpm, 3-mm mono-sector cells, −2° wedge windows, and running times of 22 h, or 44 h. The initial concentration of the total calibration mixture was $c = 2$ g/l in all cases, i.e., about 0.18 g/l for each marker component. Schlieren optical pictures of these seven density gradients (measured simultaneously in an eight-hole rotor) are given in Fig. 4.5. Horizontal 2° wedge windows, and the corresponding window holders (see Fig. 2.6) are obligatory for all static density gradient runs to compensate the steep radial refractive index gradient $n_s(r)$ connected with the density gradient $\varrho_s(r)$, and thus to prevent optical blackouts (see [18]).

Figure 4.5 (and Fig. 4.6) demonstrates the power of static density gradient experiments. By variation of the gradient composition, it is possible to cover a very broad range of sample densities. In the case of metrizamide (MA), this range can be

Table 4.1. Chemical composition and corresponding particle density $\varrho_p = \varrho_{Kratky}$ of 11 acrylic copolymer latices (with d_p values of about 200 nm) for AUC density gradient calibration (= marker particles)

Comp. No.	w_{EHA} [wt%]	w_{MA} [wt%]	ϱ_{Kr} [g/cm^3]
1	100	0	0.980
2	90	10	1.000
3	80	20	1.021
4	70	30	1.043
5	60	40	1.066
6	50	50	1.089
7	40	60	1.114
8	30	70	1.140
9	20	80	1.167
10	10	90	1.196
11	0	100	1.225

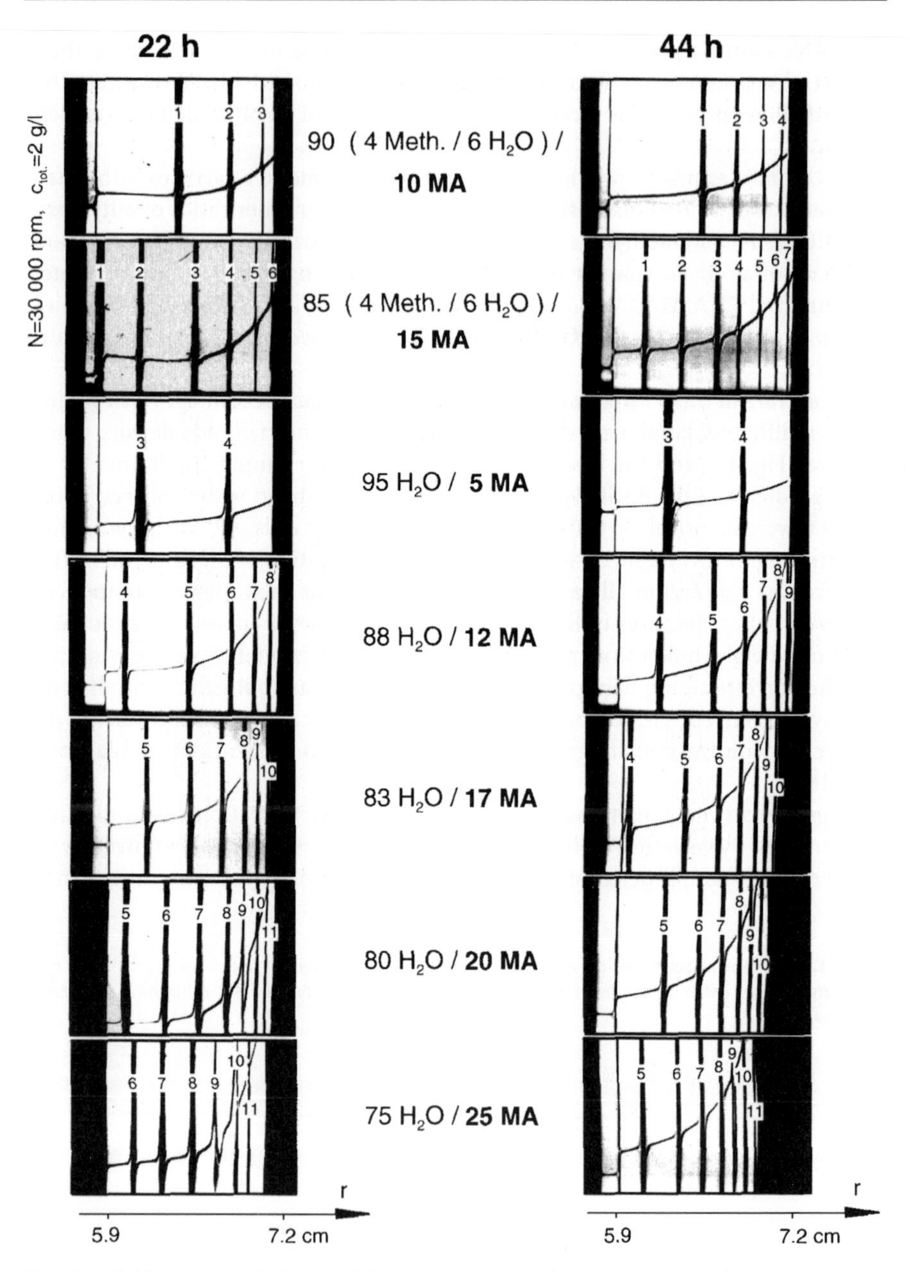

Fig. 4.5. Schlieren optical pictures (taken at 22- and 44-h running time) of seven static water/metrizamide density gradients with different compositions. Depending on the density range covered by the corresponding density gradient, different density marker particles, of the 11 marker component-containing mixture, appear at corresponding radial positions (simultaneously measured with an eight-cell multiplexer) (reprinted with permission from [3])

extended to lower values by adding light methanol, and to higher values by shifting from H_2O to heavy water D_2O (see lower part of Fig. 4.3). The density range covered is approximately from 0.85 g/cm^3 up to 1.35 g/cm^3. All 11 latices are fractionated according to their densities ρ_P, and thus appear in Fig. 4.5 as separated, narrow turbidity bands in the corresponding density gradients. This is an impressive demonstration of "density spectroscopy", and illustrates that the density gradient centrifugation proves to be very sensitive to the chemical composition of samples, which allows us to gain information about this composition.

However, the experiments in Figs. 4.5 and 4.6 were performed with the intention of density gradient *calibration* (and, incidentally, to prove the theoretical equations

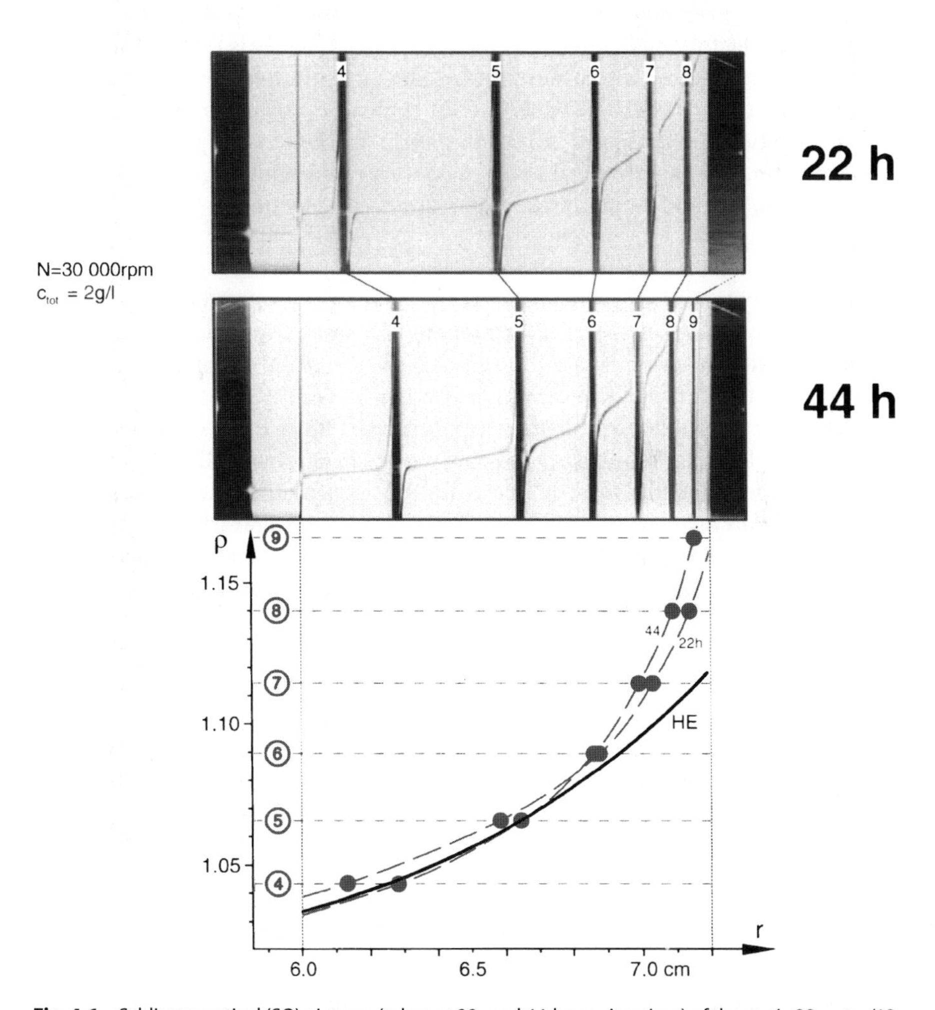

Fig. 4.6. Schlieren optical (SO) pictures (taken at 22- and 44-h running time) of the static 88 water/12 metrizamide density gradient containing the 11 latices calibration mixture. Below the pictures, the $\rho_S(r)$–r diagrams derived from the SO pictures are given. The $\rho_S(r)$ curve calculated from the Hermans–Ende theory (4.16) is also given for comparison (reprinted with permission from [3])

(4.8) and (4.16 – 4.18), especially the Herman-Ende equations). Therefore, in Fig. 4.6 the 88 H_2O/12 metrizamide density gradient Schlieren photographs, taken after 22 h and 44 h, are given again for studying the time dependence of $\varrho_s(r)$. In total, six signals of six different calibration particles can be recognized after 44 h, whereas after 22 h only five marker components are detected in the density gradient.

This exemplifies that equilibrium was not yet reached completely after 22 h (after 44 h, equilibrium was indeed reached). The corresponding particle numbers at 44 h are nos. 4–9. The three $\varrho_s(r)$ diagrams in Fig. 4.6 show the difference between the density values determined with the calibration particles (at 22 and at 44 h) and the theoretical values obtained by the Hermans–Ende theory (4.16). Whereas the agreement between the experimental findings and Hermans–Ende theory is satisfactory in the meniscus region, the discrepancy is tremendous in the bottom region of the cell where the metrizamide concentration is very high. In order to make practical use of these experiments, empirical $\varrho_s(r)$ diagrams of all seven density gradients investigated, and using all 11 markers, were generated in the same manner. The experimental points then were fitted, applying (4.18), resulting in a set of a_c, b_c and c_c parameters and seven corresponding $\varrho_s(r)$ calibration curves (see Fig. 4.7).

The parameter sets recorded for the seven differently composed water/metrizamide density gradients were assessed by means of a software program that allowed us to determine particle or polymer densities from the corresponding density gradient experiments with high accuracy (see [3]). Unfortunately, this 11-marker set is not available commercially, but it is easy to create in any chemical laboratory. Available are single markers such as PS or PMMA latices or glass beads. Even if no markers are available, however, the Hermans–Ende $\varrho_s(r)$ calculation is, as mentioned above, sufficient for most cases, and the high-resolution power exists in all density gradients. Often, the isoconcentration point procedure ((4.8) and (4.9)), or the calibration with only one marker (4.6) yield a good approximation for $\varrho_s(r)$, especially in the radius range around the marker position r_0.

Time to Reach $\varrho_s(r)$ Equilibrium

Theoretically, equilibrium is not reachable within finite time but, of course, from the practical point of view, the time can be estimated after which effectively no changes in the detected density profile $\varrho_s(r)$ of the gradient-building material will occur within the errors of measurement. In practice, this is 15–70 h if the cell is completely filled.

In the authors' experience, no theoretical approach to this problem has to date given very satisfying results (see, for example, [7–10]). Nevertheless, there are some approaches to be mentioned, e.g., [11]. Among other fruitful debates, a detailed discussion on this topic can be found in the internet ([12]). Therefore, we only give rules of thumb here. If a density gradient consisting of water and up to 15 wt% metrizamide or Nycodenz (see Sect. 4.2.2) is centrifuged in usual analytical ultracentrifugation measuring cells with column heights of about 12 mm

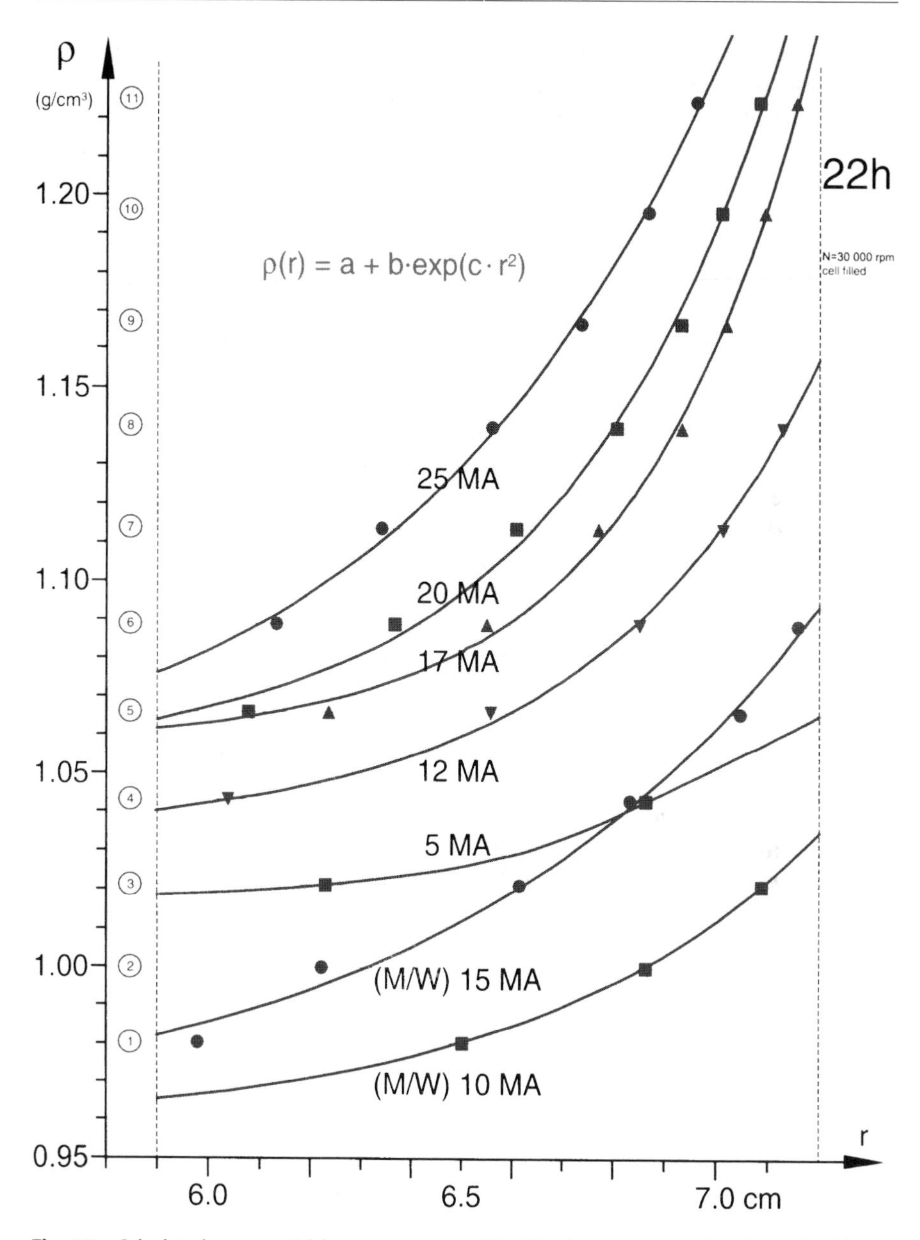

Fig. 4.7. Calculated exponential three-parameter $\rho_S(r)$ calibration curves based on the 22-h Schlieren optical pictures given in Fig. 4.5. The curves are obtained by fitting of the experimental data to (4.18) (reprinted with permission from [3])

(i.e., completely filled cells) at 30 000 rpm, then equilibrium is reached after 40 h. Aqueous CsCl density gradients may build up in time periods of 8–16 h (at speeds above 50 000 rpm), and within days at lower speed.

For standard sedimentation equilibrium runs to measure M (see Chap. 5), van Holde and Baldwin [8] found that the time to reach (near) equilibrium is proportional to the square of the column height $(r_b - r_m)^2$. In our experience, this is valid for equilibrium density gradient runs, too. This means that a considerable reduction of the long density gradient measuring time is possible by the reduction of the column height. Unfortunately, the price for this time reduction is a considerable reduction in the accessible density range $\rho_s(r)$, as follows from the Hermans–Ende equations (4.13) and (4.16). Indeed, in practice a $\rho_s(r)$ range as broad as possible is very important. Thus, a compromise for most experiments is to restrict the running time to 22 h (overnight), and not to attempt to really reach full equilibrium. Perhaps the resulting particle densities are somewhat falsified, and not really absolute ones in this case, but nevertheless a high-resolution fractionation has taken place also during these 22 h, yielding valuable analytical information. This is illustrated in Figs. 4.5 and 4.6, when one compares the 22- and the 44-h measurements. Figure 4.5 illustrates additionally that the use of the eight-cell multiplexer reduces the problem of long running times.

A very substantial reduction of running times, from hours to minutes, is possible with the technique of *dynamic* density gradients (see Sect. 4.3). However, this is possible only for rapidly sedimenting particles ($d_p > 20$ nm), and not for slowly sedimenting dissolved macromolecules (discussed in the following section). The marker calibration technique ([3]; Figs. 4.5–4.7) can also be regarded as a kind of "dynamic" density gradient method (during the time to reach equilibrium), because no equilibrium is really needed to measure absolute sample particle densities in this way.

Distribution of the Sample Within the Static Density Gradient

Whereas in the above section the formation of the density gradient and the evaluation of the resulting solution densities $\rho_s(r)$ were described, the distribution of sample material to be analyzed within the gradient will be the subject of the following section.

In this section, an equation will be derived showing that, at the stage of equilibrium, (monodisperse) sample particles are distributed around their radial equilibrium position r_0 in the density gradient following a Gaussian-like curve (see Figs. 4.9 and 4.10a). The radial derivation of such a *Gaussian* Schlieren curve is a *double* Schlieren peak, such as those shown in Figs. 4.1a and 4.8.

The equilibrium condition for the pure gradient solution given in (4.1) is assumed to be valid also in the presence of small amounts of dissolved macromolecules and dispersed particles. This is fulfilled because $c_{\text{sample}} \ll c_s$. Taking this requirement into account, (4.19) can be set up:

$$c_p s_p \omega^2 r = D_p \frac{\partial c_p}{\partial r} \, . \tag{4.19}$$

For the sake of convenience, in the following we refer solely to polymers by means of the subscript "p". In (4.19), c_p is the sample concentration, and D_p the sample's

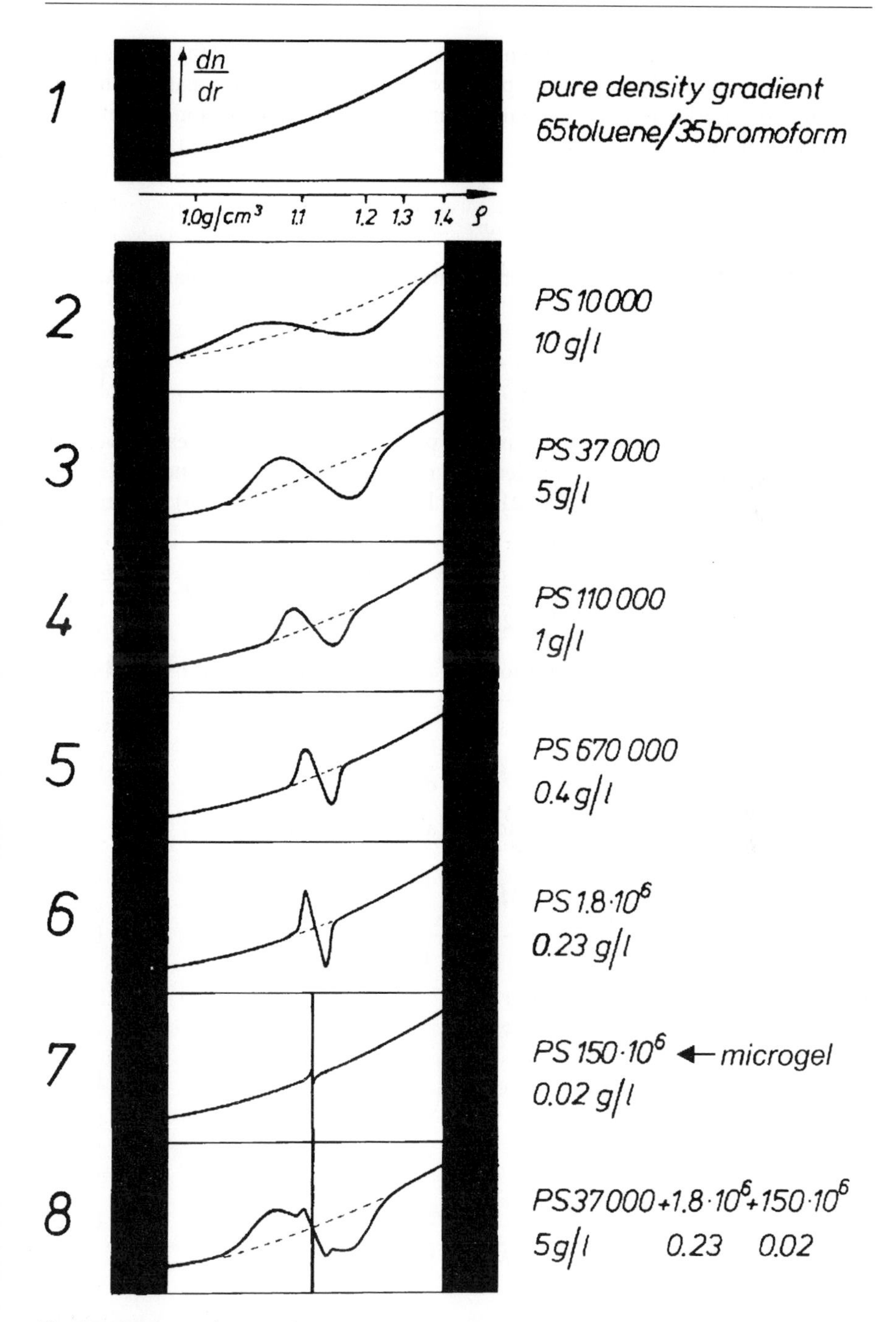

Fig. 4.8. Static 65 toluene/35 bromoform density gradients of six polystyrenes with different molar masses *M*, and of a mixture of these measured in a one-night run, using an eight-cell AUC Schlieren optics multiplexer (reprinted with permission from [20])

diffusion coefficient. Rearrangement of (4.19) gives (4.20), with the radial position r_0 at which the density of the polymer sample ϱ_p is exactly matched by the gradient solution density ϱ_s. The maximum polymer concentration is therefore given at position r_0.

$$\frac{s_p}{D_p} = \frac{1}{\omega^2 r} \cdot \frac{\partial \ln c_p}{\partial (r - r_0)} \tag{4.20}$$

In analogy to (4.3), s_p/D_p can be substituted by the Svedberg equation (1.8) to obtain (4.21):

$$\frac{\partial \ln c_p}{\partial (r - r_0)} = \frac{M\omega^2 r(1 - \bar{v}\varrho_s)}{RT}, \tag{4.21}$$

where M and $\bar{v}$ are the molar mass and the partial specific volume, respectively, of the sample to be analyzed in the density gradient. The radial dependent density of the density gradient $\varrho_s(r)$ in (4.21) may be approximated by (4.6) that may be rearranged to give (4.22):

$$\varrho_s(r) = \varrho_{s,0} + \frac{\omega^2}{\beta}\left(\frac{r^2 - r_0^2}{2}\right) = \varrho_{s,0} + \frac{\omega^2}{\beta}\frac{(r - r_0)(r + r_0)}{2} \tag{4.22}$$

With (4.4), and under the assumption that r is close to r_0, and thus $r + r_0 \approx 2r_0$ is valid, (4.23) is obtained:

$$\varrho_s(r) = \varrho_{s,0} + \frac{\partial \varrho_s}{\partial r}(r - r_0) \tag{4.23}$$

Substitution of (4.23) into (4.21), and rearrangement leads to (4.24):

$$\frac{\partial \ln c_p}{\partial (r - r_0)} = \frac{M\omega^2 r_0 \left(\frac{\partial \varrho_s}{\partial r}\right)_{r_0} (r - r_0)^2}{\varrho_{s,0} RT} \tag{4.24}$$

Integration finally leads to (4.25):

$$c_p(r) = c_{p,0} \cdot \exp\left(-\frac{(r - r_0)^2}{2\sigma^2}\right) \tag{4.25}$$

with σ^2 given by (4.26):

$$\sigma^2 = \frac{\varrho_{s,0} RT}{M\omega^2 r_0 \left(\frac{\partial \varrho_s}{\partial r}\right)_{r_0}} \tag{4.26}$$

Equation (4.25) is a typical Gaussian function created by diffusion broadening of a monodisperse sample with the maximum polymer concentration $c_{p,0}$ at the radial position r_0. The standard deviation σ in principle allows us to calculate the molar mass M (or the particle diameter d_p) of the polymer, once the equilibrium is reached. Also the broadness of every double Schlieren peak yields experimentally a value of σ, and thus a rough estimation of M (or d_p) by applying (4.26).

Schlieren optics detects the radial changes of the refractive index gradient $dn(r)/dr$ within the measuring cell. This corresponds to the derivation of the concentration over the radius $dc(r)/dr$. Therefore, when density gradient experiments are monitored by Schlieren optics (such as in Figs. 4.1a and 4.8), rather than a Gaussian-type signal (such as in Figs. 4.9 and 4.10a), the derivation of the same is obtained. This is a typical so-called double Schlieren peak. Figure 4.8 proves experimentally the expected dependency on the molar mass M, according to (4.25) and (4.26), i.e., the M dependency of the broadness of the detected double Schlieren peaks: The higher the value of M is, the narrower is the diffusion broadening. In this eight-cell multiplexer density gradient experiment, six (nearly monodisperse) differently sized polystyrene molecules and a mixture of three of these are measured (conditions: 12-mm single-sector cells, 60 000 rpm, 18 h, Philpot angle 85°, 3° horizontal wedge windows). The pure 65 toluene/35 bromoform density gradient is also shown at the top of Fig. 4.8.

Deviations of concentration distribution curves from the shape of a pure Gaussian curve (or a symmetrical double Schlieren peak) can serve as an indication of heterogeneity in the sample (see, for example, cell 8 in Fig. 4.8). Although the deviations can be very small, and the method is not of high sensitivity, it can also be concluded from (4.25) and (4.26) that for latex particles, which have extremely high M values (in other words, extremely low diffusion coefficients D_p), the double Schlieren peak will "degenerate" into a narrow *turbidity band*. In Fig. 4.8, this is illustrated by the signal caused by the polystyrene with $M_w = 150 \times 10^6$ g/mol. This polystyrene sample is simply a strongly crosslinked, 80-nm PS latex (i.e., compact particles), which generates the narrow turbidity band detected. It is also important that from the measured double Schlieren peak *area* A_{schl}, a rough estimation of the concentration c_p of the corresponding macromolecules (or of very small particles, $d_p < 30$ nm) is possible, using (3.32).

4.2.2 Gradient Materials

For isopycnic separations, the most important feature of a density gradient medium is that the maximum density of its solutions $\varrho_{s,max}$ exceed that of the sample particles ϱ_p to be analyzed. In fact, the limited range of attainable densities $\varrho_s(r)$ is the most confining restriction of density gradient materials.

Generally, gradient materials should exhibit certain properties that may be summarized as follows:

(i) The compound should be inert (if possible, no preferential solvation).
(ii) The physicochemical properties of the material and its solutions (see (4.12)–(4.16)) should be known and applicable to determine the radial concentration distribution of the heavy compound, and so $\varrho_s(r)$ by means of optical detectors (this is not necessary if marker calibration methods are used for the $\varrho_s(r)$ evaluation).

(iii) The gradient material solutions should not constrain the optical detection of the radial fractions of the sample in the density gradient.
(iv) Depending on whether a preparative or an analytical density gradient experiment is planned, the price of the material plays an important role.

There are several materials that meet the described criteria for density gradient media. The materials can be classified in five types. The most important representatives, and the accessible density ranges $\rho_s(r)$ are given in brackets below:

(i) nonionic iodinated compounds (water/metrizamide, 0.85 – 1.35 g/cm^3, see Fig. 4.3),
(ii) organic solvent mixtures (THF/DIM, 0.9 – 1.4 g/cm^3, Fig. 4.1a),
(iii) sugars and polysaccharides (water/sucrose, 1.0 – 1.15 g/cm^3),
(iv) colloidal silica (water/Percoll, 0.85 – 1.15 g/cm^3, see [16]), and
(v) alkali metal salts (water/cesium chloride, 1.0 – 1.9 g/cm^3).

The order given above reflects the importance in practical work from the authors' point of view. This is definitely different if biochemical questions are concerned. Because it is very difficult and very expensive to obtain the multipurpose metrizamide on the market at present, we replaced it by the cheaper Nycodenz, and found nearly no difference in the resulting aqueous static density gradients.

Not only binary density gradients are possible but also *ternary* ones. For example, the $\rho_s(r)$ range can be increased (see Figs. 4.3, 4.5 and 4.7) by replacing parts of water by the lighter methanol (ρ = 0.793 g/cm^3 at 25 °C) or the heavier D_2O (ρ = 1.104 g/cm^3). The Hermans–Ende evaluation fails (partly) in this case, but a marker evaluation is always possible. Beside the two mentioned organic solvent mixtures, THF/DIM and toluene/bromoform (Fig. 4.8), also other organic solvent mixtures are used in the authors' laboratory, for example, n-hexane/chloroform (see Figs. 6.39 and 6.40), DMF/bromoform, DMF/DIM, and cyclohexane/CCl_4.

Corrosion problems with centerpieces, for example, of CsCl or bromoform with aluminum centerpieces, are avoidable by using titanium centerpieces. Unfortunately, as stated above, the density range $\rho_s(r)$ of the high-resolution/fractionation density gradient technique of the AUC is restricted to 0.85 – 1.9 g/cm^3. Thus, organic/inorganic hybrid particles with densities up to 20 g/cm^3, intensively discussed at present, can not be assessed by this technique. Perhaps H_2O/D_2O analysis (see Sect. 3.5.3) will allow us to determine these density distributions in the near future, if the accuracy of measurement of this method can be improved considerably.

It is important to have in mind that possibly problems will arise if the sample incorporates or shows marked interaction with the gradient material. A dependence of the (apparent) sample's density on the experimental conditions would be the consequence, which could lead to wrong conclusions from the experiment.

In biological systems, in fact, the hydration of the sample plays an important role. It can vary according to the medium, resulting in different (apparent) densities of the same sample depending on the chosen gradient material. In organic polymer systems, often a preferential solvation of one of the two organic density gradient solvents is observed (see, for example, the different density values of dissolved polystyrene molecules, 1.04 g/cm^3 (preferential solvation of the light THF) and 1.12 g/cm^3 (preferential solvation of the heavy bromoform) in two different organic density gradient mixtures in Figs. 4.1a and 4.8). Nevertheless, also in the case of preferential solvation or other interactions, the density gradient is always a valuable analytical tool because of its high resolution power.

Interference optics is not well suited for organic density gradient detection, because it is too sensitive (steep optical gradients, optical blackouts, wedge windows to compensate, etc.). UV and Schlieren optics are a better option in this respect, but unfortunately most organic solvents for polymers, as well as metrizamide (the preferred density gradient material for polymeric latex particles), are not transparent to UV light. Thus, Schlieren optics is the most suitable, and therefore the most important density gradient detection method for polymers and polymeric latex particles. Therefore, an appeal is made to Beckman-Coulter to supply a commercially available Schlieren optics detector for the XL-A/I, and special density gradient accessories, including wedge windows, and centerpieces.

4.2.3 Experimental Procedure

There is no single, non-ambiguous procedure to perform equilibrium density gradients. Therefore, we only give a recommendation for the experimental procedure here, which was proved to be appropriate in laboratory practice: primarily, equilibrium density gradients are performed in an AUC equipped with Schlieren optics. If possible, pictures (or radial scans) should be taken every 1 h. On one hand, this allows one to identify the point in time (t_{equil}) when equilibrium is reached. On the other hand, one can obtain, via these 1-h pictures/scans, an indication of whether, for example, components of the sample are moving out of the density gradient, or whether agglomeration or related phenomena occur. Even a rough estimation of the particle size or molar mass can be gained from the velocity that the components need in order to reach their radial r_{iso} point of equal density in the gradient. In Schlieren optical experiments, mono-sector cells are used, which conveniently contain −2° horizontal wedge windows to compensate the steep radial optical refractive index gradient.

The standard rotor speed may be chosen to be 30 000 rpm for water/metrizamide (see Fig. 4.3), and 40 000 rpm for THF/DIM density gradients. The cells should be filled nearly completely in order to cover a maximum density range $\varrho_s(r_b)-\varrho_s(r_m)$.

The choice of sample concentration depends, of course, on the purpose of the experiment. If an unknown sample has to be analyzed, it is often helpful to investigate the sample at two different concentrations, one at standard, the other at increased concentration, in one run. The standard concentration may be around 0.05 g/l, whereas the higher concentration may be chosen to be around 0.5 g/l. This allows monitoring the main components of the sample, as well as possible secondary components.

Typical running times are 22 and 44 h. To be definitely in the quasi-equilibrium state, 72 h are recommended (weekend runs). The time to reach equilibrium can be shortened by applying the procedure of *overspeeding* (see, for example, [13] and Sect. 5.2), linked with the loss of experimental information obtained during the time to reach equilibrium. The possibility to measure several samples simultaneously (up to seven in an eight-cell rotor multiplexer) results in a high efficiency of the experiment (see Fig. 4.8).

4.2.4 Examples

Figure 4.5 illustrates the high-resolution fractionating power of equilibrium (= static) density gradients. Additional examples for the application of equilibrium density gradients are described in Chap. 6. Thus, we give only one further example here. In contrast to all other density gradient examples given in this book, where *Schlieren* optics is used, the following equilibrium density gradient experiment was performed in an Optima XL-A/I [14], i.e., with *interference* optics. Figure 4.9 shows the XL-A/I interference patterns of five differently composed static water/metrizamide density gradient experiments performed on the same sample. The sample consists of 30-nm polystyrene latex particles with a known density (Kratky balance) of $\varrho_P = 1.055\,g/cm^3$ (= marker particles). The metrizamide content in the gradient-forming medium consisting of water/metrizamide varied in the range 6–15 wt%. The radial position r_{iso} of the sample particles can be clearly identified in all patterns.

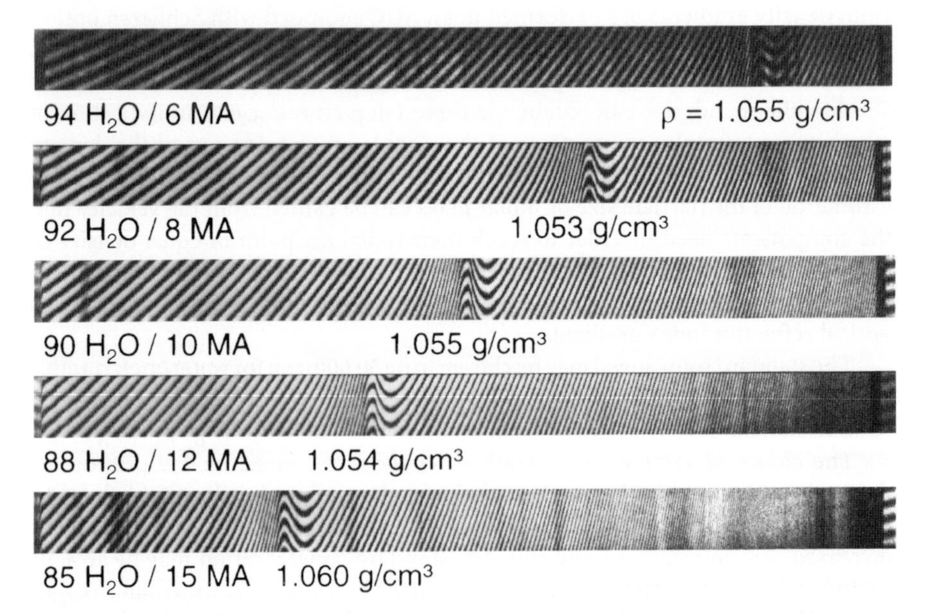

Fig. 4.9. XL-A/I interference optical pattern of 30-nm polystyrene latex particles (c = 0.25 g/l) in static water/metrizamide equilibrium density gradients with different compositions. The radial position of the sample in the different gradients can clearly be seen. The corresponding densities, calculated from the radial course of the interference fringes (see Fig. 4.10), are given (reprinted with permission from [14])

The inference optical signal is a superposition of the Gaussian peak caused by the sample particles, and the underlying exponential density gradient formed by inter-diffusion of water and metrizamide. As an example, Fig. 4.10a gives the measured fringe shift ΔJ over the radius r obtained from the inference optical pattern (such as in Fig. 4.9) of the 8-wt% metrizamide gradient after 92 h (a weekend run at 40 000 rpm). The big advantage of monitoring equilibrium density gradients by interference optics is the possibility to *directly* measure the radial density distribution $\varrho_s(r)$ via the (measurable!) absolute refractive index $n_s(r)$, which means without the need of a Hermans–Ende calculation. Interference optics allows one to determine the absolute refractive index of the water/metrizamide solution at any radial position $n_s(r)$ ([14]; cf. this is in accordance with the law of conservation of mass). From this $n_s(r)$ function, the radial concentration distribution of metrizamide $c_s(r)$, and thus the radial density distribution of the gradient solution $\varrho_s(r)$ can be calculated. The calculated $\varrho_s(r)$ function is given in Fig. 4.10b for the 92 water/8 metrizamide density gradient discussed above. This function, and the known r_{iso} position yield the ϱ_p value of the corresponding polystyrene particles. All five different polystyrene ϱ_p

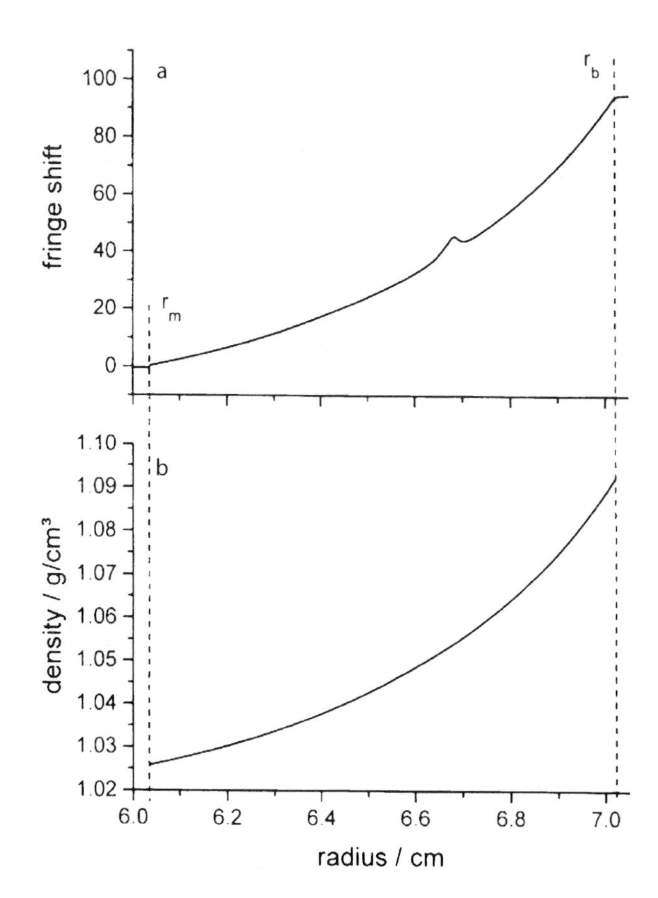

Fig. 4.10. **a** Interference fringe shift over radial distance from the axis of rotation. The static 92 water/8 metrizamide density gradient was formed after 92 h at 40 000 rpm. The applied sample consists of 30-nm polystyrene latex particles (c = 0.25 g/l). **b** The radial density distribution $\varrho_s(r)$ in the measuring cell, calculated from the radial distribution of the absolute refractive index $n_s(r)$, is measured by interference optics (reprinted with permission from [14])

values calculated in this manner for the five differently composed density gradients are given in Fig. 4.9. They agree well with the Kratky balance value of $1.055\,g/cm^3$.

However, the big drawback of performing static density gradients via interference optics in an XL-A/I is the uncertainty of detecting really all components of an investigated sample if it shows a complex composition. Especially minor secondary components and strongly diffusion-broadened components can easily be overlooked in interference "photos"/scans. The resolution power is also limited, making it impossible to identify two components of very similar density. Interference optical detection of density gradients is restricted to flat gradients (i.e., metrizamide contents $< 15\,wt\%$, see Fig. 4.9). In steep gradients, the (basic) interference fringes of the gradient itself are no more resolvable, especially in the bottom area. In practice, only Schlieren optics is suitable for density gradients of complex samples. For further details on the calculation of $\rho_S(r)$ from the interference optical XL-A/I signal and on the experiment itself, the reader should consult [14].

4.3 Dynamic Density Gradients

The main drawback of static (= equilibrium) density gradients is that the experiments are very time-consuming. Especially in an industrial laboratory, this is of very high relevance. A second major disadvantage of the static density gradient experiment, described in Sect. 4.2, is the need to apply gradient material in relatively high amounts. This runs the danger of interactions of the analyzed sample with the gradient material. As mentioned above, preferential solvation can result in misleading information if absolute particle densities ρ_P are desired.

For certain systems, these problems are overcome by a technique developed in the 1980s by Lange [15]. This technique, called *dynamic* density gradient, was then further modified by one of the authors [16]. As we will see in the next paragraph, this allows one to avoid the time-consuming formation of the density gradient equilibrium in the cell, and makes use of gradient materials that are chemically as similar as possible. The basis of dynamic density gradients is mainly the use of the synthetic boundary technique (see Sect. 3.6).

Theory of Dynamic Density Gradients

In dynamic density gradients, a pair of solvents that are chemically nearly identical, but exhibit a sufficient density difference, are overlaid (superimposed) in a synthetic boundary cell. Effectively, dynamic density gradients are performed only with the solvent pair H_2O ($\rho_{H_2O} = 1.0\,g/cm^3$) and D_2O ($\rho_{D_2O} = 1.1\,g/cm^3$), and therefore the maximum density range covered in this standard experiment is $1.0-1.1\,g/cm^3$. In the following, using again Fig. 3.25, for a start we describe the dynamic density gradient experiment without the presence of a sample, and subsequently explain the actual dynamic density gradient experiment as a fractionation tool of samples.

In a synthetic boundary cell of the valve type (see Sect. 2.3 and Figs. 2.7e, 3.24 and 4.11), pure D_2O is placed into the chamber of the measuring cell, whereas pure H_2O is put into the reservoir (= storage bin, see Fig. 3.25). Depending on the valve rubber used, the H_2O is forced through the small hole in the bottom of the storage bin into the chamber on top of the D_2O at a certain centrifugal field force (usually about 10 000 rpm). Around the well-defined radial overlaying position r_{overlay} in the cell (called r_u in Fig. 4.11), the gradient density changes from 1.0 to 1.1 g/cm^3, following a steep step-like radial function $\rho_s(r)$ in the first seconds (see Fig. 4.11). This function becomes broader with time. The standard final rotor speed of dynamic H_2O/D_2O density gradients is 40 000 rpm. Inter-diffusion of heavy water and (light) water molecules starts immediately after overlaying to equalize the radial density distribution of the solvent mixture. At the (late) end of the experiment, a homogeneous mixture of H_2O and D_2O, and therefore a unique radial density has become established throughout the measuring cell. In the meantime, a *time*-variable density gradient $\rho_s(r,t)$, often referred to as a *dynamic* density gradient, is built up due to the inter-diffusion of heavy water and (light) water molecules. Different volume fractions of heavy water and (light) water at different radial positions in the gradient zone of the cell result in different radial densities $\rho_s(r)$. Here, it has to

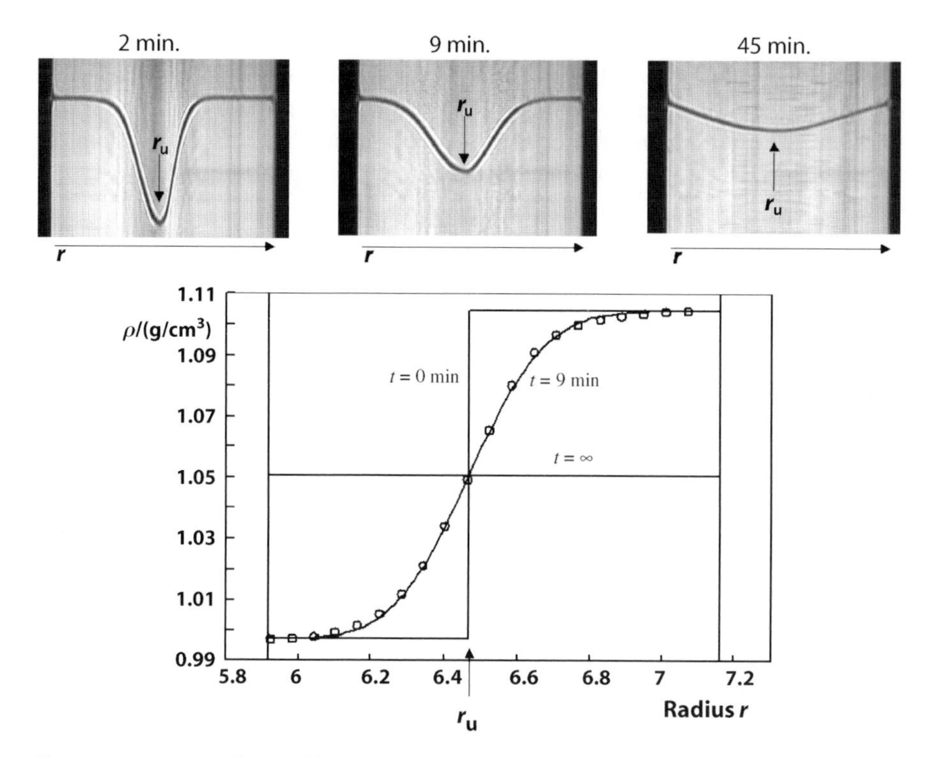

Fig. 4.11. *Upper part* Three Schlieren optical pictures of a dynamic H_2O/D_2O density gradient run, using a 12-mm mono-sector valve-type synthetic boundary cell at 40 000 rpm. *Lower part* $\rho_s(r)$ evaluation of the picture taken at 9-min running time ($r_u = r_{\text{overlay}}$)

be noted that the centrifugal forces applied are not sufficient to separate H_2O from D_2O.

Measurement of the Dynamic Density Gradient Course $\varrho_s(r)$

For the determination of particle densities, exact knowledge of the density gradient course with radius $\varrho_s(r)$ is needed (at a defined running time t). This can be measured by Schlieren optics by following the change of the experimentally registered refractive index $(dn/dr)(r)$ curve with the radius r (see Fig. 4.11, upper part). Via integration over r, the radial course of the refractive index is obtained (for the chosen running time; see (4.27)):

$$n(r) = n_{r_m} + \left(n_{r_b} - n_{r_m}\right) \cdot \frac{\int_{r_m}^{r} \frac{dn}{dr} \, dr}{\int_{r_m}^{r_b} \frac{dn}{dr} \, dr} \tag{4.27}$$

Under the assumption of ideal mixing behavior (given in the case of H_2O and D_2O), and with the Gladstone–Dale rule [17], (4.28) is obtained:

$$\frac{\varrho_s(r) - \varrho_{s,r_m}}{\varrho_{s,r_b} - \varrho_{s,r_m}} = \frac{n(r) - n_{r_m}}{n_{r_b} - n_{r_m}} \tag{4.28}$$

Combined with (4.27), we obtain (4.29), which gives the radial course of density $\varrho_s(r)$ from the radial change of the refractive index $n_s(r)$, if the densities of the solution at the bottom and at the meniscus are known. This is the case as long as no mixture of H_2O and D_2O has been formed at the corresponding radial positions r_m and r_b.

$$\varrho_s(r) = \varrho_{s,r_m} + \left(\varrho_{s,r_b} - \varrho_{s,r_m}\right) \cdot \frac{\int_{r_m}^{r} \frac{dn}{dr} \, dr}{\int_{r_m}^{r_b} \frac{dn}{dr} \, dr} \tag{4.29}$$

Therefore, the radial density gradient $\varrho_s(r)$ can be determined very simply (see lower part of Fig. 4.11, for the running time 9 min) from the area ratio between the Schlieren peak area at radial position r, $A_{schl}(r)$, and the total peak area $A_{schl,total}$ (see (4.30)) and the (known) densities at the bottom and at the meniscus (ϱ_{D_2O} and ϱ_{H_2O}).

$$\frac{A_{schl}(r)}{A_{schl,total}} = \frac{\int_{r_m}^{r} \frac{dn}{dr} \, dr}{\int_{r_m}^{r_b} \frac{dn}{dr} \, dr} \tag{4.30}$$

For the reasons given above, the dynamic density gradient centrifugation can not be evaluated properly as soon as the densities at the bottom and at the meniscus are no longer known, i.e., as soon as the diffusion has spread the H_2O/D_2O boundary (visible as a negative Gaussian Schlieren curve in Figs. 4.11 and 4.12) so strongly that D_2O molecules reach the meniscus, and H_2O reaches the bottom (which is the case in the Schlieren photograph of Fig. 4.11, taken at 45-min running time).

The considerations described above hold only if the sample fractions to be analyzed migrate fast toward the point where their densities are matched, so that no significant change of the density gradient occurs during this movement. This constraint is normally fulfilled for polymeric dispersions, i.e., for compact particles with $d_p > 30\,\text{nm}$. Dissolved macromolecules are not analyzable with dynamic density gradients because they do not move fast enough.

The density range of dynamic H_2O/D_2O density gradient experiments can be shifted [15] to either side by adding a third component to the H_2O/D_2O mixture (as for static density gradients, see Sect. 4.2.2 and Fig. 4.3). Most commonly used as shift agents are methanol ($\varrho_{\text{methanol}} = 0.793\,\text{g/cm}^3$ at 25 °C) to extend the range to lower values, and ethylene glycol or glycerin ($\varrho_{\text{ethylene glycol}} = 1.114\,\text{g/cm}^3$ and $\varrho_{\text{glycerin}} = 1.261\,\text{g/cm}^3$) for shifting to higher values. From this third component, significant additional complications arise, as mixing enthalpies, etc., have to be taken into account. Thus, the overall conclusion is that dynamic H_2O/D_2O density gradients are suited for the density range $0.85 < \varrho_s(r) < 1.25\,\text{g/cm}^3$.

To demonstrate the fractionating power of the dynamic density gradient performed in the AUC, we give one example here (another example was presented in Fig. 3.25 above). The same mixture of 11 polymeric latex particles (see Table 4.1) used to set up the marker calibration evaluation of static equilibrium density gradients (see Sect. 4.2.1 and Fig. 4.5) is investigated in a dynamic H_2O/D_2O density gradient experiment. According to Table 4.1, five components of the mixture are expected to exhibit densities in the range covered by this dynamic density gradient between 1.0 and 1.1 g/cm^3. Figure 4.12 shows four Schlieren optical pictures of the corresponding experiment, taken at 0.5-, 4-, 6- and 10-min experimental running time at the standard rotor speed 40 000 rpm.

The Schlieren photograph taken after 0.5 min (in the startup phase of the rotor at about 5000 rpm) shows the measuring cell before overlaying has occurred. The cell is half filled with D_2O, which contains the dispersed (turbid) sample ($c_{\text{total}} = 3\,\text{g/l}$, i.e., about 0.19 g/l of each component). In the next Schlieren picture (taken after 4 min), the meniscus moves to the top of the cell during the overlaying of H_2O onto the D_2O phase, because the valve opens at about 10 000 rpm. The dynamic density gradient has already formed at the H_2O/D_2O interface. In the course of the experiment, the components of the sample having densities below 1.0 g/cm^3 or above 1.1 g/cm^3 will float or sediment out of the region of the gradient. The expected five components are fractionated within the region of the density gradient, and form narrow turbidity bands. The particle densities determined, using (4.29), the 10 min photograph and the procedure described in Fig. 4.11, are 1.004, 1.021, 1.043, 1.067 and 1.089 g/cm^3. These values correspond very well

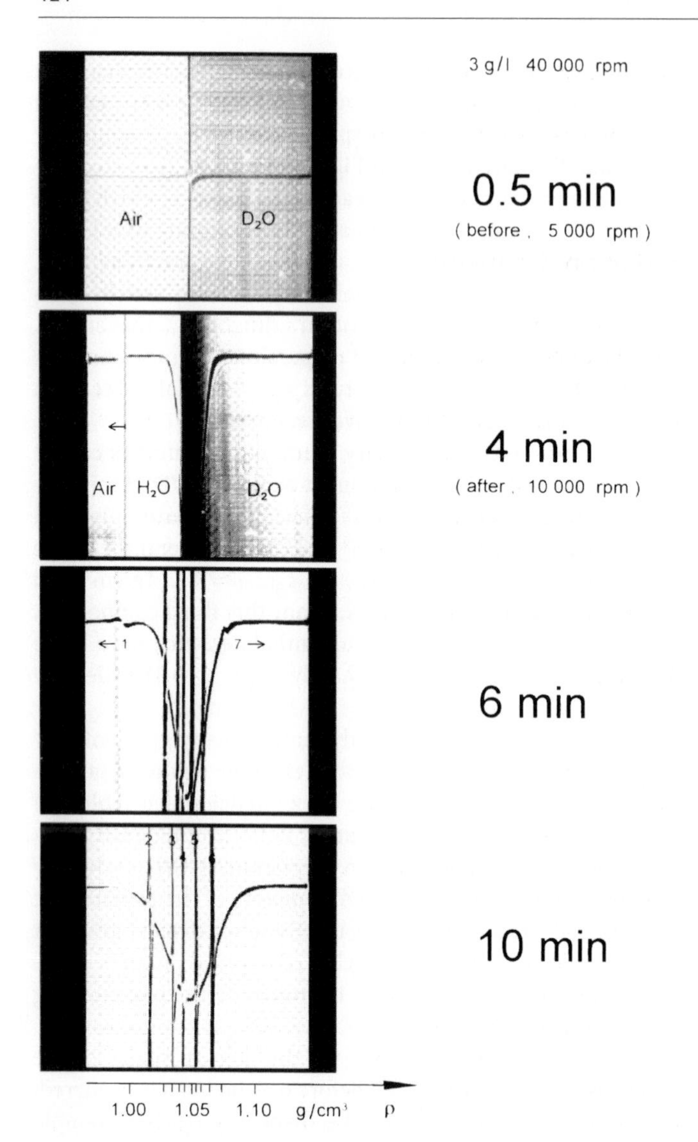

Fig. 4.12. Four Schlieren optical pictures of a dynamic H_2O/D_2O density gradient experiment performed on a mixture composed of 11 polymeric latex components with different particle densities. The pictures are taken after 0.5, 4, 6, and 10 min. Five of the 11 density fractions are recorded within this density gradient range. A 12-mm mono-sector valve-type synthetic boundary cell was applied. The $\varrho_S(r)$-axis below the 10 min photograph was calculated as described in Fig. 4.11 (reprinted with permission from [18])

with the independently measured "true" values (Kratky balance) 1.000, 1.021, 1.043, 1.066 and 1.089 g/cm^3, given in Table 4.1. This experiment is an impressive demonstration of (i) the fractionating power, and (ii) the high (absolute!) precision of the dynamic density gradient as well as (iii) the short duration of the analysis: Within only 10 min, much information on sample composition is gained. Again,

Fig. 4.12 is a good example for what we call "density spectroscopy" via AUC density gradients. It is interesting to note that this experiment in Fig. 4.12, done in an analytical ultracentrifuge, was repeated in a *preparative* ultracentrifuge, using 38-ml tubes (see Sect. 7.2 and Fig. 7.5).

4.4 Other Types of Density Gradients

Beside the two main types of density gradients (the static equilibrium, and the dynamic density gradients) described in this Chap. 4, there are other types of density gradients. Examples are the preparative dynamic density gradient (see Sect. 7.2), density gradient-related techniques such as the pH gradient (see Sect. 7.2), and the dynamic Percoll density gradient (see Sect. 4.4.1). These have in common that they are not (yet) applied broadly, and thus are not so much of practical interest. Nevertheless, as new approaches are also envisaged, it is worth considering at least the principles of alternative types of (density) gradient experiments. One example of such an exotic density gradient, the aqueous Percoll density gradient, will be discussed in more detail in the following section.

4.4.1 Percoll Density Gradient

A special type of fast *dynamic* AUC density gradient experiment, the (analytical) dynamic Percoll density gradient, was introduced by one of the authors in 1984 [16]. It allows one to analyze particle densities in the range 0.85 – 1.15 g/cm^3 within the timescale of a sedimentation velocity experiment, i.e., within 30 min. For this dynamic Percoll density gradient, simple standard cells, mono- or double-sector cells, are used, rather than synthetic boundary cells such as those used in the dynamic H_2O/D_2O density gradient described above in Sect. 4.3. The principle of this type of experiment is described in the following.

Figure 4.13 shows a TEM micrograph and the particle size distribution of Percoll. This is a dispersion of very small Percoll nanoparticles in water, supplied by Pharmacia Fine Chemicals, Uppsala, Sweden. Percoll consists of inert SiO_2 particles coated with polyvinyl pyrrolidone (PVP). The density of the Percoll particles is $\varrho_{\text{Percoll}} = 1.992\,\text{g/cm}^3$ (at 25 °C), particle sizes varying in the range 5 – 35 nm (see Fig. 4.13). The original aqueous Percoll dispersion, as supplied by Pharmacia, has a particle concentration of $c_{\text{Percoll}} = 24\,\text{wt\%}$, and a total density of $\varrho = 1.135\,\text{g/cm}^3$.

This aqueous Percoll dispersion, diluted to 12 wt% with water, is subjected to a sedimentation velocity run (see Fig. 4.14). Usually, the standard conditions are: 20 000 rpm, 3-mm mono-sector cells, and Schlieren optics. Three Schlieren pictures, after 20-, 30- and 40-min running time, are shown in Fig. 4.14. They yield an *average* sedimentation velocity coefficient $s = 180\,\text{S}$ for these Percoll particles (but there is a considerable s distribution of $40 < s < 600\,\text{S}$).

Because of the relatively *broad* particle size distribution of the Percoll particles, the Schlieren peak representing the sedimentation boundary of these Percoll particles is broadened with increasing experimental time, as the boundary moves

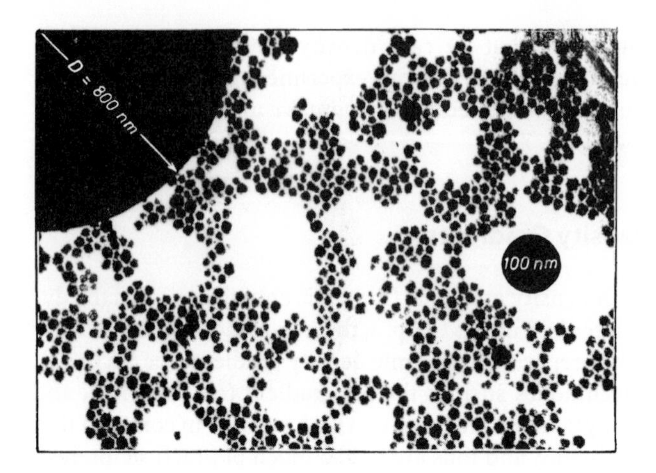

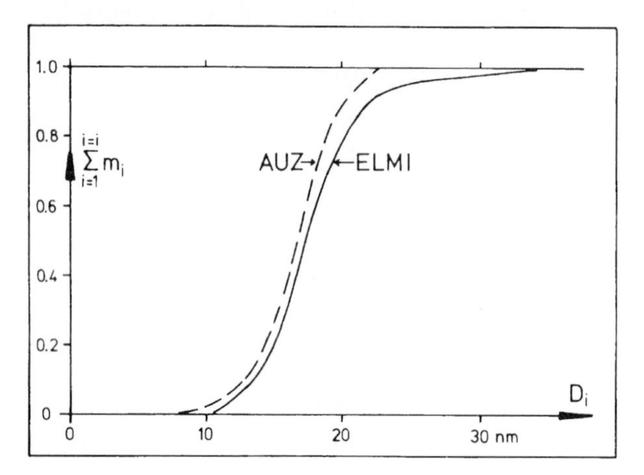

Fig. 4.13. Transmission electron microscopy (TEM) picture and particle size distribution of Percoll particles. Artificial *black spots* representing 100- and 800-nm particles are added to the TEM micrograph in order to illustrate the size ratios between the density gradient-forming Percoll particles and the analyzed polystyrene latex sample (reprinted with permission from [16])

toward the bottom of the AUC measuring cell. Within this sedimentation boundary, i.e., within the radial range of the corresponding Schlieren peak, the Percoll concentration $c(r)$ changes from zero at the upper end of the sedimentation boundary (near the meniscus) to 12 wt% at the lower end of the boundary (near the bottom) where the boundary passes into the plateau region with constant Percoll concentration. The radial concentration change $c(r)$ is associated with a radial density change $\rho_s(r)$ varying from 1.0 g/cm^3 in the pure solvent region (= pure water) to the density of the starting, 12-wt% Percoll dispersion, 1.06 g/cm^3 in the region of constant Percoll concentration. Therefore, the integration of the Schlieren optical signal in Fig. 4.14 (at a defined running time) yields directly the radial density gradient course $\rho_s(r)$ (in analogy to the dynamic H_2O/D_2O density gradient, (4.29) and (4.30) in Fig. 4.11). In the upper 20-min Schlieren photograph of Fig. 4.14, the result of this integration is visualized as a ρ axis (parallel to the r axis). This ρ axis varies with the running time t, i.e., it shifts increasingly in the direction of the cell bottom. It can be summarized that by performing a standard sedimentation

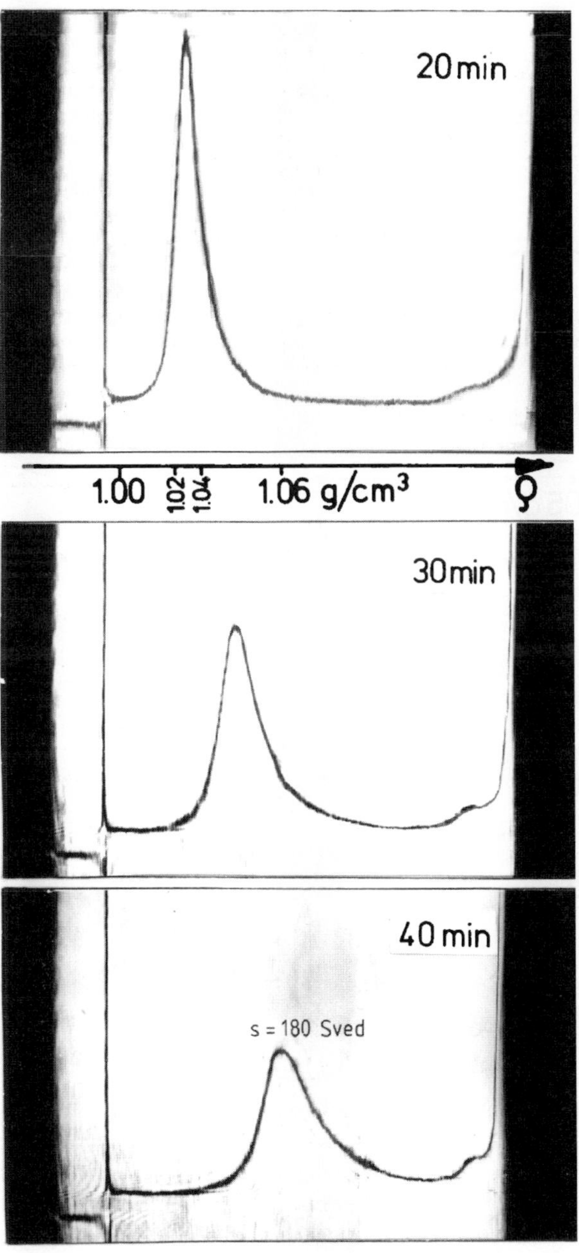

Fig. 4.14. Schlieren optical pictures of a sedimentation velocity run of a dispersion of 12-wt% Percoll particles dispersed in H_2O. Pictures taken after 20, 30, and 40 min. Due to the different sedimentation velocities of the differently sized PVP-coated SiO_2 particles ($5 < d_p < 35$ nm), the Schlieren peak is broadened in the course of the experiment. This creates a radial dynamic density gradient $\varrho_s(r,t)$ in the region of the Percoll peak ($1.00 < \varrho_s(r) < 1.06\,\text{g/cm}^3$), shown as a ϱ axis below the 20-min photograph (reprinted with permission from [16])

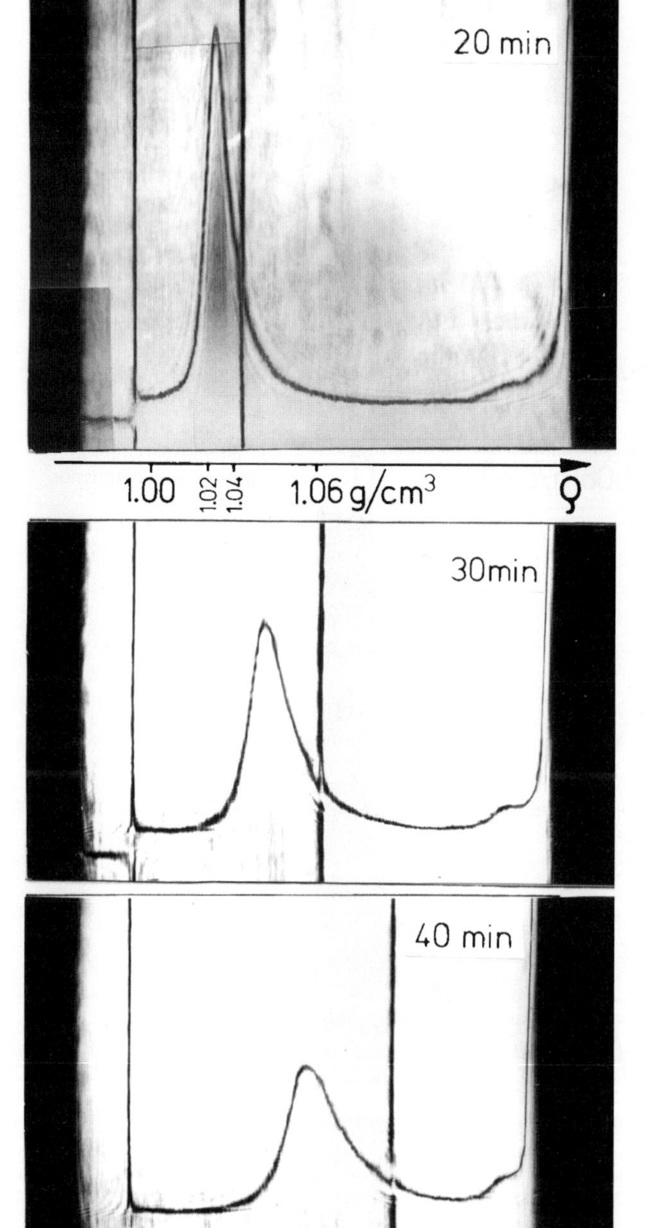

Fig. 4.15. Schlieren optical pictures of a dynamic 88 H_2O/12 Percoll density gradient experiment of a polystyrene latex sample consisting of 800-nm particles (c_{PS} = 0.01 g/l). The turbidity band caused by the rapidly sedimenting/floating polystyrene particles stays at its relative position of equal density throughout the experiment, and sediments together with the Percoll Schlieren peak toward the cell bottom (reprinted with permission from [16])

velocity run with Percoll particles, a relatively fast *dynamic* density gradient $\varrho_s(r,t)$ is established within 20–40 min in an AUC. This can be advantageous, as will be shown in the following.

In Fig. 4.15, Schlieren optical pictures of another experiment performed similarly to that described above are shown. In this case, a small amount of an aqueous polystyrene dispersion ($c_{PS} = 0.01$ g/l, consisting of big latex particles 800 nm in diameter) is added to the Percoll dispersion. In addition to the broadened Schlieren peak of Percoll, a narrow turbidity band can now be observed. This turbidity band is formed by the polystyrene particles, which move very fast, within minutes, by simultaneous sedimentation and flotation, to their isopycnic radial position r_{iso} in this dynamic Percoll density gradient.

The turbidity band remains in the region of the Percoll peak, and does not show further relative sedimentation or flotation, compared to the Percoll peak. Together with the Percoll Schlieren peak, the turbidity band sediments completely to the cell bottom. The radial position of the turbidity band within the Percoll density gradient can easily be analyzed at any given running time, and thus ϱ_P calculated. From all photographs in Fig. 4.15, the obtained density of the turbidity band is the same, i.e., $\varrho_P = 1.052\,\mathrm{g/cm^3}$. This value corresponds very well with the known density of polystyrene latex particles ($\varrho_{PS} = 1.055\,\mathrm{g/cm^3}$), a fact that can be taken as a proof of concept. The relatively limited density range of dynamic water/Percoll density gradients can be shifted to $0.85-1.15\,\mathrm{g/cm^3}$ by substitution of the H_2O by either methanol (for lower densities) or heavy water D_2O (for higher densities). For details of experiments performed on samples containing different density fractions, unknown samples, and even on living biological cells, the reader is referred to [16]. The dynamic aqueous Percoll density gradient is restricted to large particles ($d_P > 100$ nm), which sediment much faster than the Percoll particles.

References

1. Cheng PY, Schachman HK (1955) J Am Chem Soc 77:1498
2. Meselson M, Stahl FW, Vinograd J (1957) Proc Natl Acad Sci 43:581
3. Mächtle W, Lechner MD (2002) Prog Colloid Polym Sci 119:1
4. Thomas CA, Berns KI (1961) J Mol Biol 3:277
5. Ifft JB, Voet DH, Vinograd J (1961) J Phys Chem 65:1138
6. Hermans JJ, Ende HA (1963) J Polym Sci Part C Polym Symp 1:161
7. Fujita H (1962) Mathematical theory of sedimentation analysis. Academic Press, New York
8. Van Holde KE, Baldwin RL (1958) J Phys Chem 62:734
9. Rolfe R (1962) J Mol Biol 4:22
10. Rockwood D (ed) (1984) Centrifugation, a practical approach, 2nd edn. IRL Press, Oxford
11. Correia JJ, Weiss GH, Yphantis DA (1977) Biophysical J 20:153
12. See http://rasmb-email.bbri.org/rasmb_search.html
13. Hexner PE, Radford LE, Beams JW (1961) Proc Natl Acad Sci 47:1848
14. Rossmanith P, Mächtle W (1997) Prog Colloid Polym Sci 107:159
15. Lange H (1980) Colloid Polym Sci 258:1077
16. Mächtle W (1984) Colloid Polym Sci 262:270
17. Dale D, Gladstone F (1858) Philos Trans 148:887

18. Mächtle W (1999) Prog Colloid Polym Sci 113:1
19. Mächtle W (1992) In: Harding SE, Rowe AJ, Horton JC (eds) Analytical ultracentrifugation in biochemistry and polymer science. The Royal Society of Chemistry, Cambridge, p. 147
20. Mächtle W, Klodwig U (1979) Makromol Chem 180:2507
21. Lechner MD, Mächtle W, Sedlack U (1997) Prog Colloid Polym Sci 107:148
22. Lechner MD, Borchard W (1999) Eur Polym J 35:371

5 Sedimentation Equilibrium

5.1 Introduction

Sedimentation equilibrium runs in an AUC – in short, "equilibrium runs" – are long-duration runs of 10–200 h. The essential is that during this long time, equilibrium is reached between sedimentation to the cell bottom and (back-) diffusion of the dissolved (macro) molecules inside the solution column. This means that the radial concentration distribution $c(r)$ of the solute inside the measuring cell is independent of time or, in other words, the left-hand term in Lamm's equation ((1.10) or (1.15)) is zero, which means that $\mathrm{d}c/\mathrm{d}t = 0$ is valid all times.

There are two kinds of equilibrium runs in an AUC – first, the static density gradient run to measure densities ϱ_p of dissolved macromolecules and particles, and second, the sedimentation equilibrium run to measure molar masses M of dissolved macromolecules. In the case of polydisperse solutes, the sedimentation equilibrium run allows us to measure the different average molar masses, M_n, M_w, M_z (see also Sect. 3.4.2 and (3.18)), and the complete molar mass distribution $W(M)$, where

$$M_\mathrm{n} = \frac{\sum c_\mathrm{i}}{\sum c_\mathrm{i} \cdot M_\mathrm{i}^{-1}} \quad \text{is the number average molar mass,} \tag{5.1a}$$

$$M_\mathrm{w} = \frac{\sum c_\mathrm{i} M_\mathrm{i}}{\sum c_\mathrm{i}} \quad \text{is the weight average molar mass,} \tag{5.1b}$$

$$M_\mathrm{z} = \frac{\sum c_\mathrm{i} M_\mathrm{i}^2}{\sum c_\mathrm{i} \cdot M_\mathrm{i}} \quad \text{is the z-average molar mass} \tag{5.1c}$$

and c_i is the mass percentage of the component i with molar mass M_i within the whole polydisperse polymer. Density gradient runs are already broadly treated in Chap. 4, and thus this Chap. 5 deals only with the sedimentation equilibrium run to measure molar masses M and molar mass distributions $W(M)$. In addition to the sedimentation equilibrium method, there are some other AUC methods to measure M (see Sect. 5.5), but the sedimentation equilibrium method is the most important one.

Beside the AUC, there are other methods to measure molar masses and molar mass distributions, such as the scattering methods (static light scattering (SLS), small angle neutron scattering (SANS), small angle X-ray scattering (SAXS)), osmometry, viscosimetry, size exclusion chromatography (SEC), gel electrophoresis, and matrix-assisted laser desorption/ionization time-of-flight mass spectrometry (MALDI-TOF-MS). However, the AUC is one of the most powerful and versatile tools, especially to determine molar mass distributions (see example in Sect. 3.4.2), because a *fractionation* according to M_i takes place within the AUC measuring cell. In contrast to SEC and gel electrophoresis, which are relative methods, the AUC is an absolute method, which means that no calibration standards are necessary to measure molar masses. Additionally, the AUC covers the broadest range of molar masses, from about 300 up to several million Dalton (g/mol). This is possible because the rotor speed ω can be varied easily between 500 (for high M) and 60 000 rpm (for small M).

The basic equation of sedimentation equilibrium runs used to calculate M from the measured concentration distribution $c(r)$ (for monodisperse solutes in an ideal non-interacting solution, where $A_2 = 0$ is valid, and at infinite dilution $c \to 0$) follows from Lamm's equation (1.15) for $\mathrm{d}c/\mathrm{d}t = 0$, or more simply, from the net mass transport equation (1.13) for $\mathrm{d}m/\mathrm{d}t = 0$ to

$$\frac{s}{D} = \frac{(\mathrm{d}c/\mathrm{d}r)}{\omega^2 rc} \tag{5.2}$$

and further, by substituting s/D by means of the Svedberg equation (1.8), to

$$M = \frac{RT}{\left(1 - \overline{\nu} \cdot \varrho\right)\omega^2} \cdot \frac{(\mathrm{d}c/\mathrm{d}r)}{rc} \tag{5.3}$$

or, with $\mathrm{d}c/c = \mathrm{d}\ln c$ and $2r\,\mathrm{d}r = \mathrm{d}(r^2)$, to

$$M = \frac{2RT}{\left(1 - \overline{\nu} \cdot \varrho\right)\omega^2} \cdot \frac{\mathrm{d}\ln c}{\mathrm{d}r^2} \tag{5.4}$$

This final basic equation (5.4) says that a plot of the logarithm of concentration $c(r)$ versus the square of the radial distance r^2 should give a straight line (see, for example, Fig. 5.3b), with the slope being directly related to the molar mass M of the (monodisperse!) solute. In the case of a polydisperse solute in a non-ideal solution ($A_2 > 0$), this straight line will be transformed into a curvilinear line (upward-bending in the case of polydispersity/association, and downward-bending in the case of non-ideality). A careful analysis of this curvature will yield the molar mass distribution of the solute, as well as information on possible interactions, such as self-association, or on the thermodynamic quality of the analyzed solution, i.e., about the association constants K_a or the second virial coefficients A_2. This analysis is the main topic of Chap. 5.

The sedimentation equilibrium method was used extensively in the years 1950–1970 for molar mass measurements of biopolymers, especially for proteins. Nowadays for proteins, amino acid sequencing and gel electrophoresis is much faster and more precise. Nevertheless, still today only the equilibrium method allows us to study reversible associations of proteins in solution (see Sect. 5.4.2). In the years

1950–1980, molar mass determinations via the equilibrium method, using the old analytical ultracentrifuge of Beckman, the famous "Model E", was a tedious, long procedure, because of working with photo-plates, special calibrations, a complex digitization of the photo-plates in a comparator by hand, and manual graphical extrapolations. Nowadays, all these problems are eliminated with the new Beckman Optima XL-A/I machine, which is fully computerized and digitized via a CCD camera and an image processing system, delivering 100-fold more measuring data than before, and having an eight-cell rotor multiplexer. For synthetic polymers and especially synthetic polyelectrolytes, which are mostly broadly distributed, molar mass determinations via the equilibrium method have now become easier, faster, and more precise. We will demonstrate that in this Chap. 5 by presenting only measurements on synthetic polymers (with one exception in Sect. 5.4.3).

There are many publications about the sedimentation equilibrium method in the literature. For a deeper study and for further details, we recommend the following review articles: Creeth and Pain [1], van Holde and Baldwin [2], Chervenka [3], Yphantis [4], and Harding et al. [5].

5.2 Experimental Procedure

Equilibrium runs in an AUC to measure molar masses M can be done with all optical systems: Schlieren optics, interference optics, or the UV/VIS scanner. The latter is appropriate only if the solute shows absorption, and if the solvent is transparent for the corresponding wavelength λ. This is always valid for aqueous solvents (i.e., for biopolymers and polyelectrolytes), but it is not valid for the most common solvents of synthetic polymers, such as toluene, THF and DMF. These solvents show strong UV absorption, so that the dissolved polymer is "invisible". Thus, interference optics is the universal detector, and mostly used for synthetic polymers. Figure 5.1 shows the primary results (before data evaluation) of a standard equilibrium run recorded with the three different optical systems mostly used: a UV scan (done with an Optima XL-A/I), and a combined interference and Schlieren photograph (both done with an old Model E during the same run). In the new Optima XL-A/I, these "photos" are done with a digital CCD camera (see the interference "photos" made by an Optima XL-A/I in Fig. 5.2). All three scans/photographs in Fig. 5.1 show the typical result of an equilibrium run: an exponential radial course of the concentration distribution $c(r)$ within the measuring cell.

The standard measuring cell is a double-sector cell with a thickness of $a = 12\,\mathrm{mm}$, and with the corresponding window holder (UV, interference or Schlieren). Also 3- or 25-mm centerpieces (aluminum, Epon or titanium, the latter is recommended for organic and aggressive solvents) are possible. The initial concentration c_0 of the solute should be as low as possible to be near the ideal state $c \to 0$. On the other hand, c_0 depends on the sensitivity of the optical system chosen: 0.05 – 1.00 g/l for the most sensitive UV scanner, 0.5 – 10.0 g/l for the interference optics, and 2 – 40 g/l for the Schlieren optics. For non-ideal solutions, a series of concentrations $c_1, c_2, c_3, \ldots$ is recommended (three for the four-cell rotor,

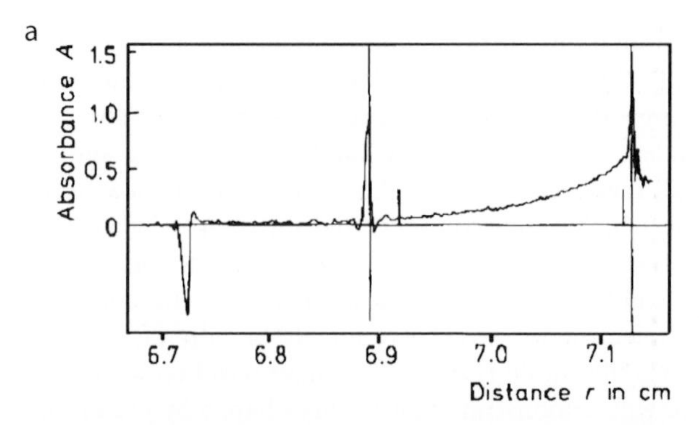

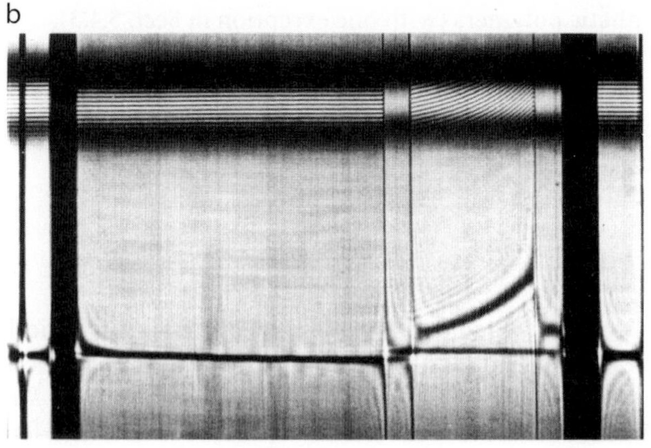

Fig. 5.1a,b. Typical primary results of standard equilibrium runs to measure molar masses. **a** A radial UV scan, and **b** a combined Schlieren and interference photograph, using 12-mm double-sector cells, 2.5-mm filling heights, and an artificial FC-43 oil bottom (reprinted from [3] and [6] with permission)

and seven for the eight-cell rotor). This allows us to carry out the necessary extrapolation $c \to 0$. Sapphire windows are here preferred for optical reasons, compared to quartz windows in the case of interference optics.

The total measuring time t_{equil} needed to reach equilibrium is proportional to the square of the cell filling height, $(r_b - r_m)^2$, according to [2]. Thus, measuring times of hours up to weeks are possible. In practice, one chooses a 2–3 mm length of the solution column (as in Fig. 5.1), in order to be close to the equilibrium state within about 24 h. Also short column equilibrium runs with filling heights of 0.5–1.0 mm and shorter running times of only 3–10 h are possible. Although these M determinations by short column runs are not very precise, especially in the case of polydisperse samples, they may be used for analyzing (mostly monodisperse) proteins. In contrast to former days, with the tedious, manual comparator evaluation of photo-plates, it is today much easier to check whether the equilibrium state is really reached. This is done by continuous automatic scans during the whole equilibrium run, until comparisons between following scans do not show any difference within the detector sensitivity. Only the last scan and its digital data are used for the evaluation of the molar mass M and the molar mass distribution $W(M)$.

Because the measuring data at both ends of the solution column are very important, it is common practice to use a small artificial bottom (about 0.5 mm in

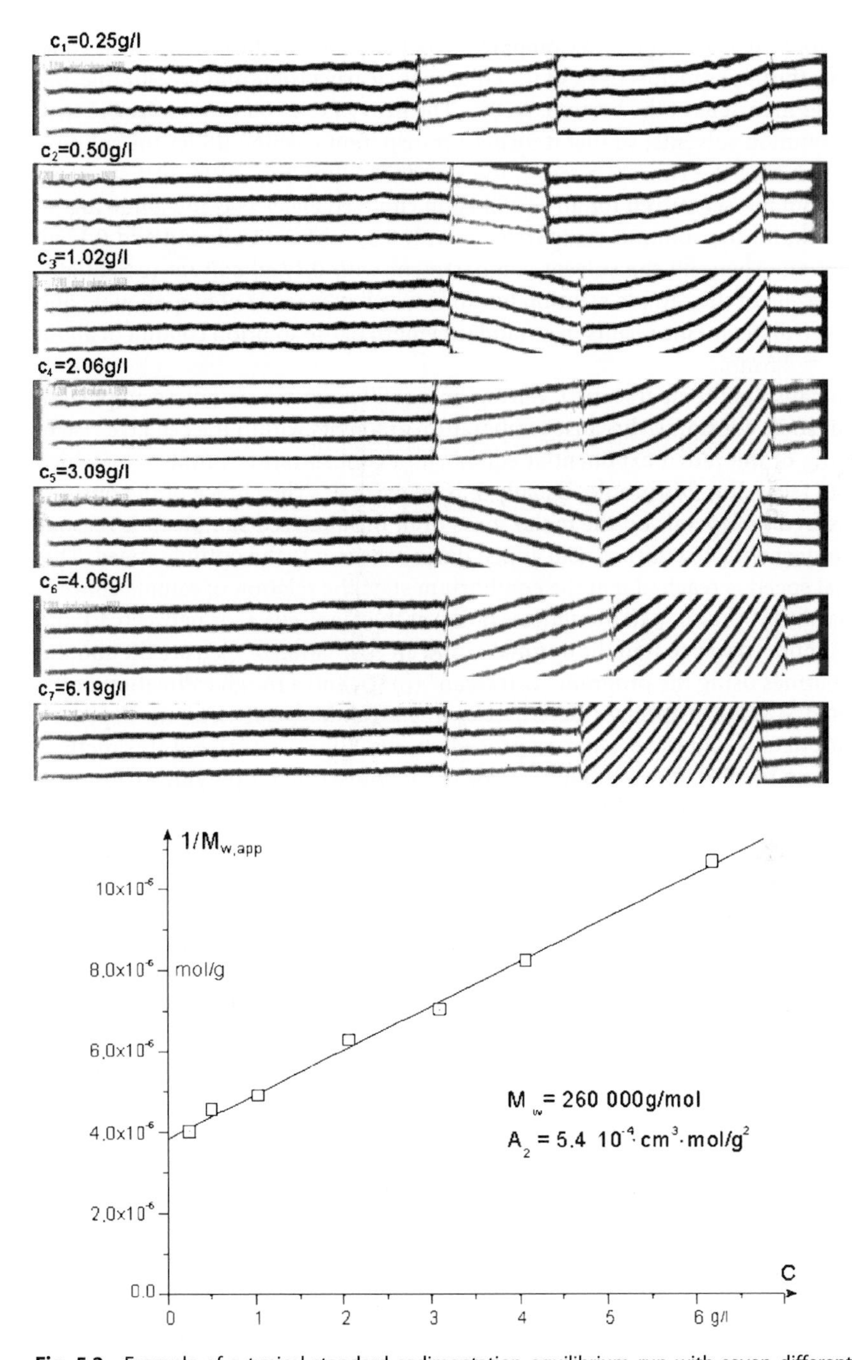

Fig. 5.2. Example of a typical standard sedimentation equilibrium run with seven different concentrations of the NBS 706 polystyrene in toluene. The *upper part* shows the seven primarily taken interference "photos" of an Optima XL-A/I, and the *lower part* the resulting (classical) evaluation, delivering M_w and A_2

height) of heavy FC-43 oil (3M Company) to better visualize the bottom area. Thus, not only the upper meniscus r_m of the column is clearly visible and measurable, but the lower meniscus at the bottom, r_b, too. The heavy FC-43 oil is not miscible with most common solvents, so that it forms a transparent cushion under the solution column during the run. As a result, the standard filling procedure of a standard 12-mm double-sector cell for an equilibrium run (see Fig. 5.1) is the following: to start with, fill the solution compartment with 0.03 ml FC-43 oil, and add 0.12 ml of solution. Then, fill the solvent compartment with 0.17 ml solvent (essentially, the dialyzate of the sample solution has to be used in the case of biopolymers and polyelectrolytes). This results in a standard length of the solution column of about $r_b - r_m = 2.5\,\text{mm}$.

The absolute temperature T during an equilibrium run is not so important but not so its constancy, because temperature gradients within the cell give rise to remixing of the radial exponential equilibrium concentration profile $c(r)$, which leads to falsifying this profile. Usually T is 20 or 25 °C. The choice of the optimal rotor speed ω depends on the molar mass M of the analyzed sample. It is reasonable to start with 10 000 rpm, and adjust during the run to the optimal speed. This optimal speed is reached if at the equilibrium state the relation of sample concentration at the bottom c_b to the concentration at the meniscus c_m is between 3:1 and 10:1. A modern way to find the optimal ω and an estimation of t_{equil} are the simulation routines using the program "Ultrascan" ([33]), and a rough estimation of the expected M value. A special case, where $c_m = 0$ is valid, we call meniscus depletion run (for details and the evaluation procedure, see Yphantis [4]), but this method is not very precise. It especially fails for broadly distributed polymers (to prove $c_m = 0$, one needs longer solution columns, $r_b - r_m$). In order to reach the equilibrium state faster, often overspeeding is used, which means choosing for the initial 3 h of an equilibrium run a rotor speed that is 1.5 times higher than the final speed.

The basic equilibrium run equations (5.3) and (5.4) at the beginning of this Chap. 5 show that for the calculation of M, beside T, ω and the measured concentration profiles, $c(r)$ or $\mathrm{d}c/\mathrm{d}r(r)$, a precise knowledge of the important buoyancy term $\left(1 - \overline{v} \cdot \rho\right)$ is needed. Often the partial specific volume $\overline{v}$ of the solute, and the solution density ρ (which is approximately the density of the pure solvent) are known from the literature or tables. If not, we have to measure it (as described in Sect. 2.6.1) with a high-precision Kratky density balance on a concentration series (5, 10, 20 g/l) of the solute in the solvent used.

To transfer the primary measured radial profiles of the absorption $A(r)$, the refractive index $n(r)$, or refractive index gradient $\mathrm{d}n/\mathrm{d}r(r)$ (see Fig. 5.1) into the secondary concentration profile $c(r)$ used for calculating M, specific optical solute/solvent constants are needed. For the absorption optics, this is the specific decadic absorption coefficient ε, and for the interference and Schlieren optics this is the specific refractive index increment $\mathrm{d}n/\mathrm{d}c$ (usually, as described in Sect. 2.6.2, $\mathrm{d}n/\mathrm{d}c$ is measured by means of a high-sensitivity or a differential refractometer, using the same concentration series as that used for the $\overline{v}$ determination).

For the absorption optics, this transfer relationship (see [6]) between the absorbance $A = \lg(I_0/I)$ and the concentration c is given by Lambert-Beer's law (2.2), $A(r) = \lg(I_0/I(r)) = \varepsilon c(r)a$, if ε is known from the literature or tables. If ε is not known, any equilibrium run can deliver ε with the help of the law of conservation of mass,

$$\int_0^1 c(X)\,\mathrm{d}X = c_0 \tag{5.5}$$

where c_0 is the initial concentration, and $X = (r^2 - r_\mathrm{m}^2)/(r_\mathrm{b}^2 - r_\mathrm{m}^2)$ is the normalized squared radius r. Combination of (2.2) and (5.5) gives the wanted

$$\varepsilon = \frac{1}{c_0 a} \int_0^1 A(X)\,\mathrm{d}X \tag{5.6}$$

As c_0, a and $\int A(X)\,\mathrm{d}X$ are obtained experimentally, ε may be easily determined from equilibrium runs with absorption optics. The parameter ε depends considerably on the selected UV wavelength λ. A presupposition for the validity of this procedure is that the total amount of solute is recorded within the $A(X)$ signal, and no (high molar mass) parts of the solute have completely sedimented to the cell bottom.

For the interference optics (and the Schlieren optics – not presented here), this transfer relationship can be derived in a similar way. Because the vertical shift of the interference fringes J is proportional to the concentration c (see Sect. 2.4.2 and (2.3)), in analogy to the absorption optics and the Lambert-Beer equation (2.2), the following relation is valid:

$$c(X) = \frac{\left(\Delta J(X) + J_\mathrm{m}\right)\lambda}{a(\mathrm{d}n/\mathrm{d}c)} \tag{5.7}$$

where $\Delta J(X) = (J(X) - J_\mathrm{m})$ is the number of fringes shifted in relation to the meniscus position of the chosen fringe at the (radial) distance X, and J_m is the unknown (absolute) number of fringes at the meniscus. Again, J_m can be calculated with the help of the equation for the conservation of mass, (5.5), via the following relation:

$$J_\mathrm{m} = J_0 - \int_0^1 \Delta J(X)\,\mathrm{d}X = \frac{c_0 a(\mathrm{d}n/\mathrm{d}c)}{\lambda} - \int_0^1 \Delta J(X)\,\mathrm{d}X \tag{5.8}$$

where $J_0 = (c_0 a(\mathrm{d}n/\mathrm{d}c))/\lambda$, calculated with (2.3), is the total number of fringes of the initial concentration c_0. Again, as c_0, a, $\mathrm{d}n/\mathrm{d}c$, λ, and $\int \Delta J(X)\,\mathrm{d}X$ are obtained experimentally, J_m and also $c(X)$ may be determined from equilibrium runs with interference optics. If c_0 of the initial concentration is not known, it is possible to determine J_0 with a synthetic boundary run of that solution (see Sect. 3.6). The accuracy of AUC M measurements via interference optics depends essentially on the accuracy of the J_m determination (see also [20]).

For nearly all measuring methods of molar masses M, also for the equilibrium run, we need to know the initial concentration c_0 of the starting solution. An M value is always only as precise as this value c_0. Usually, c_0 is determined in a relatively precise manner by weighing in the solid solute into the solvent. However, there are cases where this is not possible, for example, if there is a ready solution with an unknown concentration c_0. In this case, one can solve the problem by a synthetic boundary run of this solution within the AUC, and determine c_0 as described in Sect. 3.6.

We would like to conclude the description of the experimental procedure with a typical example of a standard equilibrium run, shown in Fig. 5.2: the primary interference "photos", taken first by means of an Optima XL-A/I CCD camera within the same run, where seven different concentrations of a polystyrene sample, dissolved in toluene, were measured simultaneously. Also shown are the resulting molar mass M_w and the second virial coefficient A_2, delivered by a complex evaluation. How this data analysis is handled will be described in the following Sect. 5.3.

5.3 Data Analysis

This Sect. 5.3 is divided into four parts. In Sect. 5.3.1, we present the well-known classical evaluation of sedimentation equilibrium runs, and the corresponding equations to calculate M of monodisperse samples, and M_w and M_z of polydisperse samples. In Sect. 5.3.2, we describe a powerful nonlinear regression evaluation method, created by Lechner in 1991, to determine the complete molar mass distribution $W(M)$, using (i) a versatile three-parameter *model* molar mass distribution $W(M, P, B, K)$, and (ii) also the M_w and M_z values of the classical evaluation, in Sect. 5.3.1, as starting parameters for the nonlinear regression. Lechner's method is particularly well suited for polydisperse samples. Section 5.3.3 shows practical applications of the classical and of the nonlinear regression evaluation methods. In Sect. 5.3.4, we outline the M-STAR evaluation method (M^*), which delivers, as does the classical evaluation, an M_w value without the need of a model. The M-STAR program can be used as a pre-evaluation program preceding any model-dependent evaluation.

5.3.1 Basic Equations for Molar Mass Averages – Classical Evaluation of AUC Equilibrium Runs

The full theoretical thermodynamic foundation of the sedimentation equilibrium is given in the famous books of Fujita [7,8]. The basic equation for the radial concentration $c(r)$, or $c(X)$ with $X = (r^2 - r_m^2)/(r_b^2 - r_m^2)$ at the sedimentation equilibrium state can be described for non-ideal solutions ($A_2, A_3, \ldots > 0$) of *monodisperse* polymers by the following relation [8]:

$$r\,\mathrm{d}r \quad \frac{\left(1-\bar{v}_2\varrho_1\right)\omega^2}{RT} = \mathrm{d}(\ln c)\left\{\frac{1}{M} + 2A_2^{se}c + 3A_3^{se}c^2 + \ldots\right\} \tag{5.9}$$

where A_2^{se}, A_3^{se}, ... are the virial coefficients of sedimentation equilibrium. This (5.9) is essentially identical with the basic (5.3) and (5.4) for $A_2^{se} = A_3^{se} = \ldots = 0$.

For non-ideal solutions of *polydisperse* polymers, the following relation holds for the ith (monodisperse) component within that polydisperse sample with in all $i = 1, 2, \ldots, q$ components [8]:

$$r\,\mathrm{d}r\frac{\left(1-\bar{v}_2\varrho_1\right)\omega^2}{RT} = \frac{\mathrm{d}\left(\ln c_i\right)}{M_i} + 2\sum_{k=1}^{q} A_{2ik}\,\mathrm{d}c_k + \ldots \tag{5.10}$$

Summation over all solute components i, and division by $\sum c_i \equiv c(r)$ at distance r yields the "point" weight-average molar mass $M_w(r)$ (see (5.1b) at a radial displacement r (or X):

$$\frac{\sum c_i M_i}{\sum c_i} = M_{w,app}(r) = \frac{\left(r_b^2 - r_m^2\right)}{\Lambda}\frac{\mathrm{d}(\ln c)}{\mathrm{d}(r^2)} = \frac{\left(r_b^2 - r_m^2\right)}{2\Lambda}\frac{(\mathrm{d}c/\mathrm{d}r)}{c \cdot r} = \frac{\mathrm{d}\ln c}{\Lambda\,\mathrm{d}X} \tag{5.11}$$

with the constant

$$\Lambda = \frac{\left(1-\bar{v}_2\varrho_1\right)\omega^2\left(r_b^2 - r_m^2\right)}{2RT} \tag{5.12}$$

The suffix "app" in $M_{w,app}(r)$ includes correction terms for non-ideality (i.e., for $A_2, A_3 > 0$). The correct $M_w(r)$ is obtained by extrapolating the $M_{w,app}(r)$ values to zero concentration $c \to 0$. As Schlieren optics measures $\mathrm{d}n/\mathrm{d}r$ (which is proportional to $\mathrm{d}c/\mathrm{d}r$), rather than $c(r)$, one has to rearrange (5.10) with respect to $\mathrm{d}n/\mathrm{d}r \sim \mathrm{d}c/\mathrm{d}r$. This gives the "point" z-average molar mass $M_z(r)$ (see (5.1c) at a radial displacement r (or X):

$$\frac{\sum c_i M_i^2}{\sum c_i M_i} = M_{z,app}(r) = \frac{\left(r_b^2 - r_m^2\right)}{\Lambda}\frac{\mathrm{d}}{\mathrm{d}r^2}\left[\ln\left\{\frac{1}{r}\frac{\mathrm{d}c}{\mathrm{d}r}\right\}\right] = \frac{2}{\Lambda}\frac{\mathrm{d}[\ln(\mathrm{d}c/\mathrm{d}X^2)]}{\mathrm{d}X} \tag{5.13}$$

The operational "point" average molar masses $M_{w,app}(r)$ and $M_{z,app}(r)$ are functions of the radial distance r (or X) at a point in the ultracentrifuge cell or, in other words, the local molar masses defined by the local slopes of the $c(r)$ curve or the $\mathrm{d}c(r)/\mathrm{d}r$ curve. To obtain the real, r-independent average molar masses $M_{w,app}$ and $M_{z,app}$ of the *whole* polydisperse polymer, one has to carry out an integration of (5.11) and (5.13), from the meniscus to the bottom of the cell (with the use of the relation for the conservation of mass, (5.5)). The results of that integration are the two following, well-known equations:

$$M_{\mathrm{w,app}} = \frac{(c_\mathrm{b} - c_\mathrm{m})}{\Lambda \cdot c_0} \tag{5.14}$$

$$M_{\mathrm{z,app}} = \frac{(r_\mathrm{b}^2 - r_\mathrm{m}^2)}{2\Lambda} \frac{1}{(c_\mathrm{b} - c_\mathrm{m})} \left[\frac{1}{r_\mathrm{b}} \left(\frac{\mathrm{d}c}{\mathrm{d}r} \right)_\mathrm{b} - \frac{1}{r_\mathrm{m}} \left(\frac{\mathrm{d}c}{\mathrm{d}r} \right)_\mathrm{m} \right] \tag{5.15}$$

Equation (5.14) is called the Lansing–Kraemer equation [9], and (5.15) the Richards–Schachman equation [10], after the scientists who first derived these in 1935 and 1959.

According to (5.14), we need only the difference $c_\mathrm{b} - c_\mathrm{m}$ to calculate $M_{\mathrm{w,app}}$, because all other values in this equation are known. This difference in concentration between the bottom of the cell and the meniscus can easily be determined directly by the UV scanner, using (2.2), or with interference optics by the (relative) number of shifted fringes at the cell bottom, $\Delta J(r_\mathrm{b}) = (J_\mathrm{b} - J_\mathrm{m})$, using (2.3):

$$c_\mathrm{b} - c_\mathrm{m} = \Delta c(r_\mathrm{b}) = \int_{r_\mathrm{b}}^{r_\mathrm{m}} \mathrm{d}c = \int_{r_\mathrm{b}}^{r_\mathrm{m}} \left(\frac{\mathrm{d}c}{\mathrm{d}r} \right) \mathrm{d}r = \frac{A(r_\mathrm{b}) - A(r_\mathrm{m})}{a\varepsilon} = \frac{(J_\mathrm{b} - J_\mathrm{m})\Lambda}{a(\mathrm{d}n/\mathrm{d}c)} \tag{5.16}$$

For Schlieren optics, integration is necessary (not presented here). To calculate $M_{\mathrm{z,app}}$ according to (5.15) is more complex, because we need, in addition to $c_\mathrm{b} - c_\mathrm{m}$, the local slopes $\mathrm{d}c/\mathrm{d}r$ at the radial positions r_b and r_m. These slopes are connected according to (5.11), as $\frac{\mathrm{d}c}{\mathrm{d}r} = \frac{2 \cdot \Lambda \cdot c \cdot r}{(r_\mathrm{b}^2 - r_\mathrm{m}^2)} \cdot M_{\mathrm{w,app}}(r)$, with the two point weight-average molar masses $M_{\mathrm{w,app}}(r_\mathrm{b})$ and $M_{\mathrm{w,app}}(r_\mathrm{m})$ at these radial positions. Introduction of these two relations into (5.15) yields

$$M_{\mathrm{z,app}} = \frac{c_\mathrm{b} M_{\mathrm{w,app}}(r_\mathrm{b}) - c_\mathrm{m} M_{\mathrm{w,app}}(r_\mathrm{m})}{(c_\mathrm{b} - c_\mathrm{m})} \tag{5.17}$$

Equation (5.17) is identical with (5.15), and is also called the Richards–Schachman equation. In contrast to the calculation of $M_{\mathrm{w,app}}$ (see (5.14)), where we need only the difference of the concentration at the bottom and at the meniscus, $c_\mathrm{b} - c_\mathrm{m}$), we now need for the calculation of $M_{\mathrm{z,app}}$ (see (5.17)) the absolute values of c_b and c_m, too. These values are delivered according to the procedure described in Sect. 5.2, using (2.2), (5.5) and (5.6) for absorption optics, as well as (2.3), (5.7) and (5.8) for interference optics.

As has been pointed out above, all M_{app} values include correction terms for non-ideality. Correct molar masses M are obtained by an extrapolation $M_{\mathrm{app}} \to M$. This is an extrapolation to infinite dilution $c \to 0$. The exact necessary mathematical extrapolation relation is, according to [8],

$$\frac{1}{M_{\mathrm{app}}} = \frac{1}{M} + 2A_2^{\mathrm{se}} \left\{ \frac{c_\mathrm{b} + c_\mathrm{m}}{2} \right\} + 3A_3^{\mathrm{se}} \left\{ \frac{c_\mathrm{b} + c_\mathrm{m}}{2} \right\}^2 + \ldots \tag{5.18}$$

However, if c is relatively low, a good approximation is

$$\frac{1}{M_{\text{app}}} = \frac{1}{M} + 2A_2^{\text{se}}c_0 + 3A_3^{\text{se}}c_0^2 + ... \quad (5.19)$$

In the following, we use only (5.19).

The (classical) determination of number-average M_n (see (5.1a)) or (z + 1)-average M_{z+1} molar masses will in principle be possible with equations that are similar to (5.14) or (5.15). Nevertheless, the applicability is very restricted, due to larger experimental errors on determining intercepts, slopes and further derivatives of the primary experimental curves, i.e., of $A(r)$, $\Delta J(r)$, and $\mathrm{d}n(r)/\mathrm{d}r \sim \mathrm{d}c(r)/\mathrm{d}r$, especially at the bottom of the cell and at the meniscus. It will be pointed out in the following Sect. 5.3.2 that the direct transformation (nonlinear regression) or the indirect (inverse Laplace) transformation of the reduced concentration profile $U_w(X) = c(X)/c_0$ or the reduced concentration gradient profile $V_w(X) = (\mathrm{d}c(X)/\mathrm{d}X)/c_0$ is more suitable for the determination of the molar mass averages M_n, M_w and M_z via an approximated three-parameter model $W(M)$ than is the case with the classical evaluation method. If the determined approximated $W(M)$ is of high quality, it allows us to calculate all needed average molar masses with (5.1), or with the integral form of this equation (see (3.18)).

5.3.2 Basic Equations for Molar Mass Distributions – Nonlinear Regression Evaluation (Lechner Method)

The determination of the molar mass distribution $W(M)$ of a polydisperse polymer via AUC equilibrium runs requires the connection of the experimentally measured, reduced concentration profile $c(X)/c_0 = U_w(X)$ (for absorption and interference optics), or the reduced concentration gradient profile $(\mathrm{d}c(X)/\mathrm{d}X)/c_0 = V_w(X)$ (for Schlieren optics) with (i) the molar mass distribution function $W(M)$, and (ii) the reduced concentration profile $U(X,M)$ or the reduced concentration gradient profile $V(X,M)$ of a monodisperse sample [8,11–13]:

$$U_w(X) = \left(\frac{c(X)}{c_0}\right)_{c_0\to 0} = \int_0^\infty W(M)U(X,M)\,\mathrm{d}M \quad (5.20)$$

$$U(X,M) = \Lambda M \frac{e^{\Lambda MX}}{e^{\Lambda M}-1} \qquad 0 < X < 1 \quad (5.21)$$

$$V_w(X) = \left(\frac{1}{c_0}\frac{\mathrm{d}c(X)}{\mathrm{d}X}\right)_{c_0\to 0} = \int_0^\infty W(M)V(X,M)\,\mathrm{d}M \quad (5.22)$$

$$V(X,M) = \Lambda^2 M^2 \frac{e^{\Lambda MX}}{e^{\Lambda M}-1} \qquad 0 < X < 1 \quad (5.23)$$

Equations (5.20) to (5.23) are valid only for ideal or pseudo-ideal solutions (where $A_2, A_3, ... = 0$ is valid). For non-ideal solutions, one has to perform a very precise extrapolation to infinite dilution $c \to 0$, or to introduce correction terms.

The problem of non-ideal solutions may be avoided by measuring pseudo-ideal solutions, i.e., using theta solvents at theta temperatures, where $A_2 = 0$ is valid. However, pseudo-ideal solutions are not available in all cases, which would restrict the application of this method, if needed.

In principle, (5.20) to (5.23) allow the direct estimation of the desired (approximated) molar mass distribution $W(M)$, and the corresponding molar mass averages M_n, M_w and M_z from the experimentally measured $U_w(X)$ or $V_w(X)$, if there are enough (and very accurate) measuring data. As mentioned above, this calculation is feasible only if a precise extrapolation toward zero concentration $c \rightarrow 0$ is possible, and has been made. The new Optima XL-A/I with its digitized absorption and interference optics allows us to carry out such a precise extrapolation to zero concentration by using a sufficient number of concentrations. Up to seven concentrations, simultaneously measured, each with about 300 measuring points, $A(r_i)$ or $J(r_i)$, deliver enough data for a reliable mathematical treatment.

There are two possible ways to obtain the desired molar mass distribution $W(M)$ from the experimentally measured concentration profiles, an "inverse" approach, and nonlinear regression. For a long time, the former option was performed by carrying out inverse Laplace transformations of the two Fredholm integral (5.20) and (5.21), and (5.22) and (5.23) [8, 11–14]. However, the disadvantage of this mathematically rigorous inverse Laplace transformation is that the operators are poorly conditioned, i.e., small changes of the experimental values $U_w(X)$ or $V_w(X)$ exhibit large changes of the distribution function $W(M)$, and, as a result, physically meaningless negative values of $W(M)$ may occur [8, 11–14]. Even the introduction of regularizing parameters basically does not solve the difficulties of the inverse Laplace transformation [13, 14]. An advantage of the Laplace transformation is that one does not need any model *MMD*.

The second way, the nonlinear regression evaluation, was first proposed by Lechner [15, 16], so we call it the Lechner method. The basic idea behind this approach was that nowadays fast computers are available in every laboratory, and so the direct estimation of $W(M)$ via nonlinear regression, and a reasonably assumed *model* $W(M)$ for the start of the regression are a good alternative to the inverse Laplace transformation. For this nonlinear regression, one supposes that $U_w(X)$ and $V_w(X)$ in (5.20) and (5.22) are functions of X and $\mathbf{k}$ where $\mathbf{k} = (k_1, k_2, ..., k_n)$ determines the model molar mass distribution function:

$$U_w(X) = \mathrm{f}\left(X, k_1, k_2, ..., k_n\right)$$

$$V_w(X) = \mathrm{g}\left(X, k_1, k_2, ..., k_n\right)$$

$$S = \sum \left(\mathrm{f}\left(X_i, k_1, k_2, ..., k_n\right) - U_w\left(X_i\right)\right)^2 = \text{minimum!}$$

$$S = \sum \left(\mathrm{g}\left(X_i, k_1, k_2, ..., k_n\right) - V_w\left(X_i\right)\right)^2 = \text{minimum!}$$

$$\mathrm{d}S/\mathrm{d}k_1 = \mathrm{d}S/\mathrm{d}k_2 = ... = \mathrm{d}S/\mathrm{d}k_n = 0$$

with $k_1, k_2, ..., k_n \equiv \mathbf{k}$ = constants of the model molar mass distribution function $W(M)$.

The nonlinear regression problem is then defined by minimizing S, which is the squared sum of the differences between the experimental points $U_w(X_i)$ and the calculated points $f(X_i, \mathbf{k})$ or $g(X_i, \mathbf{k})$. The problem is solved in the usual way by calculating n partial derivatives of S with respect to k, and setting these equal to zero. The constants may be $M_n, M_w, M_z, \ldots$, or constants of a model molar mass distribution function, e.g., $P, B, K, \ldots$ (see (5.24)). Unlike the inverse Laplace transformation problem, the nonlinear regression method of calculating $W(M)$ is well conditioned and very stable. The calculation can be performed mathematically by the multidimensional Newton method, or by the multidimensional Simplex method: the Simplex method is preferred because it needs no derivatives of the integral equation. The Newton method normally converges faster but it sometimes causes problems, especially if the experimental values are noisy. The calculation needs a model molar mass distribution $W(M)$. Lechner [17] proposed the following versatile exponential unimodal three-parameter model distribution function $W(M, P, B, K)$:

$$W(M) = \bar{c} M^{K} e^{-BM^{P}} \tag{5.24}$$

with

$$\begin{aligned}
\bar{c} &= \frac{PB^{K/P+1/P}}{\Gamma(K/P + 1/P)} = f(P, B, K) \\
M_n &= \frac{\Gamma(K/P + 1/P)}{B^{1/P}\Gamma(K/P)} = f(P, B, K) \\
M_w &= \frac{\Gamma(K/P + 2/P)}{B^{1/P}\Gamma(K/P + 1/P)} = f(P, B, K) \\
M_z &= \frac{\Gamma(K/P + 3/P)}{B^{1/P}\Gamma(K/P + 2/P)} = f(P, B, K) \\
M_{z+1} &= \ldots = f(P, B, K)
\end{aligned}$$

where P, B, K are parameters of the model molar mass distribution $W(M, P, B, K)$, Γ the Gamma function, $P = 0.2 - 0.5$ is valid for the Wesslau distribution or the log-normal distribution, $P = 1$ for the Schulz–Flory distribution, and $P = 2$ for the Maxwell distribution or the Poisson distribution.

Equation (5.24) covers nearly all types of unimodal molar mass distribution functions (Poisson, Schulz–Flory, Wesslau, log-normal, etc.) with different broadnesses and skewnesses, i.e., for most polymers created by radical or ionic polymerization. The constant P can also be considered as skewness parameter. In the following Sect. 5.3.3, we will demonstrate, by means of practical application examples, that Lechner's nonlinear regression method works very well. We will also compare average values of M_w and M_z from the same equilibrium run, but calculated with the two different evaluation methods, these being the classical one and the nonlinear regression method. Lechner's method is also applicable for multi-modal distribution functions $W(M)$, by modeling $W(M)$ by an addition of several model functions such as (5.24), with different P, B, K parameter sets for each component (see [16]).

5.3.3 Application of the Basic Equations – Classical and Nonlinear Regression Evaluation

To analyze an equilibrium run such as the one shown in Fig. 5.2, Lechner [15, 16] proposed the following five-step procedure, which combines the classical evaluation (steps 1 and 2, see theoretical consideration in Sect. 5.3.1) and the nonlinear regression evaluation (steps 3–5, see Sect. 5.3.2) in one computer program (this program is available on request from the author). Steps 1 + 2 deliver classically M_w, M_z and A_2 (which are used as starting parameters for the nonlinear regression), and steps 3–5 deliver the approximated three-parameter model molar mass distribution $W(M, P, B, K)$, and the resulting M_n^0, M_w^0, M_z^0 and P.

Step 1
Calculation of the reduced concentration profile $c(X)/c_0$ and the relative squared distance $X = (r^2 - r_m^2)/(r_b^2 - r_m^2)$ from the primary experimental values $A(r)$, or $\Delta J(r)$ or $(dn/dr)(r)$. There are only small differences in this calculation between absorption, interference, and Schlieren optics.

Step 2
Smoothing the experimental $c(X)/c_0$ measuring points (about 300 in the XL-A/I) by a continuous line using a special regression spline procedure [18]: for classical evaluation from these smoothed data, we (i) determine at meniscus and bottom c_m, c_b and the slopes $(dc/dr)_m$, $(dc/dr)_b$, and (ii) calculate, with (5.14) and (5.15), $M_{w,app}$ and $M_{z,app}$ for every concentration c_0; (iii) these M values are extrapolated according to (5.19) to infinite dilution $c_0 \to 0$. The extrapolated values are M_w and M_z. A_2 follows from the (beginning) slope of the $1/M_{app}$ versus c_0 extrapolation line. Subsequently, in step 5, these values are used as starting values in the nonlinear regression.

Step 3
Here, the nonlinear regression starts with the calculation of interpolated values of $c(X)/c_0$ for a dataset of 41 equal values of pre-selected equidistant X values $(0, 0.025, 0.050, 0.075, ..., 0.975, 1)$ from 0 to 1 with a Lagrange or spline interpolation procedure (we prefer the latter). This interpolation is absolutely necessary, as values of $c(X)$ are not measured for the same X values at different concentrations.

Step 4
Extrapolation of $c(X)/c_0$ to zero concentration $c_0 \to 0$ for constant X values by linear regression for lower-quality solvents or by quadratic regression for higher-quality solvents: the precision of this extrapolation is very important.

Step 5
Calculation of the approximated three-parameter model molar mass distribution $W(M, P, B, K)$ and the resulting parameters M_n^0, M_w^0, M_z^0 and P by the non-

linear regression evaluation according to the Simplex method (by minimizing $S = \sum[f(X_i, k_1, k_2, ..., k_n) - U_w(X_i)]^2$.

By means of the measuring example in Fig. 5.2, an interference optics equilibrium run of the calibration polystyrene NBS 706 (already used for *MMD* measurement via sedimentation velocity runs in Sect. 3.4.2, Fig. 3.12), in the following we demonstrate this five-step evaluation in detail. This polystyrene NBS 706 is well characterized and distributed worldwide by the National Bureau of Standards (NBS), Washington, DC. The given NBS data are: unimodal middle-broad molar mass distribution, M_n = 136 000 g/mol (osmometry), M_w = 257 800 g/mol (light scattering), M_w = 288 100 g/mol (sedimentation equilibrium), and via SEC, $M_z{:}M_w{:}M_n = 2.9{:}2.1{:}1$ (i.e., M_z = 355 000 – 400 000 g/mol, according to NBS).

The solvent, toluene, was appropriate and a thermodynamically good one (i.e. A_2 is high). Seven different concentrations were measured simultaneously at 25 °C. Rotor speed was 7500 rpm, and total running time 72 h (this was a long-weekend run; 20 h would be enough to reach equilibrium). The AUC was a Beckman Optima XL-A/I with an eight-cell rotor, and an interference optics multiplexer with modulable laser, λ = 675 nm; 12-mm double-sector cells were used. The filling height was about 3 mm (an artificial FC-43 oil bottom was used). As values for $\bar{v}$, ρ solution and d*n*/d*c*, we used $\bar{v} = 0.917\,\text{cm}^3/\text{g}$, $\rho_{\text{solution}} = \rho_{\text{solvent}} = 0.8622\,\text{g/cm}^3$, and d*n*/d*c* (675 nm, 25 °C) = $0.107\,\text{cm}^3/\text{g}$.

The evaluation of the interference "photos" of Fig. 5.2, i.e., the radial fringe shift $\Delta J(r)$ relative to the meniscus, was done inside the Optima XL-A/I by an imaging processing system. These data are accessible as ACSII files. The original Beckman programs are not usable for the five-step evaluation, so we used Lechner's program. Figure 5.3a shows the relative radial fringe shifts $\Delta J(r)$ of all seven concentrations extracted from Fig. 5.2 using the Lechner algorithm.

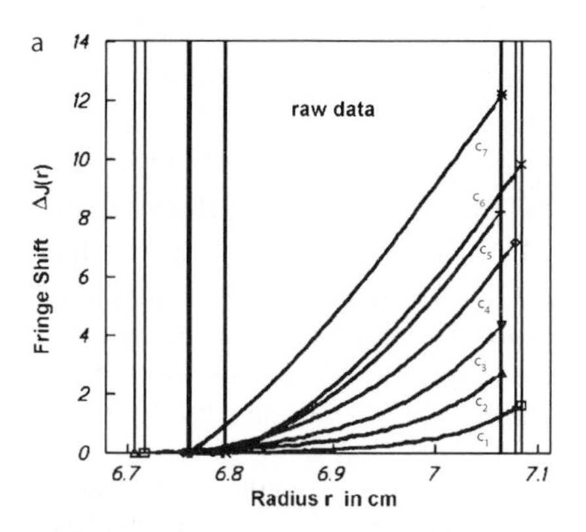

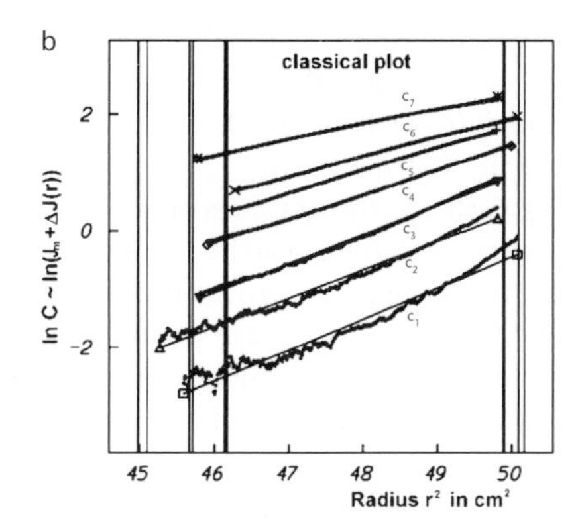

Fig. 5.3. **a** Relative shift $\Delta J(r)$ of interference fringes as a function of the rotor radius r for seven concentrations of polystyrene NBS 706 in toluene. A *continuous line* (by spline regression) is drawn through the individual measuring points, and extrapolated to the meniscus and to the cell bottom. **b** Classical plot of ln $c(r)$ as a function of r^2 to obtain M from the slope. The *vertical lines* in **a** and **b** are the positions of r_m and r_b in the different cells

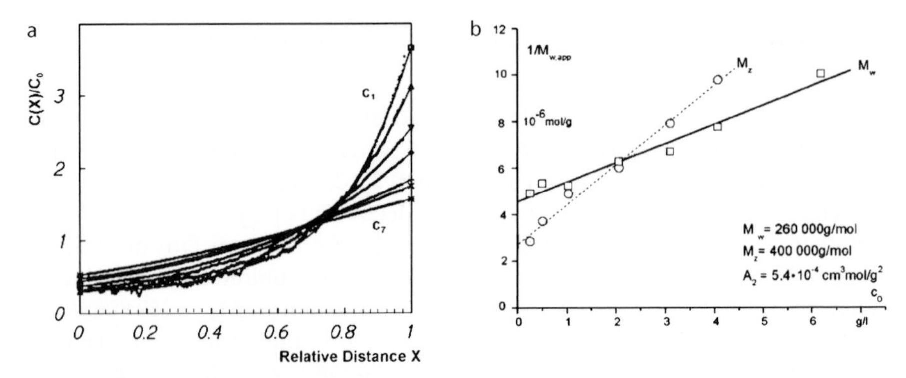

Fig. 5.4. a Reduced concentration profiles $c(X)/c_0$ as a function of X, for polystyrene NBS 706 in toluene. **b** $1/M_{\text{app}} - c_0$ extrapolation plots to obtain M_w, M_z and A_2 for $c_0 \to 0$ (classical evaluation)

In step 1, we will now transfer these relative fringe shifts $\Delta J(r)$ via (5.7) and (5.8) into the required radial (absolute) concentration profiles $c(r)$ or $c(X)$; then, $\ln c(r)$ is plotted (for visualization of linearity) as a function of r^2 (see Fig. 5.3b), and the quotient $c(X)/c_0$ is also plotted as a function of X (Fig. 5.4a).

Now in step 2, by a spline regression smoothing, we calculate the continuous lines through these measuring points, also presented in Fig. 5.4a (and Fig. 5.3a). The extrapolated values of these lines, both the $c(X)/c_0$ values and their slopes $(\mathrm{d}c(X)/\mathrm{d}r)/c_0$, at the meniscus r_m and at the bottom position r_b, are introduced into (5.14) and (5.15) to calculate $M_{w,\text{app}}$ and $M_{z,\text{app}}$. In Table 5.1, the individual numerical values of all seven concentrations measured are listed, together with the extracted $\Delta J(r_m) = (J_b - J_m)$ values and the calculated J_m values. All $M_{w,\text{app}}$ values are of high quality.

This is not valid in the case of the $M_{z,\text{app}}$ values at higher concentrations (cf. $M_{z,\text{app}}$ values in Table 5.1 in brackets), because the estimation of the slopes is not precise enough (there is some compensation in bending between polydispersity and non-ideality in the $\ln c(r)$ versus r^2 curves of Fig. 5.3b). Also listed in Table 5.1 are the (falsified!) $M_{\text{ideal,app}}$ values, which represent the straight optimal lines in Fig. 5.3b, i.e., the M values of polystyrene NBS 706 according to the basic (5.4), if we assume NBS 706 would be monodisperse and $A_2 = 0$. The bending of the measured curves in Fig. 5.3b proves that these assumptions are incorrect. The reciprocals of $M_{w,\text{app}}$ and $M_{z,\text{app}}$ are plotted in Fig. 5.4b, according to (5.19), as function of c_0, and then extrapolated via a regression line to infinite dilution $c_0 \to 0$. Intercepts and beginning slopes yield $M_w = 260\,000$ g/mol, $M_z = 400\,000$ g/mol, and $A_2 = 5.4 \times 10^{-4}\,\text{cm}^3\,\text{mol/g}^2$. These three values are the final result of the classical evaluation within step 2, also listed in Table 5.1. They will be used below as starting parameters for the nonlinear regression in step 5. The errors of measurement in these M values are about $\pm 8\%$.

In step 3, we start with the nonlinear regression by a spline interpolation of the $c(X)/c_0$ values at the 41 pre-selected fixed radial X positions. These pre-selected fixed X positions are shown above the X axis of Fig. 5.4a as small vertical lines.

In step 4, the interpolated $c(X)/c_0$ values at these fixed X positions are plotted as a function of c_0 (see Fig. 5.5), and then extrapolated to zero concentration $c_0 \to 0$ by quadratic regression to determine $c(X)/c_{0,0}$. For the sake of clarity, we show only ten of the total 41 regression lines in Fig. 5.5.

In step 5, these extrapolated $c(X)/c_{0,0}$ values at the 41 pre-selected X positions are plotted as a function of X (see Fig. 5.6a), and visualized as "experimental" points (open circles in the figure). They represent the reduced concentration profile at infinite dilution. Now, as shown also in Fig. 5.6a, through these "experimental" points a continuous, optimally fitted curve, based on the $W(M, P, B, K)$ function, is calculated by the nonlinear regression evaluation with the Simplex method,

Table 5.1. M_{ideal}, M_w, M_z and M_w^* values of polystyrene NBS 706 determined via an AUC sedimentation equilibrium run in toluene

c_0 g/l	$I_b - I_m$ –	I_m –	$M_{\text{app,ideal}}$ g/mol	$M_{\text{w,app}}$ g/mol	$M_{\text{z,app}}$ g/mol	$M^*_{\text{w,app}}$ g/mol
0.27	1.61	0.18	204 000	249 000	353 000	228 000
0.51	2.72	0.29	188 000	219 000	270 000	222 000
1.03	4.36	0.60	191 000	204 000	205 000	207 000
2.07	7.18	1.49	160 000	160 000	167 000	160 000
3.10	8.20	2.64	150 000	142 000	(127 000)	142 000
4.07	9.84	3.7	129 000	122 000	(102 000)	124 000
6.19	12.20	6.27	100 000	94 000	–	94 000
			$M_{\text{app,ideal}}$ g/mol	$M_{\text{w,app}}$ g/mol	$M_{\text{z,app}}$ g/mol	$M^*_{\text{w,app}}$ g/mol
Extrapolation to $c_0 \to 0$ yields:			220 000	260 000	400 000	246 000

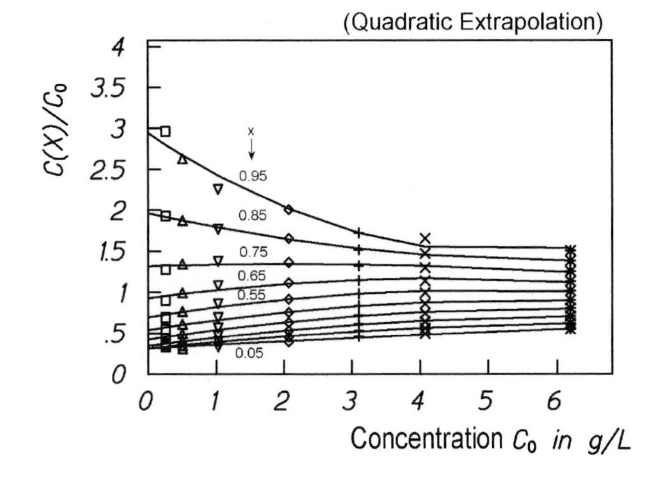

Fig. 5.5. $c(X)/c_0$ as a function of c_0 for different pre-selected X values. The *lines* indicate the quadratic extrapolation to $c_0 \to 0$; polystyrene NBS 706 in toluene

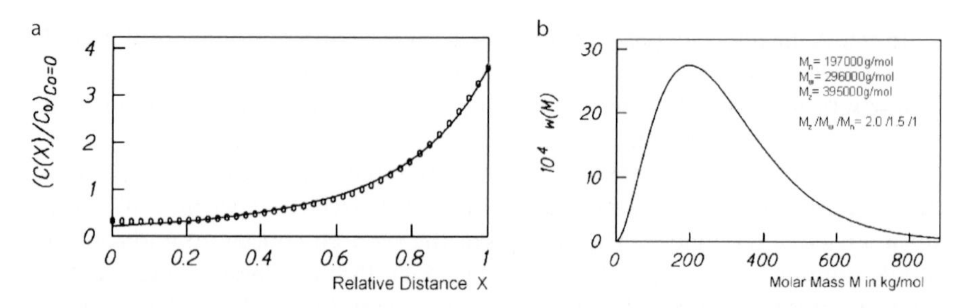

Fig. 5.6. **a** The reduced concentration profile $c(X)/c_{0,0}$ as a function of X, for polystyrene NBS 706 in toluene. An optimally fitted line, by nonlinear regression through the 41 "experimental" points, is shown. **b** Best-fit three-parameter model molar mass distribution $W(M,P,B,K)$, representing the optimally fitted line in **a**

using M_w and M_z of the classical evaluation as starting parameters. This yields the fitting parameters M_n^0, M_w^0, M_z^0 and P (chosen from some reasonably preselected P values), and the fitted three-parameter model molar mass distribution $W(M,P,B,K)$ itself. The latter is shown in Fig. 5.6b.

The resulting parameters of this nonlinear regression evaluation are $M_n^0 = 197\,000$ g/mol, $M_w^0 = 296\,000$ g/mol, $M_z^0 = 395\,000$ g/mol, and $P = 1$. There is a reasonable agreement if we compare these nonlinear regression values and also our classical evaluation values, $M_w = 260\,000$ g/mol and $M_z = 400\,000$ g/mol, with the NBS data ($M_n = 136\,000$ g/mol, $M_w = 257\,800/288\,100$ g/mol, and $M_z = 355\,000/400\,000$ g/mol). Several other, analogous comparisons in the literature [6, 15, 16, 18], based on absorption, interference, and Schlieren optics, prove that Lechner's nonlinear regression method is a reliable tool to estimate approximated molar mass distributions $W(M,P,B,K)$ and the corresponding M_n^0, M_w^0, and M_z^0 values via sedimentation equilibrium runs. The M values given above, determined by a sedimentation equilibrium run, agree well with the corresponding M values calculated in Sect. 3.4.2 (see Fig. 3.12) from a sedimentation velocity run via a scaling law.

During the *MMD* determination via the s run in Fig. 3.12a, a strong fractionation of NBS 707 took place. Thus, this *MMD* is more precise than the one obtained via the equilibrium run shown in Fig. 5.6b, where only a weak fractionation took place. Especially high molar mass components are better detected in s runs than in equilibrium runs. This becomes evident if one compares the two *MMD*s: the high molar mass tail in Fig. 3.12d is longer (up to 1300 kg/mol) than the one shown in Fig. 5.6b (up to 800 kg/mol). Nevertheless, the low molar mass tail, the maximum M, and the averages, M_n, M_w, M_z, are very similar. This result demonstrates that *MMD* determinations via equilibrium runs are a valuable analytical tool. However, the precision can be improved in the future by more precise measuring data and an improved data analysis.

5.3.4 M-STAR, a Special Data Analysis to obtain M_w

The first two steps (= classical evaluation) in Lechner's computer program to obtain M_w and M_z for polydisperse samples are model-free, i.e., they do not require any prior assumption of the nature of the system – monodisperse, polydisperse, ideal, non-ideal, non-interacting, interacting (e.g., self-associating, preferential solvation). A similarly powerful and interactive computer program to obtain M_w in another way is M-STAR (M^*), first proposed by Creeth and Harding [19] and modernized in 1997 by Cölfen and Harding [20]. In the following outline, we refer to the latter form. For further interesting details, we recommend to consult the original paper [20] (the M^* program includes the classical evaluation, too).

The M-STAR program exists in two forms, as M-STARA for absorption optics, and as M-STARI for interference optics. It takes into consideration the *whole* solute distribution $c(r)$ or $c(X)$, i.e., from the meniscus to the bottom, rather than only a selected dataset (such as in some commercial computer programs, [21]). M-STARA and M-STARI are therefore recommended, such as are Lechner's steps 1 and 2, as a first analysis program of equilibrium data. These programs are therefore particularly well suited if heterogeneity (polydispersity or interaction phenomena) or non-ideality are suspected. All these phenomena are indicated by a strong curvature (= bending) in the ln $c(r)$ versus r^2 plots (see Fig. 5.3b). This curvature often makes it difficult to determine the whole cell weight-average molar mass M_w, particularly if the regions near the cell bottom are not well defined (noisy data). This is often the case with absorption optical records, and especially with extremely broadly distributed polymers, such as polysaccharides, which show very steep, scarcely resolvable concentration gradients $dc(r)/dr$ in the bottom region.

To solve these problems was the intention of Creeth and Harding [19] when they developed their M^* procedure. $M^*(r)$ or $M^*(X)$ is an operational "point" average molar mass, similarly to $M_{w,app}(r)$ and $M_{z,app}(r)$ in (5.11) and (5.13), defined in analogy to the classical Lansing–Kraemer relation (5.14) by

$$M^*(r) = \frac{r_b^2 - r_m^2}{\Lambda} \cdot \frac{c(r) - c_m}{c_m\left[r^2 - r_m^2\right] + 2\int\limits_{r_m}^{r} r\left[c(r) - c_m\right]\,dr} \tag{5.25}$$

The only parameter in (5.25) that requires extrapolation is c_m (in contrast to Lechner's method, which additionally requires extrapolation of c_b). This c_m extrapolation can be achieved without difficulty with absorption optics, by simple linear extrapolation of $A(r)$ to the meniscus, because usually the radial concentration gradient is flat in the meniscus region. By contrast, interference optics requires a more subtle extraction for c_m, i.e., for J_m, because interference optics directly gives only the solute concentration relative to the meniscus. There are some possible ways to obtain J_m [1,20]. One of these, and perhaps the most reliable, is described in Sect. 5.2 by using (5.8), i.e., via the equation for the conservation of mass and an integration of $\Delta J(r)$.

It is a special advantage of the M^* method that, in (5.25), no differentiation process is necessary that is sensitive to noisy data. The M^* function itself has some useful identities (see [20]), the most important of which is that at the cell bottom r_b

$$M^*\left(r_b\right) = M_{w,app} \tag{5.26}$$

is valid, i.e., $M^*(r_b)$ is identical with the apparent weight-average molar mass over the *whole* distribution of the solute in the cell, from r_m to r_b. The function $M^*(r)$ is calculated using (5.25). It becomes increasingly stable with progressive integration (done with the usual trapezoid rule) from r_m to r_b.

The identical (5.26) provides a basis for obtaining $M_{w,app}$ without involving a concentration extrapolation to the cell bottom (but rather a less severe extrapolation of $M^*(r) \to M^*(r_b)$ to the cell bottom, which is less sensitive to errors). The extrapolation of $M^*(r)$ or $M^*(X)$ to the cell bottom is usually simple, and can be done by applying a straight extrapolation line in most cases. This extrapolation, done in a plot $M^*(r)$ versus r^2, or better, versus $X = (r^2 - r_m^2)/(r_b^2 - r_m^2)$, is possible even if the raw dataset is of poor quality, a feature often present with absorption optics. It should be noted here that one characteristic feature of the $M^*(r)$ function is that it starts off very noisy near the meniscus (because of small concentration increments), but then becomes more precise toward the cell bottom, thereby resulting in a good determination of $M_{w,app}$ via extrapolation (see Fig. 5.7). The use of $M^*(r)$ in this way can be thought of as an accumulator, starting (noisy) at the cell meniscus, and steadily homing in on the true $M_{w,app}$, as all information about the solute distribution is gathered by the time the cell bottom is reached. A further advantage of the M^* program is that one needs no absolute units of the concentration c in (5.25), because c stands both in the numerator and in the denominator. Relative units, such as fringe numbers J or absorption units A, are sufficient. This is ideal for samples with unknown concentration.

We will finish this M^* Sect. 5.3.4 with an application example, presented in Fig. 5.7 and in Table 5.1. For this application example, we use the measuring data of the sedimentation equilibrium measurement we evaluated in Sect. 5.3.3 with the Lechner method (by classical and nonlinear regression evaluation), namely, for polystyrene NBS 706 with seven concentrations in toluene (shown in Figs. 5.2–5.6, and Table 5.1). This allows us to make an interesting comparison between the two evaluation methods, i.e., the M^* method and Lechner's method.

Figure 5.7 presents the plot M^* versus X (see [22]) of polystyrene NBS 706. As expected, it is clearly visible that the noise within the seven different curves decreases with increasing concentration c_0 and with increasing X. The extrapolation $M^*(X \to 1)$, or $M^*(r \to r_b)$, to yield $M_{w,app}$, was done with straight lines through the seven curves in the bottom region $0.8 < X < 1$. The resulting numerical values are listed in Table 5.1. The comparison of the $M_{w,app}$ values yielded by the Lechner method and the M^* method show an excellent agreement, within only 2% (with exception of the lowest concentration $c_1 = 0.27$ g/l, which was a little too low, indicated by noisy measuring data).

This good agreement was also expected because (i) the molar mass distribution of polystyrene NBS 706, with M_w:M_n = 2.1:1, is not very broad, (ii) the measurement itself was a good one, i.e., the measuring data are not very noisy (see Figs. 5.3 and 5.4), and (iii) the Lechner method facilitates a good extrapolation of c_b. The full power of the M^* method will be visible only if the $c(r)$ curves are much steeper near the bottom region, as in Fig. 5.3a, and/or if the curvature of the ln $c(r)$ versus r^2 plot is much stronger, as in Fig. 5.3b.

5.4 Examples

In this Sect. 5.4, we will present further application examples of sedimentation equilibrium runs to measure M. In Sect. 5.4.1, two absorption optics examples, a nearly monodisperse polyelectrolyte (NaPSS 35 000) and again polystyrene NBS 706, are shown. In Sect. 5.4.2, the association of monofunctionalized poly(ethylene oxide) is studied, and a new AUC buoyant density method to determine $\overline{v}$ is presented, all done using the UV scanner and the original Beckman evaluation program. In Sect. 5.4.3, an example with the new fluorescence detector of Laue ([28]; see Sect. 2.4.4) is shown.

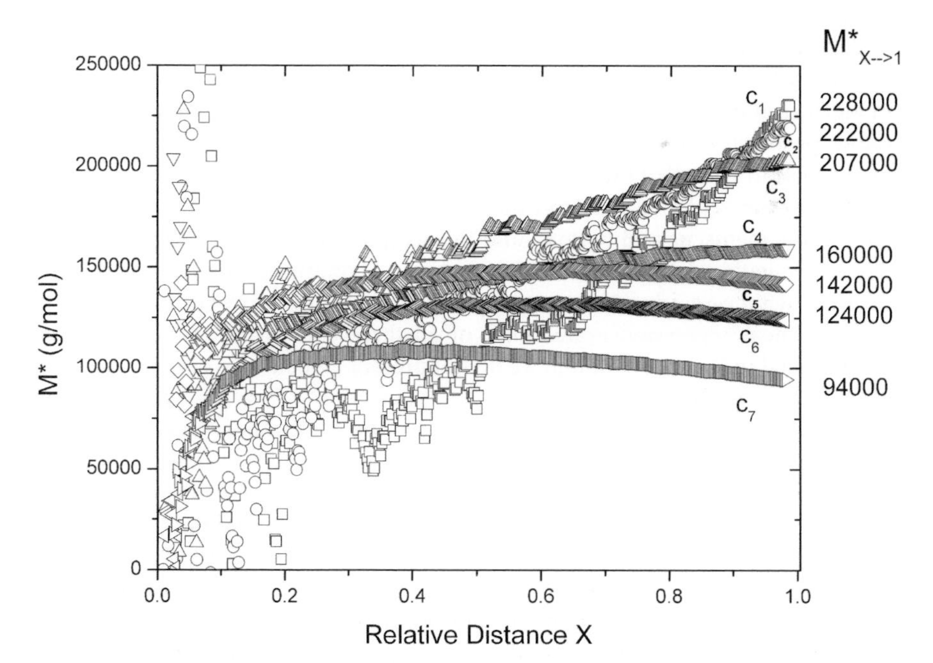

Fig. 5.7. M^* plotted as a function of X (see [22]), for polystyrene NBS 706 in toluene at seven different concentrations. The same measuring data are used as in Figs. 5.2 and 5.3a

5.4.1 Absorption Optics Examples of a Polyelectrolyte and Calibration Polystyrene NBS 706

Absorption optics is preferably used for aqueous systems (biopolymers and polyelectrolytes), and for organic solvents that are UV-transparent, because it is possible to measure at very low concentrations (0.03 – 0.50 g/l). Thus, we are near the ideal state of infinite dilution $c_0 \rightarrow 0$, and therefore it is often enough to conveniently measure only one concentration (see example in Sect. 5.4.2) to yield reasonably precise molar masses M, rather than measuring a series of concentrations and carrying out the tedious extrapolation to zero concentration.

The first example we discuss now in Fig. 5.8 is a nearly monodisperse polyelectrolyte, sodium polystyrene sulfonate NaPSS 35 000, supplied and well characterized by Polymer Standard Service, Mainz, Germany: molar mass of the peak maximum measured by SEC is $M_p = 35\,000$ g/mol and $M_w/M_n = 1.10$. The solvent used was aqueous 0.1 M NaCl, and the three concentrations $c_0 = 0.05$, 0.10, 0.15 g/l are very low. For dialyzed solutions, we measured $\bar{v} = 0.626\,\text{cm}^3/\text{g}$ and $\rho_s = 1.001\,\text{g/cm}^3$ at 25 °C. The equilibrium run was done with a Beckman Optima XL-A/I using the four-hole rotor and the UV scanner at $\lambda = 260$ nm. The filling height of the 12-mm double-sector cell was 2.0 mm, the rotor speed 6000 rpm, and the running time 140 h (this was a long-weekend run, 24 h would be enough to reach the equilibrium state). Sedimentation velocity runs of polyelectrolyte NaPSS 35 000 and other NaPSS samples were already the subject of Sects. 3.3.5 and 3.4.1.

The original XL-A/I UV scan of the lowest concentration $c_0 = 0.05$ g/l is shown in Fig. 5.1a above. There are four vertical lines in this scan. They indicate the four radius positions of the meniscus, bottom, first and last experimental values used for evaluation (note that the XL-A/I absorption values $A(r)$ very near the meniscus or the bottom region scatter too much, so they were eliminated). The vertical lines of the meniscus and bottom, and the accepted A values are plotted again in Fig. 5.8a, together with the values of the other two concentrations 0.10 and 0.15 g/l.

Now, again we carry out the five-step evaluation proposed by Lechner (classical and nonlinear regression evaluation), of all three concentrations together and summarized in Fig. 5.8, although we do not present all details (as in Figs. 5.3–5.6) for polystyrene NBS 706 in toluene.

In step 1, we transfer the absorption values $A(r)$ via Lambert–Beer's law, (2.2), into the required radial (absolute) concentration profile $c(X)$. Because the specific decadic absorption coefficient ε was not known, it was calculated with (5.5) and (5.6) using the procedure described in Sect. 5.2, and the law of the conservation of mass. The resulting numerical value (average over all three concentrations) was ε (25 °C, 260 nm) $= 182\,\text{m}^2/\text{kg}$. If $c(X)$ is determined, then we can calculate the "experimental" points of the reduced concentration profile $c(X)/c_0$.

In step 2, again by a special spline regression smoothing, the optimally fitted lines through the "experimental" points are calculated. Figure 5.8a shows these lines in $A(r)$ form (note that the $c(X)/c_0$ plots and the corresponding lines, as in Fig. 5.4a, are not shown in Fig. 5.8). For classical evaluation, we determine the values and slopes of these lines at the meniscus and bottom, calculate $M_{w,app}$ and

$M_{z,app}$ with (5.14) and (5.15), and extrapolate the $1/M$ values in Fig. 5.8c with (5.19) to $c_0 \to 0$. The result of this classical step 1 + step 2 evaluation is $M_w = 36\,200$ g/mol, $M_z =$ (roughly) 43 000 g/mol, and $A_2 =$ (roughly) 0.2×10^{-4} cm^3 mol/g^2.

In step 3, a spline interpolation, to obtain $c(X)/c_0$ values at 41 pre-selected X values, is done, and in step 4 these values (see Fig. 5.8b) are extrapolated to $c_0 \to 0$. The 41 extrapolated values $c(X)/c_{0,0}$ are plotted in Fig. 5.8d as a function of X. In step 5, through these "experimental" points, representing the reduced concentration profile at infinite dilution, a continuous optimally fitted line is calculated (shown in Fig. 5.8d) by the nonlinear regression evaluation with the Simplex method. This nonlinear regression evaluation yields the three-parameter model molar mass distribution $W(M, P, B, K)$, presented in Fig. 5.8e, and the resulting fitting parameters $M_n^0 = 32\,800$ g/mol, $M_w^0 = 35\,200$ g/mol, $M_z^0 = 37\,600$ g/mol, and $P = 1.34$. So, the broadness distribution coefficient is $M_w^0/M_n^0 = 1.07$. All these AUC values, the classical and the nonlinear regression ones, agree well with the values given by Polymer Standard Service. The errors of measurement are about

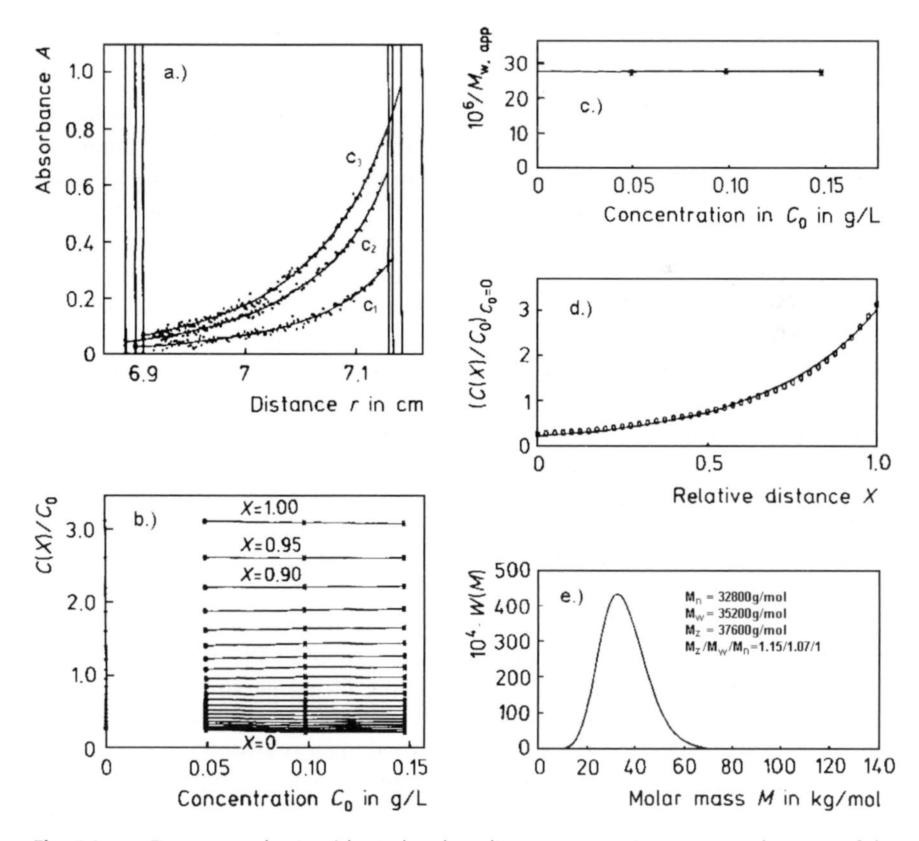

Fig. 5.8a–e. Five-step evaluation (classical and nonlinear regression) to measure the *MMD* of the polyelectrolyte NaPSS 35 000 in aqueous 0.1 M NaCl, done with XL-A/I absorption optics at $\lambda = 260$ nm, analogous to Figs. 5.3–5.6 (reprinted from [6] with permission)

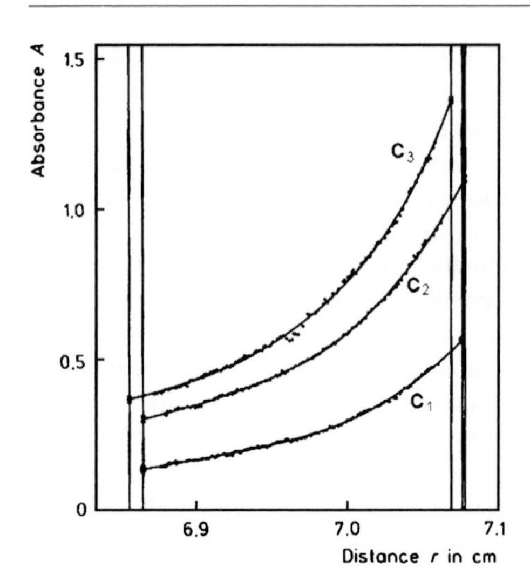

Fig. 5.9. Radial XL-A/I absorption profiles $A(r)$ at $\lambda = 260\,\text{nm}$ of three concentrations of polystyrene NBS 706 in the (pseudo-ideal) theta solvent cyclohexane at 35 °C (reprinted from [6] with permission)

$\pm 5\%$ in M, and $\pm 10\%$ in A_2 (the latter is higher because the concentrations used for absorption optics are very small).

The second absorption example we discuss here in this Sect. 5.4.1 is again polystyrene NBS 706, already discussed in Sect. 5.3.3, where we used interference optics and the non-UV-transparent solvent toluene. For this polystyrene, we now use the organic solvent cyclohexane with an UV-transparent window around $\lambda = 260\,\text{nm}$, where polystyrene shows strong absorption. Cyclohexane is a theta solvent for polystyrene at 35 °C, i.e., at this temperature polystyrene/cyclohexane solutions are pseudo-ideal and $A_2 = 0$ should be valid. We measured this second sample at this theta temperature with the same equipment (an Optima XL-A/I) and nearly the same conditions as those for the NaPSS 35 000 sample discussed above. Concentrations c_0 were 0.10, 0.20, and 0.30 g/l, filling height 2.0 mm, rotor speed 6000 rpm, running time 22 h, $\overline{v} = 0.931\,\text{cm}^3/\text{g}$, and $\varrho = 0.7647\,\text{g/cm}^3$. The result of this measurement is shown in Fig. 5.9.

Figure 5.9 shows, similarly to Fig. 5.8a, the primary measured absorption profiles $A(r)$ of the three polystyrene/cyclohexane solutions, and the optimally fitted lines through the measuring points. The classical step 1 + step 2 evaluation gives $M_\text{w} = 246\,000\,\text{g/mol}$, $M_\text{z} = 360\,000\,\text{g/mol}$, $A_2 \approx 0.5 \times 10^{-4}\,\text{cm}^3\,\text{mol/g}^2$, and $\varepsilon\,(260\,\text{nm}, 35\,°\text{C}) = 213\,\text{m}^2/\text{kg}$. The nonlinear regression step 3–step 5 evaluation yields the three-parameter model molar mass distribution $W(M, P, B, K)$, which is very similar to that in Fig. 5.6b. The resulting fitting parameters are $M_\text{n}^0 = 113\,000\,\text{g/mol}$, $M_\text{w}^0 = 241\,000\,\text{g/mol}$, and $M_\text{z}^0 = 373\,000\,\text{g/mol}$. The comparison with (i) the NBS values, (ii) the values measured via s run-MMD in Sect. 3.4.2 (Fig. 3.12d), and (iii) the values measured via equilibrium run with interference optics in toluene (see Sect. 5.3.3, Table 5.1, and Fig. 5.6b) is satisfactory, with exception of M_n. The reason for this discrepancy may be some UV-absorbing impurities in the solvent cyclohexane that were not correctly compensated for. The second virial coefficient A_2 is not zero, as expected in a theta solution, but not far removed from this value.

5.4.2 Association Equilibrium Example and AUC Buoyant Density Method to Determine $\overline{v}$

In this section, first, we outline the "buoyant density method" of Schubert et al. [23] to determine $\overline{v}$ via sedimentation equilibrium runs in *solvent mixtures* (based on [27]; similar $\overline{v}$ measurements in *Nycodenz/water mixtures* were already done in [34]), and second, we study a monomer/dimer association equilibrium of mono-functionalized poly(ethylene oxide), abbreviated PEO. Both the $\overline{v}$ determination and the association study will be explained by means of a recent paper of Schubert et al. [24].

Supramolecular chemistry utilizes non-covalent bonds to build up large, ordered structures. Among the interactions used, hydrogen bonds (beside metal-ligand interactions) play an important role. The first step of such association processes is the building up of dimers. Schubert et al. [24] studied such first steps by using hydrogen bond interactions. They quantified the dimer formation between suitable hydrogen bond-forming blocks by sedimentation equilibrium runs. As blocks, they used PEO chains with a polymerization degree of 70 ($M \approx$ 4000 g/mol). At one end of the chain, a dye-labeled group, which does not form hydrogen bonds, is linked covalently, which means that no dimerization is possible. This first compound was called the precursor. In a second step, a hydrogen bond-forming group was linked, also covalently, to the other end of the PEO chain to obtain the final compound, which is able to form dimers (for details, see [24]). The dye labeling allows UV scanner detection, and dimer monitoring via sedimentation equilibrium analysis in the AUC.

All equilibrium runs were performed using a Beckman Optima XL-A/I, an An-50-Ti rotor, 12-mm titanium double-sector centerpieces, and polyethylene gaskets. The rotor speed was always near 40 000 rpm, temperature 20 °C, filling height 4 mm, with concentrations of 0.1 – 0.3 g/l. The absorption profiles were recorded at 330 nm using the original Beckman XL-A/I software. The evaluation of these profiles for the existence of single monomers, or dimers, or both, was performed with the computer program DISCREEQ by Schuck [25, 26].

$\overline{v}$ Determination with the Buoyant Density Method

For every determination of M with equilibrium runs, we need, according to the basic equation (5.4) or (5.9), the partial specific volume $\overline{v}$ of the solute within the buoyancy term $(1 - \overline{v}\varrho)$. The measurement of $\overline{v}$ of an unknown solute is usually done with the Kratky density balance, using a concentration series as described in Sect. 2.6.1. For such concentration series, however, a large amount of solute substance, approximately 200 mg, is needed. Commonly, such high amounts of solute are not available, only nanograms. In such cases, it is often possible to estimate $\overline{v}$ by means of the so-called buoyant density method, via equilibrium runs within the AUC. Schachman and Edelstein [27] were the first to propose this method, and they employed it for the $\bar{v}$ determination of proteins using the "aqueous" solvent pair H_2O/D_2O. The idea behind this approach was to measure, without prior knowledge of $\bar{v}$, the *relative* $M_{eff} = M(1 - \bar{v}\varrho) = 2RT[\mathrm{d}\ln c(r)]/(\omega^2\,\mathrm{d}r^2)$ of the same solute with a defined M in two solvents of different densities, ϱ_1 and ϱ_2, and assuming the same $\bar{v}$ of the solute in these two solvents (there are several restrictions, which will not be discussed here). Applying two times the basic equation (5.4) with its two unknowns, M and $\bar{v}$, on two equilibrium measurements in two solvents with different but known densities will yield the two unknowns. The same idea forms the basis of our H_2O/D_2O particle density analysis described in Sect. 3.5.3, but in that case we used sedimentation velocity runs, rather than sedimentation equilibrium runs.

Schubert et al. [23] transferred this Schachman–Edelstein idea from biopolymers in "aqueous" solvent pairs to synthetic polymers in organic solvent pairs, and used for their PEO derivatives the solvent pair THF/propylene carbonate ($\varrho = 0.899$ and $1.204\,\mathrm{g/cm^3}$, respectively, at 20 °C). Both are good solvents for PEO. Schubert et al. further extended the Schachman–Edelstein idea by using not only two solvents and two equilibrium runs to measure M_{eff}. Rather, they used, in addition to the two pure solvents, several different mixtures of these, i.e., additional solvents with different densities ϱ. This procedure delivers more precise values of the two unknowns $\bar{v}$ and M. Similar measurements to determine $\bar{v}$ of detergent micelles were done by Lustig [34].

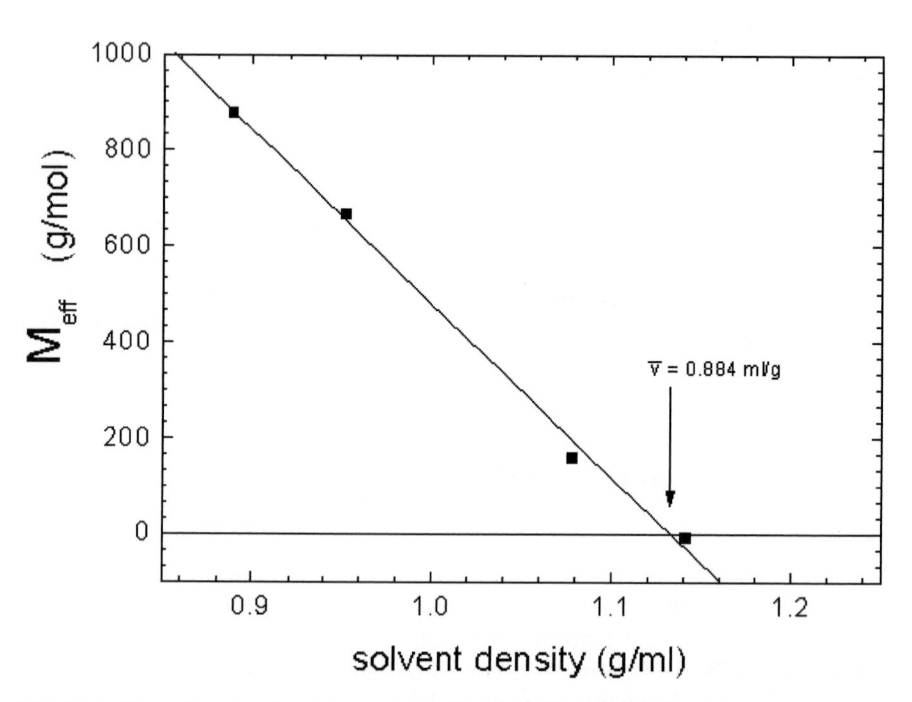

Fig. 5.10. Determination of $\bar{v}$ with the AUC buoyant density method, i.e., $M_{eff} = M(1 - \bar{v} \cdot \varrho)$ measurements of the same solute via equilibrium runs in a solvent pair and its different mixtures: PEO derivative precursor in THF/propylene carbonate (reprinted from [24] with permission)

A plot of M_{eff} as a function of ϱ, such as that in Fig. 5.10, will yield $\overline{\nu}$ at that ϱ position where $M_{\text{eff}} = 0$ is valid, i.e., $(1 - \overline{\nu}\varrho) = 0$ or $\overline{\nu} = 1/\varrho$. This ϱ position is found by interpolation via a regression line in the M_{eff} versus ϱ plot (see Fig. 5.10, where this regression line is a straight line). In this manner, Fig. 5.10 delivers for the PEO derivative precursor $\overline{\nu} = 0.884\,\text{cm}^3/\text{g}$ and $M = 4035\,\text{g/mol}$. This is in agreement with a Kratky density balance measurement on a similar compound (where a bigger amount of substance was available), and with M_{theor}, following from the known chemical structure of the precursor (see [24]). The straight line in the M_{eff} versus ϱ plot in Fig. 5.10 suggests independence of both M and $\overline{\nu}$ vis-à-vis the solvents used. Thus, the $\overline{\nu}$ value of the precursor, measured with the buoyant density method, is presumably of high reliability in the present example. Still, this is not always valid. Preferential solvation of one of the solvent components may falsify the determined $\overline{\nu}$ value.

***M* Measurements to Study the Self-Association**

To study the self-association of the final PEO compound bearing a hydrogen-bonding group, via M measurements by sedimentation equilibrium runs in pure THF, the $\overline{\nu}$ value of the precursor in THF, $0.884\,\text{cm}^3/\text{g}$, was used. The result of such a measurement is shown in Fig. 5.11, for a concentration $c_0 = 0.32\,\text{g/l}$.

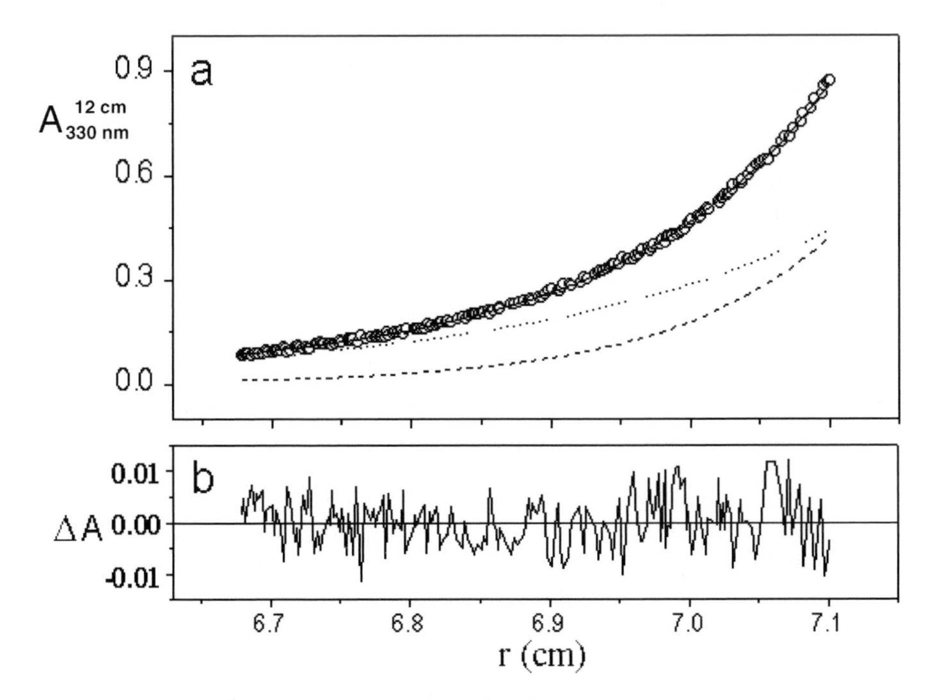

Fig. 5.11. a Equilibrium run of a PEO derivative (final compound) bearing a hydrogen-bounding group: absorption $A(r)$ as a function of radius r (THF, 0.32 g/l, 40 000 rpm, $\lambda = 330$ nm), $A(r)$ measuring points, fitted with a curve (*continuous line*) where the presence of 55% monomers and 45% dimers is assumed. The *dotted line* is the calculated contribution to $A(r)$ of the monomers, and the *hatched line* that of the dimers. **b** Residuals $\Delta A(r)$ of the fit (reprinted from [24] with permission)

In contrast to the analogous measurement of the precursor, the measured $A(r)$ profile of the final PEO compound is not very well fitted by a single exponential, but it is well fitted by a superimposition of two single exponentials, consisting of 55-wt% monomer and 45-wt% dimer. This 55/45 composition yields the smallest residuals $\Delta A(r)$ of the fit obtained by using Schuck's program DISREEQ [25,26]. The resulting residuals are shown in Fig. 5.11b. Schubert et al. [24] attached also a metal–ligand interaction group to the PEO chain. For that PEO derivative they obtained, in contrast to the hydrogen-bonding group, a complete dimerization.

The above PEO example demonstrates that (i) the sedimentation equilibrium analysis is a powerful method to detect and quantify the formation of macromolecular association complexes via supramolecular building blocks by non-covalent bonds, and that (ii) the buoyant density method is a valuable auxiliary tool for determining $\overline{v}$ at very low compound concentrations c_0, so that only nanograms of substance are necessary.

5.4.3 The New Fluorescence Detector of Laue Used for Green Fluorescent Protein

The characteristics of the new fluorescence detector of Laue [28], described in Sect. 2.4.4, are that (i) it is a specific detector for macromolecules with inherent or chemically attached specific fluorescent groups, and that (ii) it allows one to measure extremely low solute concentrations, up to 100-fold lower than is the case with absorption optics. The latter was demonstrated by Laue [28] by means of a comparison experiment based on absorption optics versus fluorescence optics, which is presented in Fig. 5.12. Both detectors were incorporated into a Beckman Optima XL-A/I, and for both the same macromolecule was chosen: the well-known (monodisperse!) green fluorescent protein (GFP).

The experimental conditions are presented in the legend of Fig. 5.12 (for details, see the original paper, [28]). The most important difference between the two experiments was the GFP concentration $c_0 = 11.4 \times 10^{-6}$ mol/l $= 0.307$ g/l for the absorption optics, and 0.435×10^{-6} mol/l $= 0.0117$ g/l for the fluorescence optics. The resulting molar masses, resulting from nonlinear least-square fitting, are $M = 27\,000 \pm 1200$ g/mol (absorption), and $M = 27\,700 \pm 2200$ g/mol (fluorescence). These agree excellently, within 1% (absorption) and 3% (fluorescence), with the known sequence molar mass of GFP, $M = 26\,900$ g/mol. These M values, and the residuals in the upper part of Fig. 5.12 (signal-to-noise ratio of 223 for absorption, and 64 for fluorescence) show that the precision of parameters obtained with the fluorescence detector will be somewhat less than those obtained with the absorption detector.

The new fluorescence detector of Laue is likely to become a powerful tool in every kind of AUC analysis. It is now commercially available (see Sect. 2.4.4).

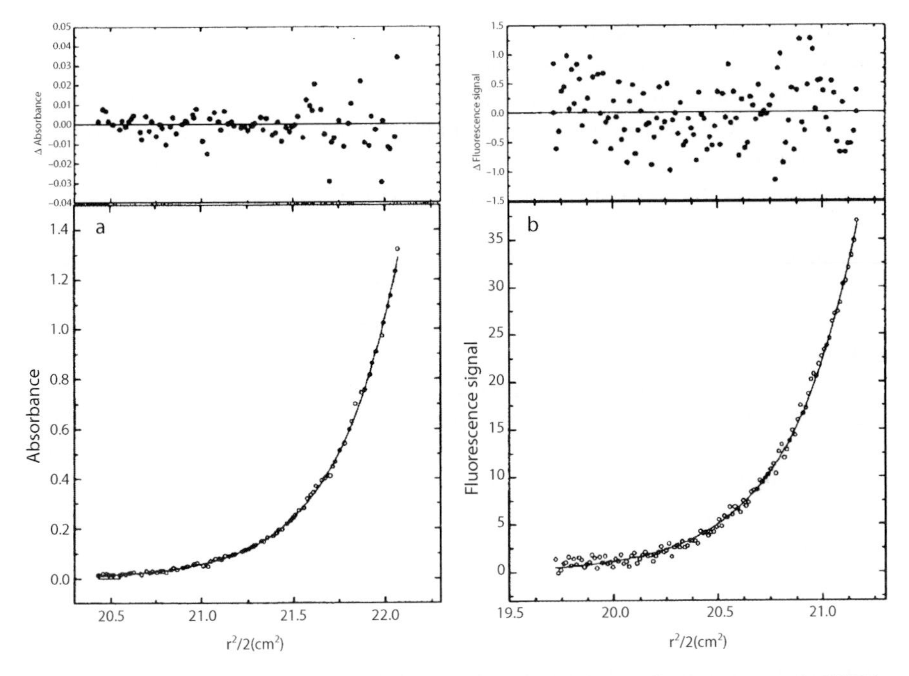

Fig. 5.12a,b. Comparison of sedimentation equilibrium data, for green fluorescent protein (GFP) in 20 mM Tris-HCl, pH 8.1, 150 mM NaCl, 2.5 mM $CaCl_2$ at 20 °C and 30 000 rpm. **a** Absorption optics data $A(r)$ vs. $r^2/2$ for $c_0 = 11.4 \times 10^{-6}$ M = 0.307 g/l. Nonlinear least-square fitting yields M = 27 000 ± 1200 g/mol. **b** Fluorescence optics data lg I_0/I vs. $r^2/2$ for $c_0 = 0.435 \times 10^{-6}$ M = 0.0117 g/l. Nonlinear least-square fitting yields M = 27 700 ± 2200 g/mol (reprinted from [28] with permission)

5.5 Further AUC Methods to Measure Molar Masses

Beside the dominating sedimentation equilibrium method described above, there are four other AUC methods to measure molar masses M: (i) the Svedberg method, (ii) the sedimentation velocity method, (iii) the density gradient method, and (iv) the approach-to-equilibrium (or Archibald) method. In the following, these four methods are outlined.

Svedberg Method

The Svedberg, or sedimentation-diffusion method is already described in Sect. 1.1. It is the oldest AUC method to measure absolute (average) molar masses, and it is not an equilibrium method, but rather a velocity method. In separate measurements, the sedimentation coefficient s and the diffusion coefficient D of the solute are determined, and subsequently M is calculated using the Svedberg equation (1.8), $M = RT/(1 - \bar{v}\varrho) \times (s/D)$. To yield the correct weight-average molar mass M_w, one has to combine the weight average of s, s_w, and the z average of D, D_z (at infinite dilution $c \to 0$). D_z is delivered by *DLS* measurements.

Sedimentation Velocity Method

This is only a relative method but nevertheless also a velocity method, and it has been described in Sects. 3.4.1 and 3.4.2 above. It is a powerful method to measure, beside average molar masses, complete molar mass distributions $W(M)$, if a scaling relation $s = KM^{a}$, as in (3.11), is known. This is not the case for most polymers, but does apply for some industrially important polymers such as polystyrene, polymethyl methacrylate, etc. It is tedious work to create such a scaling relation by preparative fractionation of the (polydisperse) polymer, and separate measurements of M and s of the single fractions.

Density Gradient Method

This is an absolute, and an equilibrium method (see Sect. 4.2.1 and (4.26)), but very inaccurate and seldom used for the measurement of M. Nevertheless, the density gradient method delivers M as valuable "byproduct" information of every density gradient run of macromolecules, because the broadness σ of every density gradient double Schlieren peak (see Fig. 4.8) yields a rough estimation of M. Especially for macromolecules with extremely high $M > 10^7$ g/mol, and minimal amounts of substance, the density gradient method is often the only allowing one to roughly estimate M.

Approach-to-Equilibrium (or Archibald) Method

This procedure is often referred to as the Archibald method, in honor of the man who first described it [29]. In principle, it depends upon the fact that the conditions for sedimentation equilibrium are fulfilled at both ends of the solution column at all times – that is, no solute can pass through the air–solution meniscus, or out through the bottom of the cell. This is valid during every kind of AUC run, for equilibrium runs, and for sedimentation runs, too. The net flow of the solute is zero at r_{m} and r_{b}. Thus, the Archibald equation [29] to measure M,

$$M = RT(1-\bar{v}\varrho)\omega^2 \cdot \frac{(\mathrm{d}c/\mathrm{d}r)_{\mathrm{m}}}{c_{\mathrm{m}} r_{\mathrm{m}}} = RT(1-\bar{v}\varrho)\omega^2 \cdot \frac{(\mathrm{d}c/\mathrm{d}r)_{\mathrm{b}}}{c_{\mathrm{b}} r_{\mathrm{b}}} \tag{5.27}$$

is similar to the basic equilibrium equation (5.3) for this transient state (this "transient" state between sedimentation (run) at the beginning, and equilibrium (run) reached only much later is responsible for the name "approach-to-equilibrium" method). From a measurement of both the concentration c and the concentration gradient $\mathrm{d}c/\mathrm{d}r$, at either r_{m} or r_{b}, the molar mass M of the solute can be calculated. This is possible with interference optics and Schlieren optics. In practice, one uses only the meniscus data, because the bottom data are not very precise, and often falsified by fast-sedimenting impurities.

The advantage of the Archibald method is that only very short measuring times of 2 h, or less, are required. The disadvantages are (i) that it depends directly upon a single measurement at one of the ends of the solution column, where the uncertainty of the data is highest, and (ii) that in particular for broadly distributed synthetic polymers during the Archibald run, a fast fractionation takes place, so that M measured at the meniscus is smaller than M measured at the bottom. This difference will increase with running time, because Archibald runs are usually done at high rotor speeds. Via a tedious extrapolation $t \rightarrow 0$ of the measuring data, these disadvantages can be compensated for.

Nowadays, the Archibald method is seldom used, and the sedimentation equilibrium method, especially in combination with an eight-cell-multiplexer, is preferred. Still, perhaps a renaissance of the Archibald method will occur in the near future, if someone writes a new, intelligent evaluation program (see Sect. 7.3, "improvement of data analysis") such as was the case for equilibrium runs, described in Sect. 5.3. An initial step was done in this direction in [35]. In combination with the new computerized analytical ultracentrifuges, with their powerful data acquisition, fast Archibald determination of M during every sedimentation/equilibrium run should be possible, producing M as "byproduct" information (if $\overline{v}$ is known). Good general references for the approach-to-equilibrium method are [1,29–32].

Closing Remarks Concerning M Measurement via AUC

For the M and MMD measurements of an unknown sample via AUC, the following three-step strategy in planning the corresponding experiments is recommended:

(i) An s run of the unknown sample, proceeding the final equilibrium run, is always helpful. It delivers basic information on whether the sample is narrowly or broadly distributed, and whether M is low or high. This allows one to plan the subsequent equilibrium run.

(ii) If an $s_0 = KM^a$ scaling law is known (this is seldom the case!), the above s run yields the complete and most precise MMD. Furthermore, if this is a high-precision s run (and an excellent data evaluation program exists for this), it can deliver an M value via the Archibald procedure. If also D of the sample is known (via a synthetic boundary run, or more precisely, via DLS), then M follows from the Svedberg equation (1.8).

(iii) The equilibrium run itself should be done with UV optics if possible, because only one (very low) concentration is necessary. If one has to use interference optics, a series of 3–7 (mostly higher) concentrations is required to allow one to carry out the extrapolation $c \rightarrow 0$ (and if A_2 is desired). Schlieren optics is not recommended here, because it is not sensitive enough. If the sample is narrowly distributed (as for proteins), 0.5–1.5 mm short column equilibrium runs of 1–5 h (or meniscus depletion runs lasting 10–40 h) are recommended. If the sample is broadly distributed (and if a precise MMD and A_2 are desired), 2–4 mm long column equilibrium runs of 10–50 h and a c series are necessary. To find the optimal rotor speed ω, and for an estimation of the run time t_{equil}, the use of a simulation routine, such as [33], is helpful. $\overline{v}$ of the sample (and dn/dc, in the case of interference optics) has to be known for calculating M and MMD.

References

1. Creeth JM, Pain RH (1967) Prog Biophys Mol Biol 17:217
2. van Holde KE, Baldwin RL (1958) J Phys Chem 62:734
3. Chervenka CH (1969, 1973) A manual of methods for the analytical ultracentrifuge. Spinco Division of Beckman Instruments, Inc, Palo Alto, California
4. Yphantis DA (1964) Biochemistry 3:297
5. Harding SE, Rowe AJ, Horton JC (eds) (1992) Analytical ultracentrifugation in biochemistry and polymer science. Part II. Equilibrium methods. The Royal Society of Chemistry, Cambridge, pp. 231–330
6. Lechner MD, Mächtle W (1992) Makromol Chem Rapid Commun 13:555
7. Fujita H (1962) Mathematical theory of sedimentation analysis. Academic Press, New York
8. Fujita H (1975) Foundation of ultracentrifugal analysis. Wiley, New York
9. Lansing WD, Kraemer EO (1935) J Am Chem Soc 57:1369
10. Richards EG, Schachman HK (1959) J Phys Chem 63:1578
11. Scholte ThG (1968) J Polym Sci 6:91 ibid. 111
12. Scholte ThG (1970) Eur Polym J 6:51
13. Greschner GS (1984) Eur Polym J 20:475
14. Wiff DR (1973) J Polym Sci Polym Symp 43:219
15. Lechner MD, Mächtle W (1991) Makromol Chem 192:1183
16. Lechner MD (1992) In: Harding SE, Rowe AJ, Horton JD (eds) Analytical ultracentrifugation in biochemistry and polymer science. The Royal Society of Chemistry, Cambridge, p. 295
17. Lechner MD (1978) Eur Polym J 14:61
18. Lechner MD, Mächtle W (1991) Prog Colloid Polym Sci 86:62
19. Creeth JM, Harding SE (1982) J Biochem Biophys Methods 7:25
20. Cölfen H, Harding SE (1997) Eur Biophys J 25:333
21. McRorie DK, Voelker PJ (1993) Self-associating systems in the analytical ultracentrifuge. Beckman Instruments Inc, California
22. Cölfen H, Schilling C (2004) Max Planck Institute of Colloids and Interfaces, Potsdam/Golm, and Nanolytics GmbH, Dallgow, Germany, created this M^* vs. X plot with a new unpublished M-STARI program
23. Tziatzios C, Precup AA, Weidl CH, Schubert US, Schuck P, Durchschlag H, Mächtle W, van den Broek JA, Schubert D (2002) Prog Colloid Polym Sci 119:24
24. Tziatzios C, Precup AA, Lohmeijer BGG, Börger L, Schubert US, Schubert D (2004) Prog Colloid Polym Sci 127:48
25. Schuck P (1994) Prog Colloid Polym Sci 94:1
26. Schuck P, Legrum B, Passow H, Schubert D (1995) Eur J Biochem 230:806
27. Edelstein SJ, Schachman HK (1967) J Biol Chem 242:306
28. MacGregor IK, Anderson AL, Laue TM (2004) Biophys Chem 108:165
29. Archibald WJ (1947) J Phys Colloid Chem 51:1204
30. Schachman HK (1957) In: Colowick SP, Kaplan NO (eds) Methods in Enzymology, vol 4. Academic Press, New York, p. 32
31. Trautman R (1964) In: Newman DW (ed) Instrumental methods of experimental biology. The Macmillan Company, New York, p. 211
32. Tanford C (1963) Physical chemistry of macromolecules. Wiley, New York
33. Demeler B, van Holde KE (2004) Anal Biochem 335(2):279
34. Lustig A, Engel A, Tsiotis G, Landau EM, Baschong W (2000) Biochim Biophys Acta 1464:199
35. Schuck P, Millar DB (1998) Anal Biochem 259:48

6 Practical Examples of Combination of Methods

As mentioned at the end of the introductory Chap. 1, colloidal and synthetic polymer samples from both industrial and scientific fields are becoming increasingly complicated and complex on the nanometer scale. To solve such analytical problems, the AUC is a powerful tool (i) with its different measuring methods, (ii) with its different fractionation principles based on size considerations and chemical heterogeneity, and (iii) with its possibility to transfer a sample from the aqueous into an organic medium. In this Chap. 6, we will demonstrate by means of practical application examples that in many cases the key to the solution of such complex analytical problems is the *combination* of different measuring techniques. In Sect. 6.1, we will combine different AUC measuring techniques to study the grafting reaction on the surface of core/shell latices (Sect. 6.1.1), the systematic buildup of crosslinking inside of polymer latices (= microgels; Sect. 6.1.2), and ion exchange within and between carboxylated latices (= model polyelectrolytes; Sect. 6.1.3). In Sect. 6.2, we will combine AUC measuring techniques with other analytical techniques, such as sedimentation field flow fractionation (SFFF; Sect. 6.2.1), and electron microscopy (EM; Sect. 6.2.2). This application of different analytical techniques (e.g., AUC, DLS, FFF, and EM) on the same sample having a complex composition and different distribution characteristics is called *global analysis*. In our opinion, this global analysis will become ever more important in the future. In the following Sects. 6.1 and 6.2, examples of the authors' works are presented, whereas the shorter Sect. 6.3 is a compilation of literature examples concerning AUC and nanoparticles.

6.1 Combination of Different AUC Methods

6.1.1 Core/Shell Particles

In order to obtain specific properties of a polymer, a simple polymer dispersion (= latex) consisting, for instance, of polybutylacrylate (PBA), can be modified by another polymer, e.g., by grafting the styrene/acrylonitrile-copolymer (SAN) onto the surface of the PBA particles in the form of a grafting shell. This is often done in the development of rubber-modified high-impact materials. Thus, the SAN-grafted *soft* PBA particles are introduced into a *hard* continuous matrix of SAN. The grafting of the SAN macromolecules onto the PBA-core particles causes a very good bonding between the rubber particles and the matrix. In the following, we will study this grafting reaction systematically by combining the AUC methods for aqueous and organic density gradients, particle size distribution measurements, and sedimentation runs. For details of this study, see [1] (as another example of core/shell particles, we will present high-impact polystyrene in Sect. 6.2.2).

Figure 6.1 shows the particle size distribution and the aqueous static density gradient of our starting PBA-core dispersion. The particle size distribution in the upper part of the figure shows that this dispersion is nearly monodisperse, with an (average) diameter of 241 nm. In the lower part of the figure, the 92 H_2O/8 metrizamide density gradient of these particles shows a narrow turbidity band with a uniform particle density of 1.038 g/cm^3. In our first example, a styrene/acrylonitrile copolymer, SAN, has been grafted as a shell onto the surface of these PBA particles. The core-to-shell relation was intended to be 1:1 by weight. Because SAN has a higher density of 1.08 g/cm^3, we expect an increase of the total particle density by this grafting to approx. 1.06 g/cm^3.

Figure 6.2 shows the result of this experiment. In this density gradient, we simultaneously measured both the ungrafted and the grafted dispersion. The grafting

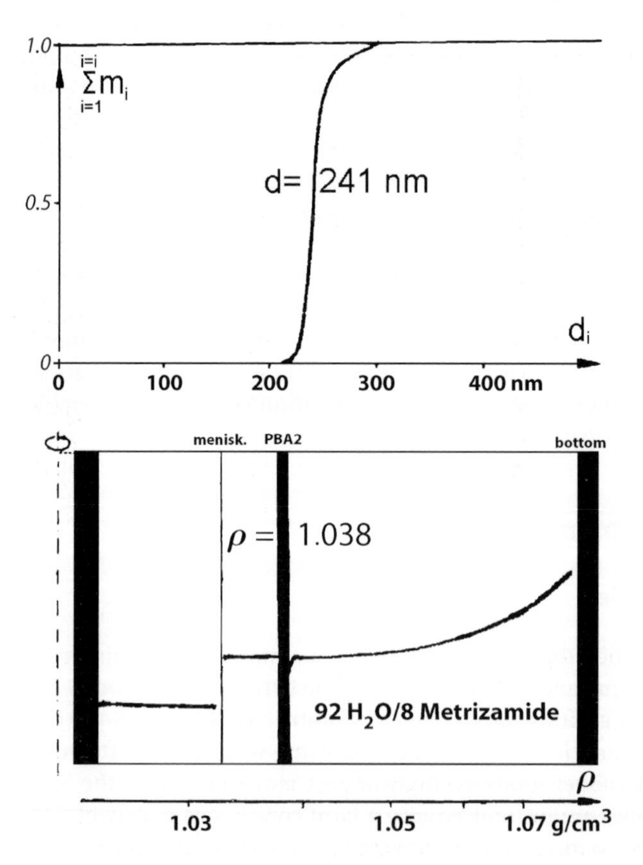

Fig. 6.1. AUC particle size distribution and aqueous static 92 H_2O/8 metrizamide density gradient of a nearly monodisperse PBA dispersion (reprinted from [1] with permission)

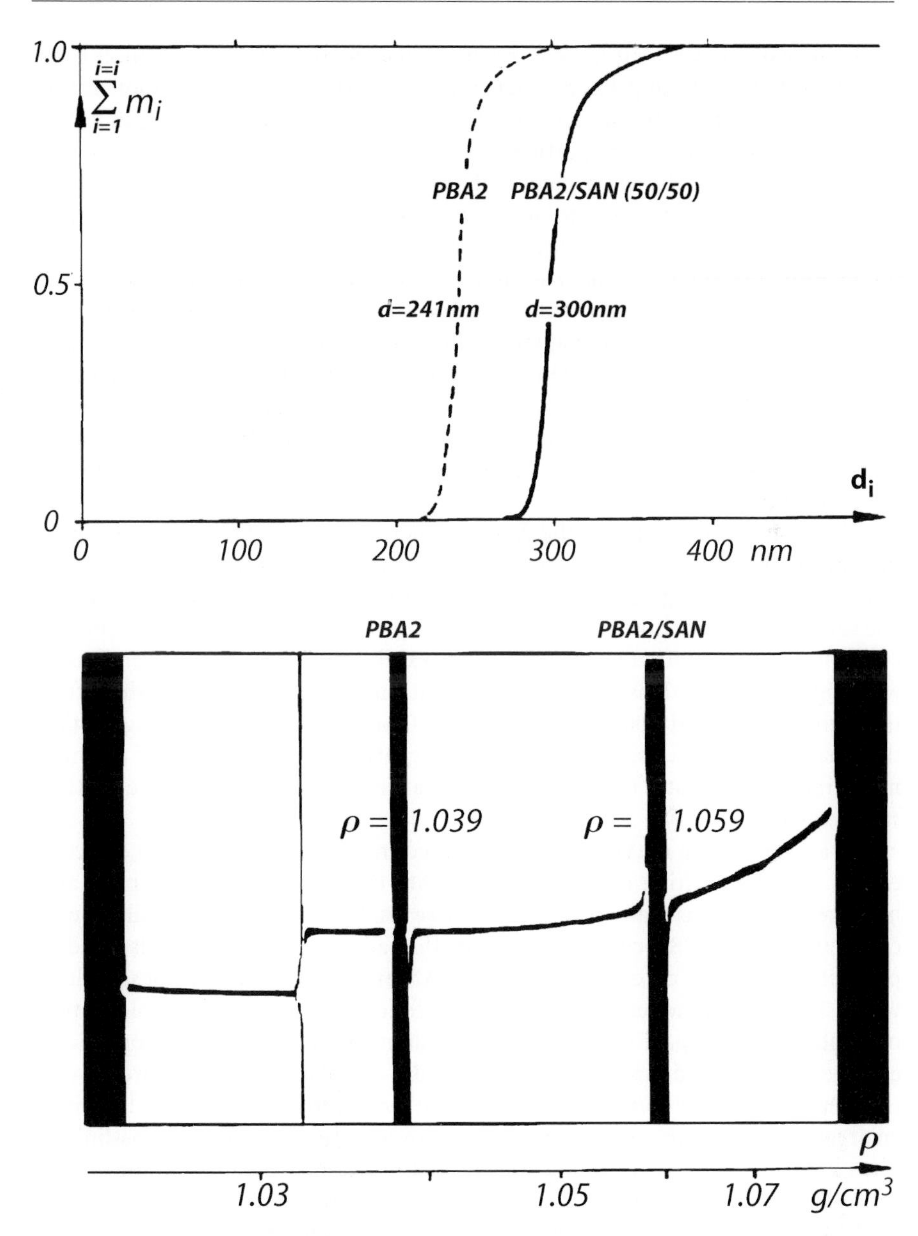

Fig. 6.2. AUC particle size distributions and aqueous 92 H_2O/8 metrizamide density gradient of the ungrafted, and the 1:1 SAN grafted, nearly monodisperse PBA dispersion (reprinted from [1] with permission)

band is also very narrow, and shows the expected particle density of 1.059 g/cm^3. Also the particle size distribution of the grafted dispersion is narrow again. The diameter rose to 300 nm, an increase of 25% that was expected theoretically.

In the case above, we grafted a nearly *monodisperse* starting latex. So, it is no surprise that this grafting was a uniform one. Now, an interesting question arises: do we obtain a uniform grafting, too, if we carry out the experiment with a PBA starting latex having a *broad* particle size distribution?

Figure 6.3 shows the experiment that gives the answer. By mixing four (nearly) monodisperse polybutylacrylate latices with diameters of about 100, 200, 300, and 400 nm, in a mixing ratio 40:30:20:10 wt%, we obtained a starting latex with the very broad, four-modal particle size distribution indicated by a dotted line in Fig. 6.3. This mixture was again grafted with SAN, 1:1. The result was the solid-line particle size distribution in Fig. 6.3. It is again distinctly four-modal, but the diameters and the weight ratios have changed. One sees immediately that the diameter increase of the 100-nm particles exceeds 25%, whereas the increase of the 400-nm particles is much smaller. Additionally, the mass percentage of the smallest particles increased from 40 to 50%. This means that smaller particles are grafted to a higher extent than bigger ones!

This result is confirmed by the density gradient shown in Fig. 6.3, where again both dispersions, the ungrafted and the grafted one, are measured simultaneously. The ungrafted starting particles, 100, 200, 300 and 400 nm, are all superimposed in *one* single band, und show exactly the same, known PBA density of 1.038 g/cm^3. By contrast, the four grafted latex components exhibit *four* discrete, different bands with higher particle densities. Their different radial positions – from the left to the right, the grafted 400-, 300-, 200- and 100-nm particles – show again that the smaller the particles, the more they are grafted! The evaluation of all these measurements indicates that the degree of grafting is proportional to the particle *surface area.*

As a result of the measurements presented in Fig. 6.2 (and Fig. 6.3), we concluded that the grafting of the SAN shell onto the PBA core was complete and homogeneous. Is this really true? Are all these SAN macromolecules really bound covalently onto the PBA-core surface, or are they all (or partly) merely adsorbed or precipitated onto the core surface, because SAN macromolecules are not soluble in water? The AUC allows us to answer such questions, because it is the sole instrument able to investigate systems where simultaneously dispersed microparticles exist beside dissolved macromolecules. In order to obtain this answer, we have to transfer the PBA/SAN core/shell particles from the originally aqueous medium into an organic medium, in which PBA and SAN are soluble if they are not crosslinked. Thus, crosslinked particles would swell in that medium, depending on the degree of crosslinking.

In a first experiment, presented in Fig. 6.4, this transfer is simply realized by a 1:300 dilution of the highly concentrated aqueous dispersion with the organic (80 tetrahydrofuran/20 di-iodomethane) density gradient medium, which is a good solvent for uncrosslinked PBA and SAN.

For comparison, we present in Fig. 6.4 both the aqueous and the organic density gradient of the PBA/SAN core/shell dispersion with the 241-nm core. Since prior to the SAN grafting the starting PBA particles were completely and densely

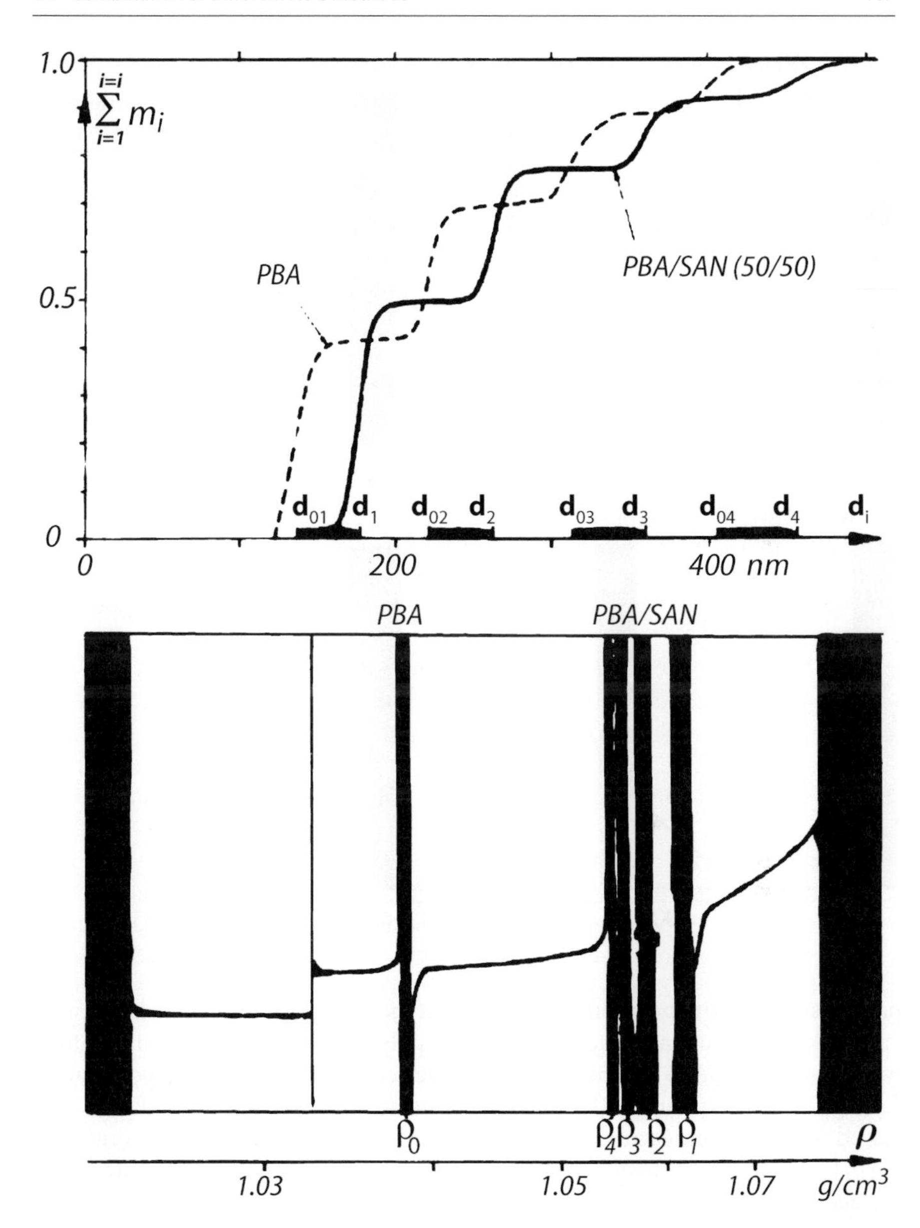

Fig. 6.3. AUC particle size distributions and aqueous 92 H_2O/8 metrizamide density gradient of the ungrafted, and the 1:1 SAN grafted, broadly distributed 40:30:20:10 wt% PBA dispersion mixture (reprinted from [1] with permission)

crosslinked, we do not expect the double Schlieren peak of dissolved PBA macromolecules at their known density position of 1.015 g/cm^3 in this organic density gradient. As expected, we see the turbidity band of (weakly swollen) crosslinked

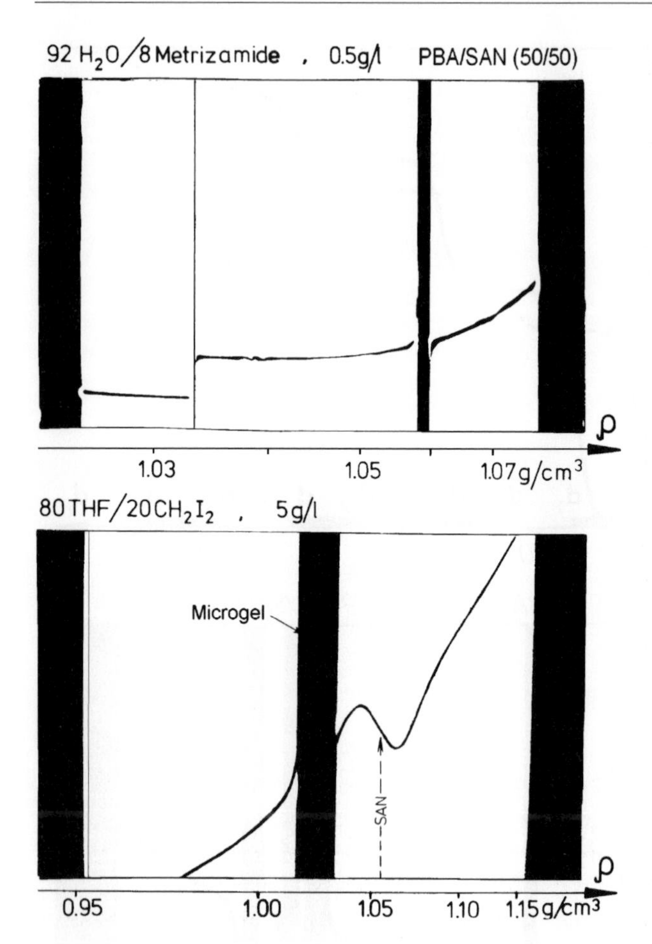

Fig. 6.4. Aqueous and organic AUC density gradients of the 1:1 SAN grafted PBA dispersion 241 nm (reprinted from [1] with permission)

particles at a density position suggesting that the particles cannot consist of pure crosslinked polybutylacrylate, but rather of polybutylacrylate covalently grafted with 40–50 wt% SAN. On the other hand, we see at a higher-density position of 1.050 g/cm^3 (known for SAN) a small, separate double Schlieren peak of pure dissolved SAN macromolecules: these have evidently separated from the core/shell particle surfaces by diffusion, because they were not covalently bound onto these surfaces.

In a second experiment, for the quantitative estimation of the mass percentage and the molar mass of the ungrafted SAN macromolecules, we transfer, again by dilution, the original aqueous dispersion into the pure organic solvent tetrahydrofuran (which is completely miscible with water), and carry out a sedimentation run, shown in Fig. 6.5. We use an unusually high concentration of 80 g/l to detect also very small amounts of free, ungrafted SAN macromolecules. Figure 6.5 shows that during the first minutes of this sedimentation run, the very fast turbidity front of the crosslinked PBA/SAN microgel particles sediment completely to the cell

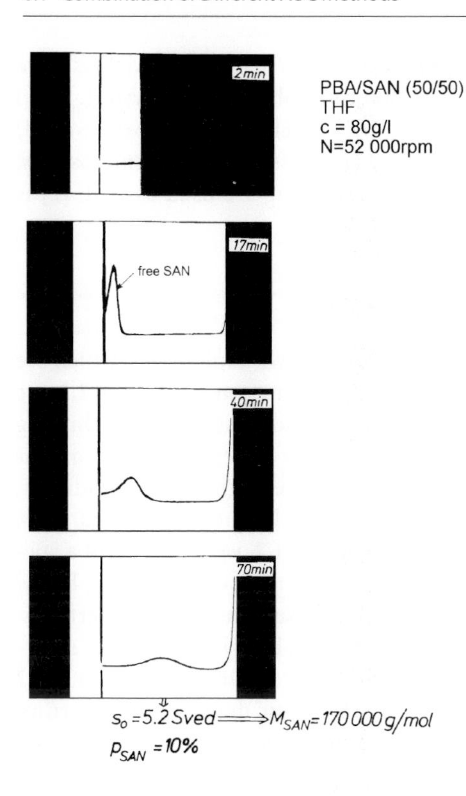

Fig. 6.5. Sedimentation run of the 1:1 SAN grafted PBA dispersion 241 nm in the organic solvent tetrahydrofuran at a high concentration of 80 g/l (reprinted from [1] with permission)

bottom (where they remain visible as a thin layer). Later, the slowly sedimenting Schlieren peak of the free, ungrafted SAN macromolecules becomes evident. The sedimentation coefficient of this Schlieren peak, $s_0 = 5.2\,\text{S}$, yields a molar mass of about 170 000 g/mol and its area a concentration of ~ 8 g/l for these free, ungrafted SAN macromolecules. Since we know the overall concentration of the whole dispersion, namely, 80 g/l, this means that ~ 10 wt% of the planned 50 wt% SAN is not covalently grafted onto the surfaces of the PBA-core particles. If the real degree of grafting were much lower, these rubber particles would yield a poor rubber-modified high-impact material (see also Sect. 6.2.2, high-impact polystyrene).

The sedimentation run experiment of Fig. 6.5 demonstrates an advantageous feature of the AUC, which is unique: the possibility to measure at super-elevated sample concentrations! Because of the fractionation capability inside the AUC measuring cell, it is possible to detect and analyze a minor component, beside a dominating major component, if there is a strong difference in the sedimentation velocities. In the first minutes of the experiment, the fast major component has sedimented completely to the cell bottom, and subsequently we can look into the "clear serum" to detect the slow minor component at an optimal concentration. If the sedimentation run is done with the synthetic boundary technique (see Sect. 3.6), we can detect very small amounts of low molar mass components, beside dominating big particles (see, for example, Fig. 3.26).

If the reader is interested in a more sophisticated core/shell particle analysis, we recommend the original paper ([2]), where 300-nm polystyrene/poly(tert.-butyl acrylate) particles and the polymer with the *inverse* structure (PTBA core/PS shell) are also extensively characterized by AUC. Since transmission electron microscopy, solid state NMR, and static and dynamic light scattering are used as well, [2] is a good example for a global analysis.

6.1.2 Characterization of Microgels and Nanogels

In the foregoing Sect. 6.1.1, we already became familiar with microgels, which are internally crosslinked polymer latex particles. In the first part of Sect. 6.1.2, we will demonstrate that the AUC is an excellent tool to study this inner crosslinking systematically. Microgel dispersions are important raw materials for the manufacturing of a wide range of industrial products, such as high-impact modifiers, filling material of chromatographic columns, adhesives, lacquers, ion exchangers, and super-absorbents. In the second part of this Sect. 6.1.2, we will present a special form of microgels, which we call *nanogels*, because their diameters are below 50 nm. These nanogels belong to the nowadays so intensely studied nanoparticles. We will demonstrate that the AUC allows an interesting look into the inner structure of these extremely small particles. Again, we will combine in this Sect. 6.1.2 the different AUC methods for aqueous and organic static density gradient runs, particle size distribution measurements, and sedimentation runs in aqueous and organic media.

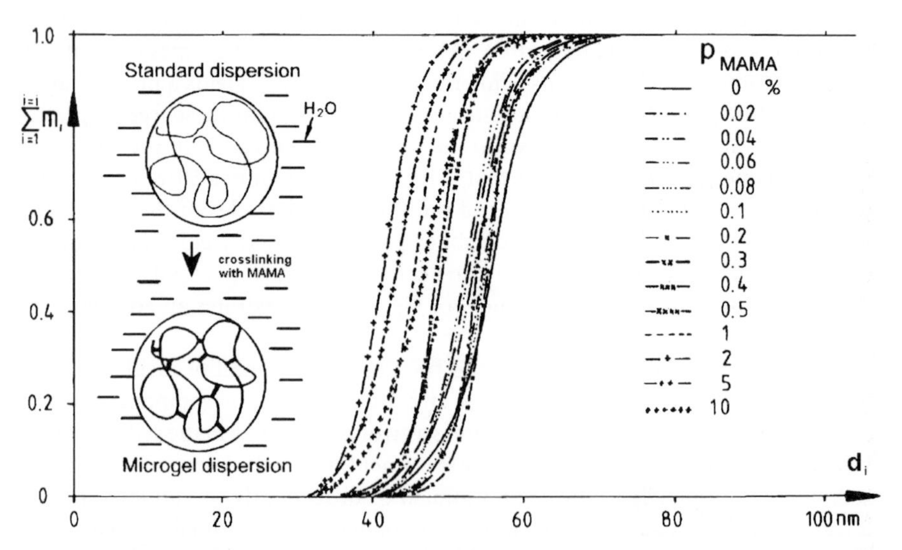

Fig. 6.6. AUC particle size distributions of 14 PBMA dispersions with different contents of the crosslinker MAMA, measured with the turbidity detector in H_2O at 15 g/l (reprinted from [3] with permission)

Microgels

Figure 6.6 shows the AUC particle size distributions of 14 polybutyl methacrylate (PBMA) dispersions. All these 14 particle size distributions have nearly the same shape, and nearly the same average diameters in the range 50–60 nm. However, inside of these particles there are significant differences: because of different amounts p of the crosslinker, methallyl methacrylate (MAMA), from $p = 0$ to 10 wt%, in addition to the main monomer BMA, these 14 dispersions vary internally between uncrosslinked, partly crosslinked, and completely crosslinked. In the following, we will study these *inner* differences by means of the AUC. For details of these studies, and additional light scattering measurements, the reader is referred to [3]. Examples for AUC analysis of microgels can also be found in [2] and [4].

To start our analysis of these 14 latices, we will discuss the aqueous density gradients shown in Fig. 6.7. For the sake of clarity, we present only five of the total 14 samples (the same is valid for Figs. 6.8 and 6.9). These aqueous density gradients show the expected result: only narrow turbidity bands of chemically uniform particles, with nearly the same known particle density of PBMA, 1.050 g/cm^3. Only the two dispersions with the highest crosslinker content, 5 and 10 wt% MAMA, show a higher particle density, because the density of MAMA is higher than that of PBMA.

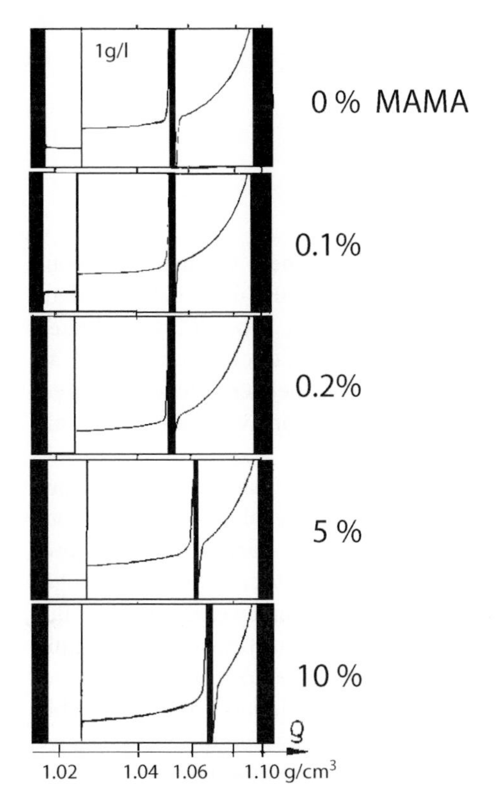

Fig. 6.7. Aqueous 90 H_2O/10 metrizamide density gradients of five PBMA dispersions with different contents of the crosslinker MAMA (reprinted from [3] with permission)

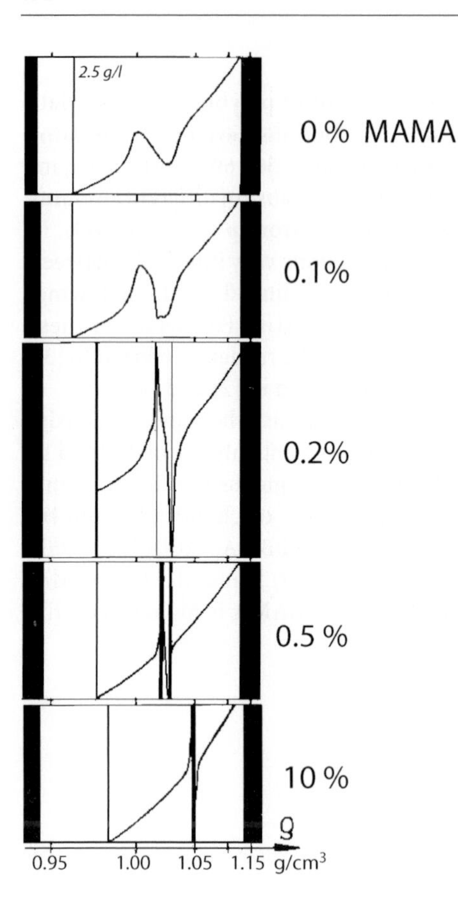

Fig. 6.8. Organic static (80 tetrahydrofuran/20 di-iodomethane) density gradients of five PBMA dispersions with different contents of the crosslinker MAMA (reprinted from [3] with permission)

Now, we again transfer the 14 dispersions from their originally aqueous medium by dilution into the organic (80 tetrahydrofuran/20 di-iodomethane) density gradient medium, which is a good solvent for PBMA, too. All uncrosslinked particles would be completely dissolved into single macromolecules, and all crosslinked particles would only swell, depending on their crosslinking degree. Figure 6.8 shows the result of this transfer experiment for five of the 14 organic density gradients.

The sample with 0 wt% MAMA (upper part of figure) shows only the double Schlieren peak of completely dissolved macromolecules (with M about 800 000 g/mol). The sample with 10 wt% MAMA (lower part of figure) shows only the narrow turbidity band of highly and completely crosslinked microgel particles (with M about 80×10^6 g/mol), which must be only weakly swollen. Between these two extremes, we find an interesting continuous transition: in sample 0.1 wt% we see, beside the dominating macromolecules, the first 5 wt% of microgel (= crosslinked part of the original particles) as a small deviation within the double Schlieren peak. In sample 0.2 wt%, the situation is already roughly 50:50, i.e., we see a 50-wt% double Schlieren peak and, superimposed on this, a broad transpar-

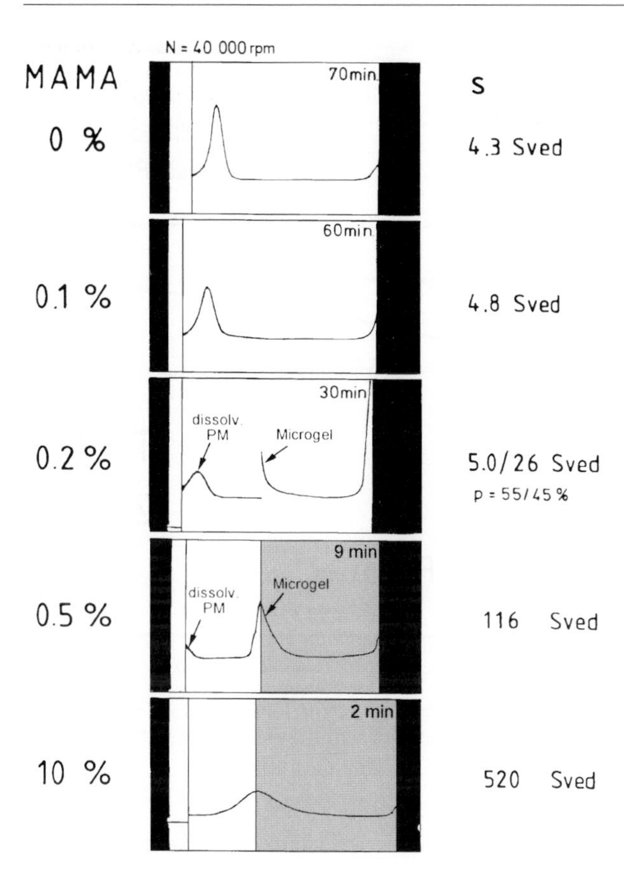

Fig. 6.9. Sedimentation runs of five PBMA dispersions with different contents of the crosslinker MAMA in the organic medium tetrahydrofuran at a concentration of 6 g/l (reprinted from [3] with permission)

ent 50-wt% band of highly swollen microgel, which shows no turbidity. In sample 0.5 wt%, we see a smaller transparent microgel band with a weak turbidity.

We will now quantify these initial results by means of sedimentation runs in the organic medium pure tetrahydrofurane (THF). Figure 6.9 presents these sedimentation runs (only one typical photograph per run, taken at different times). In the samples 0 wt% and 0.1 wt%, we see only the slow Schlieren peak of dissolved macromolecules with $s = 4.3$ and 4.8 S. In sample 0.2 wt%, we see two peaks: first, a slow peak of macromolecules with 5.0 S and a mass percentage of $p = 55$ wt%. Second, we see a fast peak of highly swollen microgel with 26 S and $p = 45$ wt%. Furthermore, in sample 0.5 wt%, we see 5 wt% of slow macromolecules and 95 wt% of fast, swollen microgel particles with 116 S, together with a weak turbidity. In sample 10 wt%, we see only very slightly swollen, fast microgel with 520 S, together with a higher turbidity.

Figure 6.10 is a summary of all these measurements on 14 microgel dispersions in the organic medium pure tetrahydrofuran. The upper diagram shows the mass percentage of microgel p_{gel} evaluated from the different Schlieren peak areas, and plotted as a function of the crosslinker content p_{MAMA} in wt%. The lower

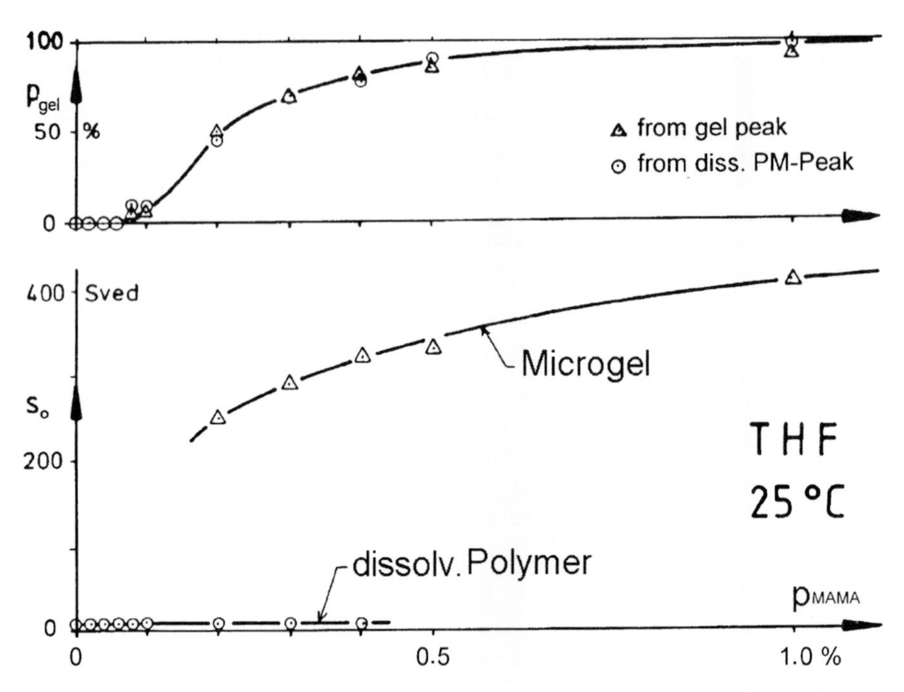

Fig. 6.10. Gel content p_{gel} and sedimentation coefficients s_0 of macromolecules and microgel parts of 14 PBMA dispersions dissolved in tetrahydrofuran as a function of the MAMA crosslinker percentage p_{MAMA} (reprinted from [3] with permission)

diagram presents the different sedimentation coefficients s_0 of the microgels, as well as of the dissolved macromolecules, also as a function of p_{MAMA}. All these measurements together show an interesting transitional region, ranging from pure macromolecules to pure, completely crosslinked microgels, with a crosslinker content between 0.1 and 0.5 wt% MAMA. Within this region, we find *both* microgel and macromolecules. Again, this composition range is a very interesting one for some applications, e.g., for pressure-sensitive adhesives.

In Fig. 6.10, the s_0 values of the microgels are plotted only in the low percentage range of crosslinker, $p_{MAMA} = 0-1$ wt%. In order to discuss the swelling ratio q of the microgels in tetrahydrofuran, we use in Fig. 6.11 the full range, $p_{MAMA} = 0-10$ wt%.

In Fig. 6.11a, one sees not only the s_0 values of the swollen microgel particles in THF, but also the s_0 values for non-swollen particles in H_2O. From the ratio $s_{0,THF}/s_{0,H_2O}$, we calculate the swelling ratio by volume q of these particles in THF, using (6.1) derived in [3], where d is the particle diameter, $\varrho_p(q = 0)$ the density of the non-swollen particle, and ϱ and η the density and viscosity of H_2O and THF:

$$q = \frac{V_{\text{swollen}}}{V_{\text{non-swollen}}} = \frac{d^3_{\text{THF}}}{d^3_{\text{H}_2\text{O}}} = \frac{\left[\left(\varrho_p(q=0) - \varrho_{\text{H}_2\text{O}}\right) \cdot \eta_{\text{THF}} \cdot s_{0,\text{THF}}\right]^{-3}}{\left[\left(\varrho_p(q=0) - \varrho_{\text{THF}}\right) \cdot \eta_{\text{H}_2\text{O}} \cdot s_{0,\text{H}_2\text{O}}\right]^{-3}} \tag{6.1}$$

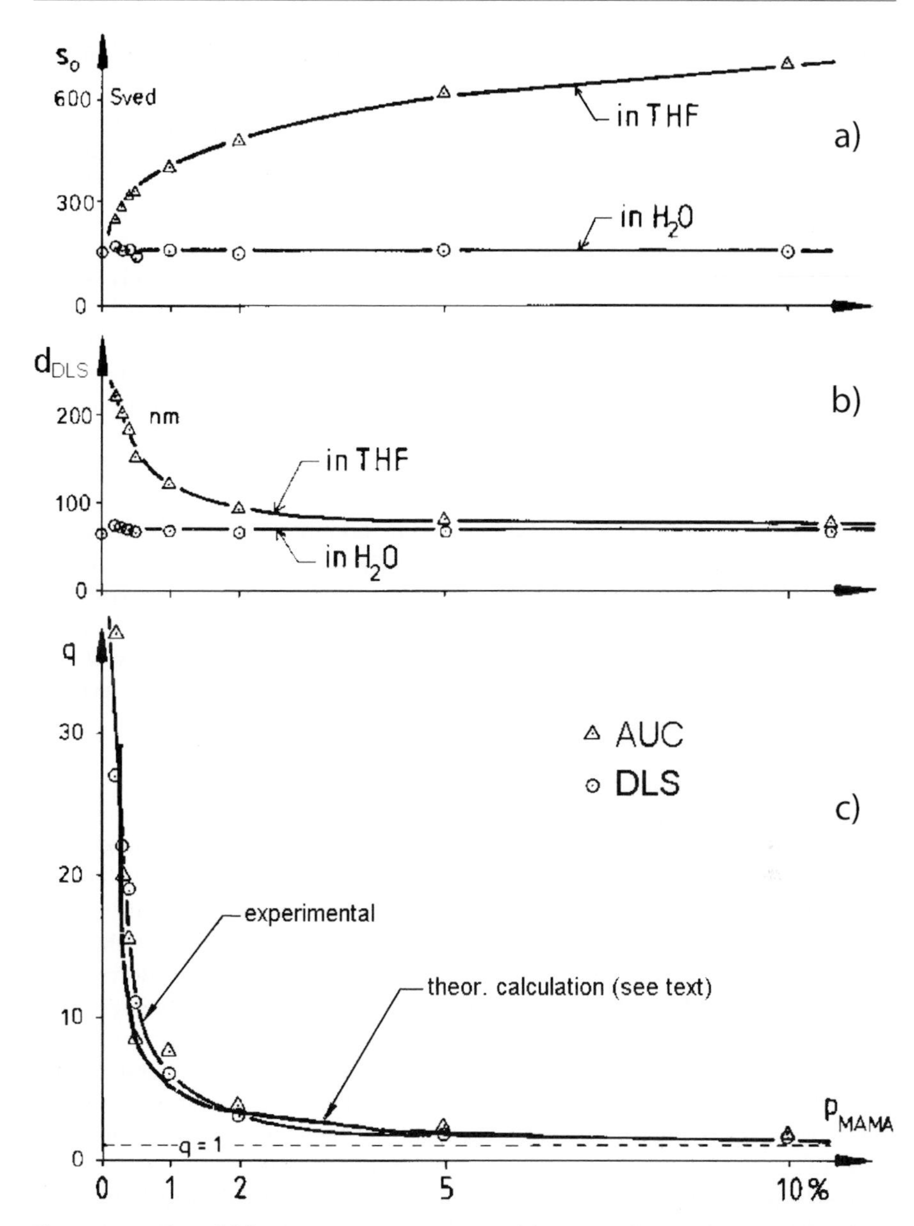

Fig. 6.11a–c. Plots of different measuring parameters of the microgel parts of 14 PBMA dispersions, in H_2O and dissolved in tetrahydrofuran, as a function of the crosslinker percentage p_{MAMA}. The parameters are **a** sedimentation coefficients s_0 in THF and in H_2O, **b** dynamical light scattering diameters d_p(DLS) in THF and in H_2O, **c** swelling ratios q calculated from the above AUC and DLS data (reprinted from [3] with permission)

The q ratios calculated in this manner from the AUC data are indicated as triangular points in Fig. 6.11c.

Additionally, q was measured independently with dynamic light scattering (DLS). This was performed, as indicated in Fig. 6.11b, by measuring the hydrodynamic (z-average) diameters d_{DLS} of the two microgel particle types separately, i.e., those swollen in THF and non-swollen in H_2O. The resulting DLS swelling ratios q calculated with the first part of (6.1) are indicated in Fig. 6.11c as circular points (similar AUC q determinations on microgels via s runs are described in [11]).

The agreement between both methods, AUC and DLS, is good. There is a large variation of the swelling ratio q, between $q = 35$ for the most weakly crosslinked microgel particles with 0.2 wt% MAMA, and $q = 1.5$ for the most strongly crosslinked particles with 10 wt% MAMA. This means that even the 10 wt% MAMA particles are weakly swollen, proving that THF is a good solvent for PBMA.

Nanogels

Nanogels are a new class of substances. They are also microgels, but their particle diameter is so small ($d_p < 50$ nm) that they nearly do not show any turbidity, and so the AUC turbidity detector is not sensitive enough to detect these nanogels. This is why Schlieren, interference, or UV optics detectors have to be used instead. Nanogels can also be considered as compact, globular, intramolecular crosslinked macromolecules (M range 1 to 20×10^6 g/mol), and therefore can be detected in the AUC like dissolved macromolecules.

Figure 6.12 shows the nearly identical particle size distributions of four aqueous nanogel dispersions measured with Schlieren optics ($c = 2$ g/l). These nanogels are very small, crosslinked polystyrene latices containing one crosslinker monomer per 80, or per 20 styrene monomers, corresponding to weakly or strongly crosslinked particles (and to a mass percentage of crosslinker of 1.3 or 5.0 wt%, respectively). Additionally, the two D_8 samples are completely deuterated, the two H_8 samples not deuterated. All four nanogel dispersions show nearly the same broad, unimodal particle size distribution. The 50% diameter is always about 25 nm (confirmed by DLS measurements, which yield constant z-average diameters around 33 nm). This means that, seen outwardly, these four latices are not distinguishable! This was intended, because small percentages of deuterated particles within "identical", non-deuterated particles should be used as tracer particles for neutron scattering experiments. This leads to the interesting question: is the AUC able to show the differences in the *inner* structure of these very small, 25-nm nanogels, which must exist? Are they weakly or strongly crosslinked, deuterated, or not deuterated?

Figure 6.13 will give a first answer to this question: we see an analysis of all four nanogel dispersions in aqueous static H_2O/metrizamide density gradients with different amounts of metrizamide, between 8 and 15 wt%, respectively. The two non-deuterated H_8 nanogels show the expected, known polystyrene particle density $\varrho_p = 1.053$ g/cm^3, but the two deuterated D_8 nanogels show an increased and – this is important – *uniform* particle density around 1.109 g/cm^3. This uniformity, and the fact that the measured density increase due to deuteration, $\Delta\varrho = 0.056$ g/cm^3,

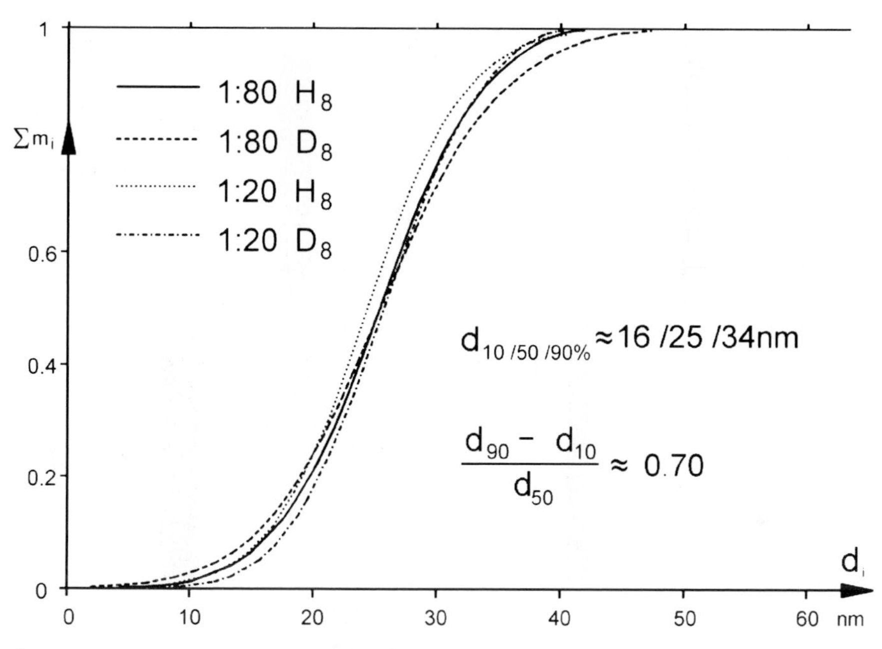

Fig. 6.12. AUC particle size distributions of four polystyrene nanogel dispersions, deuterated and non-deuterated, weakly and strongly crosslinked, measured with Schlieren optics in H_2O at 2 g/l

is the theoretically expected one, are the proof that the deuteration was successful, complete and homogeneous (compare also with the famous experiment on non-deuterated and deuterated DNA by Meselson and Stahl, described in Sect. 1.1.2, and Figs. 1.4 and 1.5).

In the following experiment, these four nanogels will be transferred again from their originally aqueous medium into the organic medium, pure tetrahydrofuran, where they will undergo swelling. Figure 6.14 shows the result of this transfer experiment, obtained by sedimentation runs at 3 g/l of all four nanogels in tetrahydrofuran.

The Schlieren photographs presented (now horizontally arranged) show, as for the aqueous medium, fast Schlieren peaks in every sample, with high sedimentation coefficients s = 39, 63, 69, and 101 S. From the peak areas follows a reproduction rate p_{gel} of nearly 100 wt% of the weighed-in material. These values of 100 wt%, and additionally the high s values, prove that these peaks belong to compact nanogel particles or, more precisely, to globular, internally crosslinked and swollen macromolecules (with M of about 6×10^6 g/mol). These peaks do not belong to linear dissolved polystyrene macromolecules! We also calculated the volume swelling degree q of the four nanogels, using (6.1), and the measured s values in THF, s_{THF}, and in water, s_{H_2O} (and checked these q values by DLS). As a result of the calculation, we get q = 7.7, 6.0, 3.8, and 4.0. This gives a clear answer to our introductory questions – the AUC allows us to detect differences in the

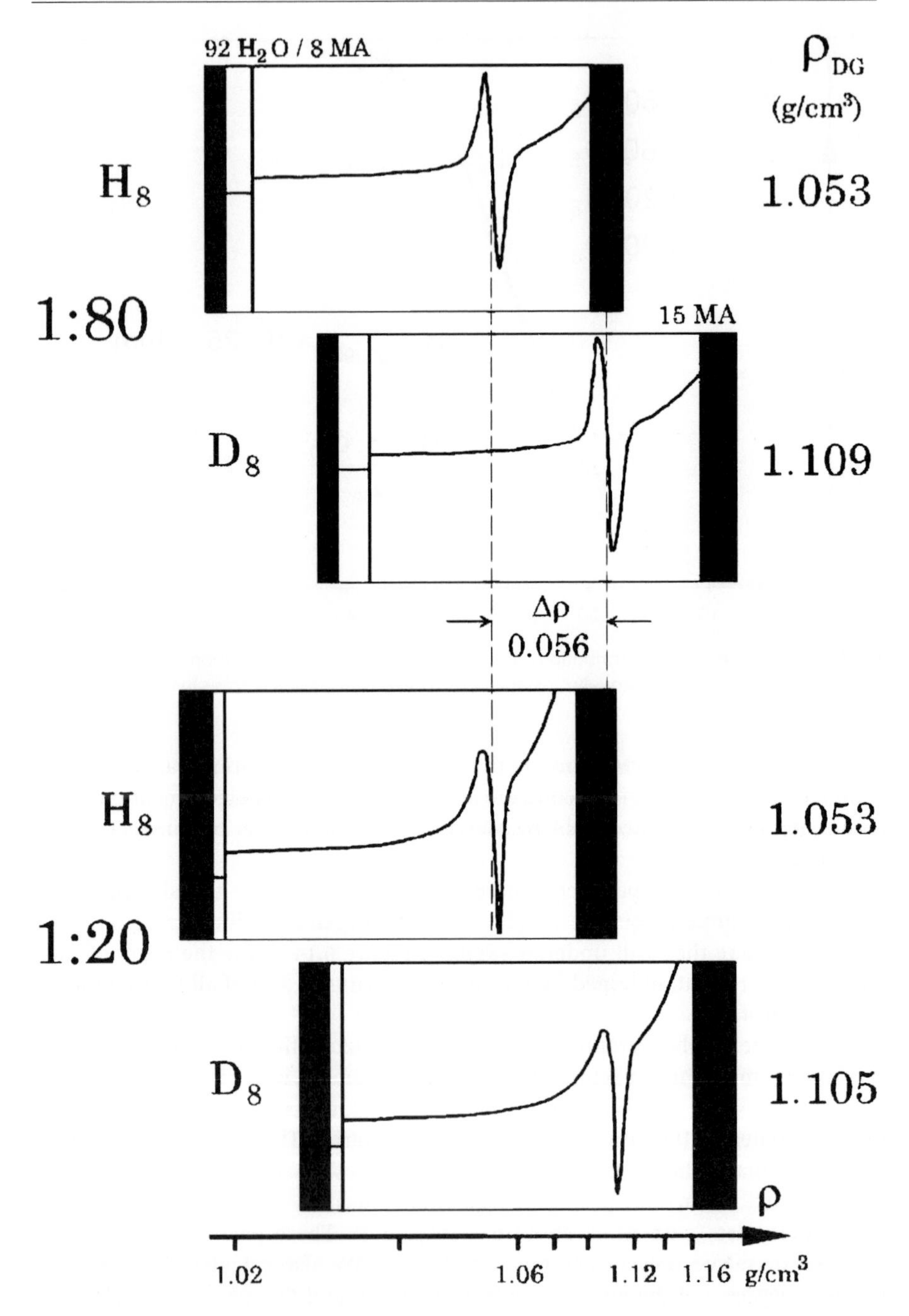

Fig. 6.13. Aqueous static H_2O/metrizamide density gradients of four polystyrene nanogel dispersions at 1 g/l

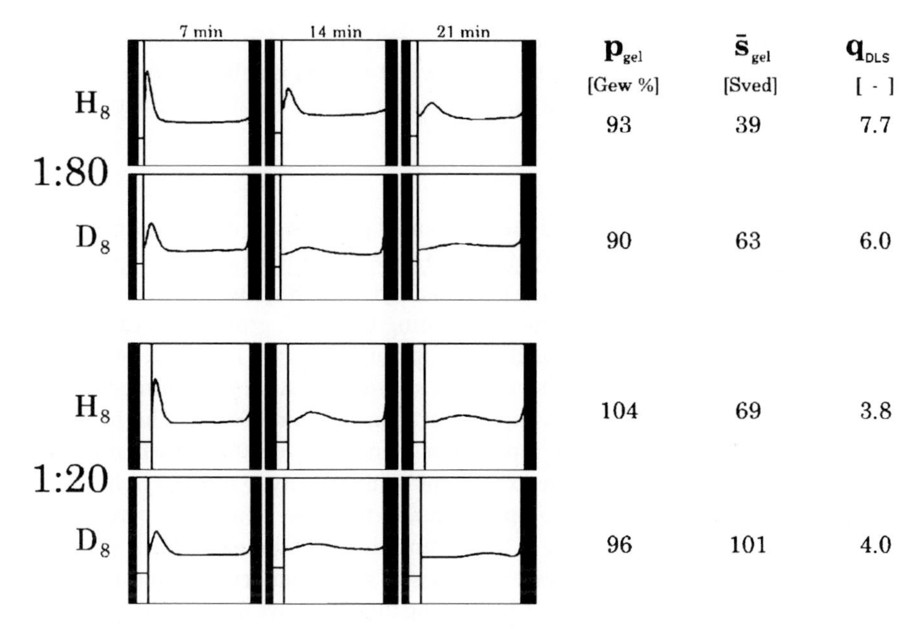

Fig. 6.14. Sedimentation runs of four polystyrene nanogel dispersions in the organic medium pure tetrahydrofuran at 3 g/l

inner structure of the four species discussed above: we differentiate (i) between deuterated and non-deuterated nanogels via ϱ measurements, and (ii) between weakly (1:80) crosslinked structures of the first two *nanogels*, having high degrees of swelling of $q = 7.7$ and 6.0, and strongly (1:20) crosslinked structures of the last two nanogels, having lower degrees of swelling of $q = 3.8$ and 4.0.

6.1.3 Ion Exchange in Carboxylated Latices

Crosslinking polymer latices by formation of metal salts during drying and film formation is widely used to produce durable coatings, strong adhesives, binders for heat-resistant materials and preservatives for floor covering. However, this ion exchange in carboxylated latices, which we now will discuss in Sect. 6.1.3, is also interesting in fundamental research, because such latex particles actually represent submicron ion exchanger beads, which are excellent models for studying analogous processes in regular (linear) polyelectrolytes. The polyelectrolyte character of carboxylated latices comes from the incorporation of electrical charges via copolymerization of $COO^- H^+$ group-bearing monomers into the macromolecule chains. All latex particles studied in this section are microgel particles, too, because they all were completely crosslinked with 2 wt% methallyl methacrylate (MAMA). This was done to preserve their identity during the different ion exchange processes. In this section, we are presenting measurements in aqueous media carried out by applying two AUC techniques, these being particle size distribution measurements, and aqueous static density gradients. In the following, it will demonstrated that especially density gradients are a unique analytical tool to study ion exchange (for details of these studies, see [4]). The present section is subdivided according to three aspects: (i) "softer" latices, (ii) "harder" latices, and (iii) kinetics of 2 $H^+ \leftrightarrow Zn^{2+}$ ion exchange.

"Softer" Latices

Figure 6.15 shows the integral and the differential particle size distributions of the "soft", "mother" latex (78 n-BA/20 MAA/2 MAMA). This is an acrylate copolymer latex with the two main components n-butyl acrylate (n-BA) and methacrylic acid (MAA), where MAA is bearing the dissociating COO^-H^+ groups. We call this a "soft" latex because its glass transition temperature $T_g = 11\,°C$ is below room temperature. The particle size distribution, shown in Fig. 6.15, is unimodal and narrow, with particle diameters ranging between 200 and 300 nm. The inserted density gradient Schlieren photograph shows a single narrow turbidity band. This means that all latex particles have exactly the same density, $\rho = 1.108\,g/cm^3$, and therefore, the same copolymer composition and the same equivalent percentage of H^+ ions. With respect to these H^+ ions, and the corresponding MAA groups, now samples of the mother latex were treated with stoichiometric amounts of MgO, CaO, ZnO, or PbO (in powder form), by stirring the respective mixtures at 85 °C for 3 h. The metal oxides are listed on the right-hand side of Fig. 6.15. By this treatment, the methacrylic acid will be transformed into the corresponding metal salts. The resulting metallized latices are hereinafter referred to as "daughter" latices.

Our aim was to replace all light H^+ ions by the heavier metal ions (Me^{++}) having increasing atomic masses A_{Me} (see Fig. 6.15). The question was whether it

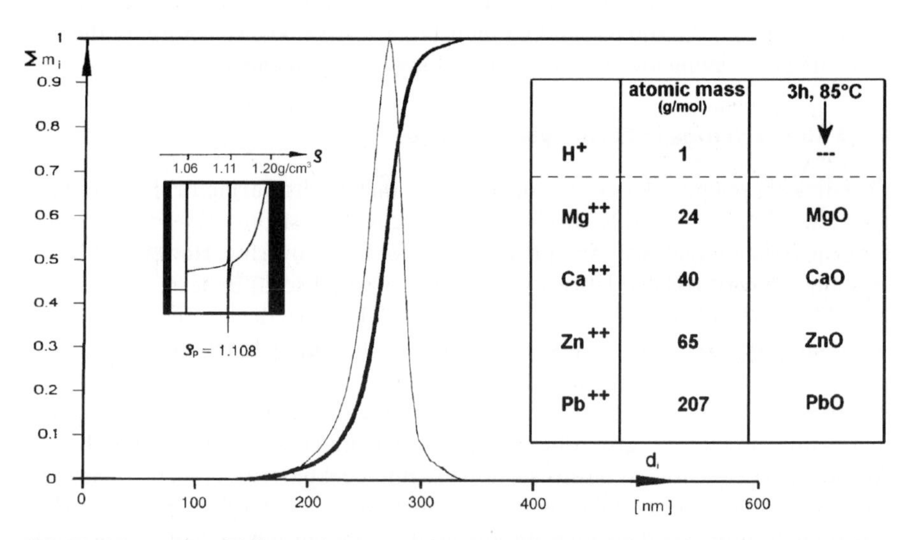

	atomic mass (g/mol)	3h, 85°C
H^+	1	---
Mg^{++}	24	MgO
Ca^{++}	40	CaO
Zn^{++}	65	ZnO
Pb^{++}	207	PbO

Fig. 6.15. Integral and differential AUC particle size distribution of the "soft" ($T_g = 11\,°C$) mother latex (78 n-BA/20 MAA/2 MAMA). *Inserts*: density gradient Schlieren photograph (*left*); table of metal oxides with which the latex was treated, and atomic masses A_{Me} of metal ions (*right*; reprinted from [4] with permission)

would be possible to achieve a complete exchange, i.e., not only would the H^+ ions at the particle surface be replaced, but also those located deep inside the particle. A sketch of this complex situation is given in Fig. 6.16.

Two neighboring H^+ ions are to be replaced by one bivalent metal ion Me^{++}, which in turn forms an ionic crosslink. We expected that, by this ion exchange, (i) the particle volume and the particle size distribution would not change, and (ii) the particle density would increase proportionally to the atomic masses A_{Me} of the heavier metal ions.

Figure 6.17 shows the particle size distributions of the four "soft" daughter latices and, for comparison, again that of their mother latex. Within experimental error limits, the distributions are all the same. Of course, these errors are larger than usual, because the errors in the ϱ_p determination are greater. The relative shape of the five different PSDs, characterized by $(d_{90} - d_{10})/d_{50}$, is exactly the same. Evidently, neither was the colloidal state disturbed by this metallization, nor did any agglomeration take place. The corresponding results of our density gradient measurements on these latices are summarized in the schematic diagram of Fig. 6.18.

All daughter latices show narrow turbidity bands in these density gradients, as does the mother latex, but the daughter particles show increased particle densities. This is a very important experimental result, because it means that all particles of each kind have exactly the same metal content. From the insert on the left-hand side of Fig. 6.18, it can be seen that $\Delta\varrho$ is indeed a linear function of A_{Me}. As a surprising result, we found that within each daughter latex particle, all particles are completely loaded with the respective metal ions. This means that a complete

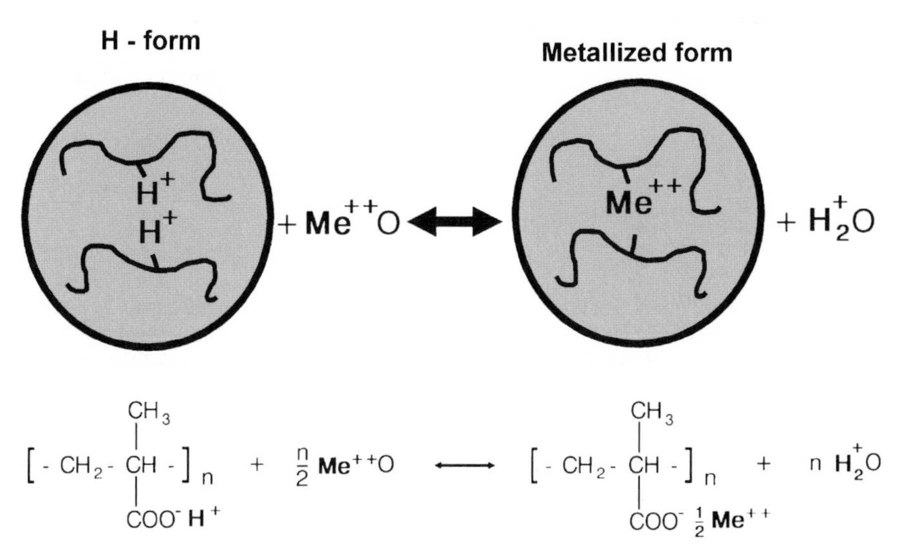

Fig. 6.16. Schematic diagram of the molecular situation within a latex particle during metal salt crosslinking by 2 $H^+ \leftrightarrow Me^{2+}$ ion exchange (reprinted from [4] with permission)

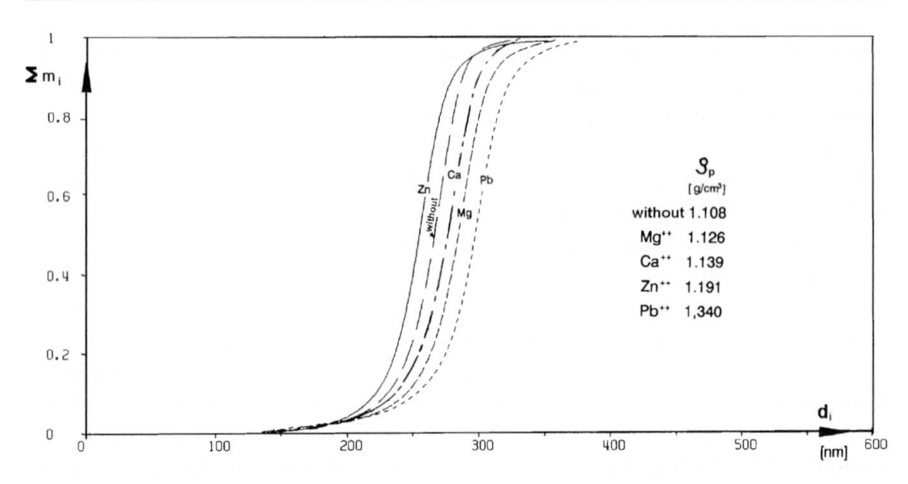

Fig. 6.17. Integral AUC particle size distributions of the "soft" mother latex 78 n-BA/20 MAA/2 MAMA, and its four metallized daughter latices (reprinted from [4] with permission)

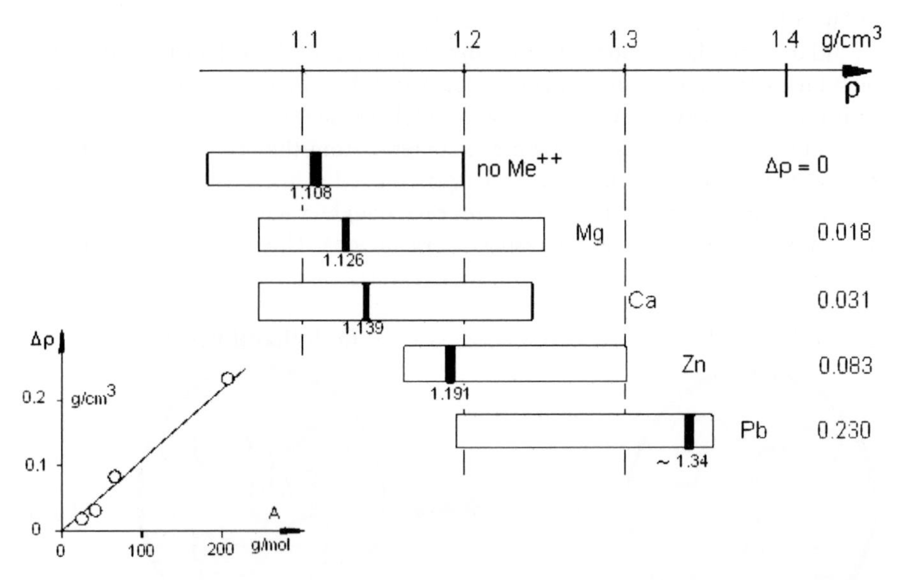

Fig. 6.18. Schematic diagram of the density gradient results on the "soft" mother latex 78 n-BA/20 MAA/2 MAMA, and its four metallized daughter latices. *Inserts*: density increase $\Delta\varrho$ by metallization versus atomic mass A_{Me} of metal ions (*left*); degree of density increase $\Delta\varrho$ by metallization (*right*; reprinted from [4] with permission)

hydrogen–metal ion exchange must have taken place, not only at the particle surfaces but also *inside* the particles.

"Harder" Latices

The complete $2\,H^+ \leftrightarrow Me^{2+}$ ion exchange in the "soft" mother latex documented above was perhaps not really very surprising, because we treated the mother latex with metal oxides above its glass transition temperature of T_g = 11 °C. To now study "harder" latices with systematically increasing T_g, we synthesized such "harder" latices by replacing, in our starting mother latex, parts p of the "soft" monomer n-BA by the "hard" monomer methyl methacrylate (MMA). Thus, the seven new mother latices have the copolymer composition (78–p) n-BA/20 MAA/2 MAMA/p MMA, with p = 0, 10, 20, 30, 40, 50, and 60 wt%.

Figure 6.19 shows the particle size distribution curves of these seven "harder" mother latices. All seven latices have narrow, unimodal, and nearly identical particle size distributions. The corresponding results of the density gradient measurements on these latices are summarized in the schematic diagram in the center of Fig. 6.20. The monomer composition of the corresponding latices is listed on the left-hand side; the corresponding T_g is specified on the right-hand side. All latices in Fig. 6.20 show narrow turbidity bands, implying they all are chemically homogeneous. The particle densities show a linear increase with the MMA content. The latter in turn results in a stepwise increasing "hardness", as indicated by the measured T_g values.

By treating these seven "harder" mother latices again with ZnO under the conditions described above, seven daughter latices were obtained. As the distance between the treatment temperature (85 °C) and the corresponding T_g was different for each latex (the two latices containing 50 and 60 wt% MMA were treated even below the T_g of the dry polymer!), we expected that only partial 2 $H^+ \leftrightarrow Zn^{2+}$ ion exchange would have taken place in the five latices having T_g values below 85 °C.

What really happened? Figure 6.21 shows the particle size distributions of the seven daughter latices. As in the case of the "soft" latices, the daughter latices have (within the experimental error limits) the same narrow, unimodal particle size distribution, essentially identical to that of the mother latices in Fig. 6.19. In Fig. 6.22, the corresponding results of the density gradi-

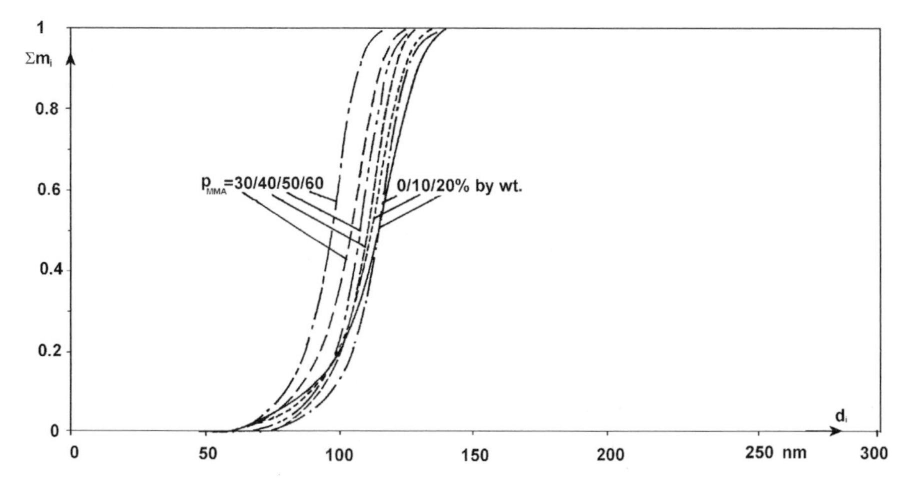

Fig. 6.19. Integral AUC particle size distributions of the seven "harder" mother latices (78–p) n-BA/20 MAA/2 MAMA/p MMA; $0 \leq p \leq 60$ wt% (reprinted from [4] with permission)

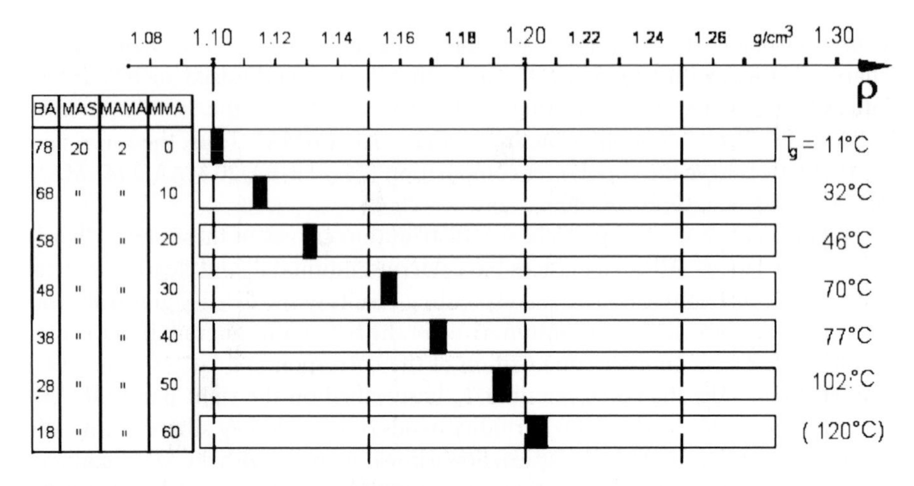

Fig. 6.20. Schematic diagram of the density gradient results on the seven "harder" mother latices (78–*p*) n-BA/20 MAA/2 MAMA/*p* MMA; $0 \leq p \leq 60$ wt%. *Inserts*: corresponding molecular composition (*left*); measured glass transition temperature T_g (*right*; reprinted from [4] with permission)

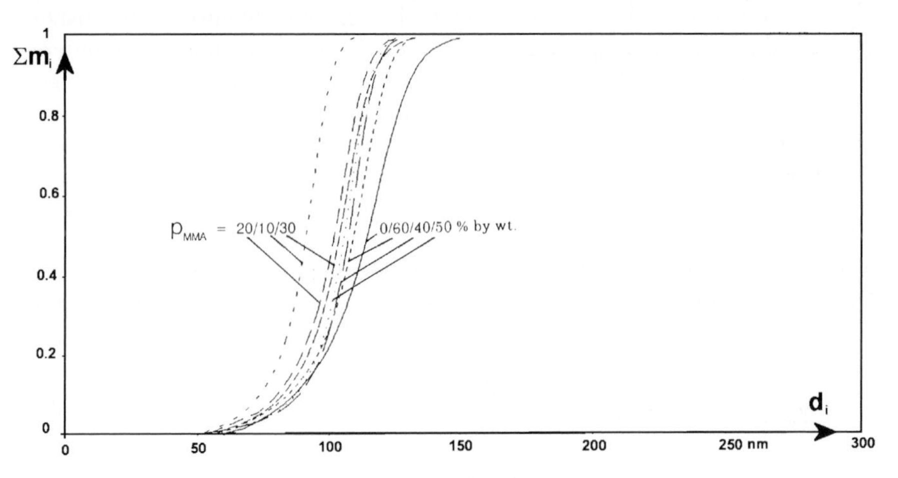

Fig. 6.21. Integral AUC particle size distributions of the seven (78–*p*) n-BA/20 MAA/2 MAMA/*p* MMA Zn-metallized daughter latices; $0 \leq p \leq 60$ wt% (reprinted from [4] with permission)

ent measurements on the daughter latices (closed bars), together with those of the mother latices (open bars), are plotted in a diagram similar to that in Fig. 6.20.

As in the case of the "soft" latices, all daughter latices show narrow turbidity bands (as did the mother latices), and particle densities increased by $\Delta\rho$ compared to the mother latices. The $\Delta\rho$ values, listed on the right-hand side of Fig. 6.22, are approximately identical for all pairs of mother–daughter latices, the mean value being $\Delta\rho = 0.077\,\text{g/cm}^3$. This means that all daughter latices are completely loaded with Zn^{2+} ions, even the two "hardest" ones. Thus, contrary to our expectation, in

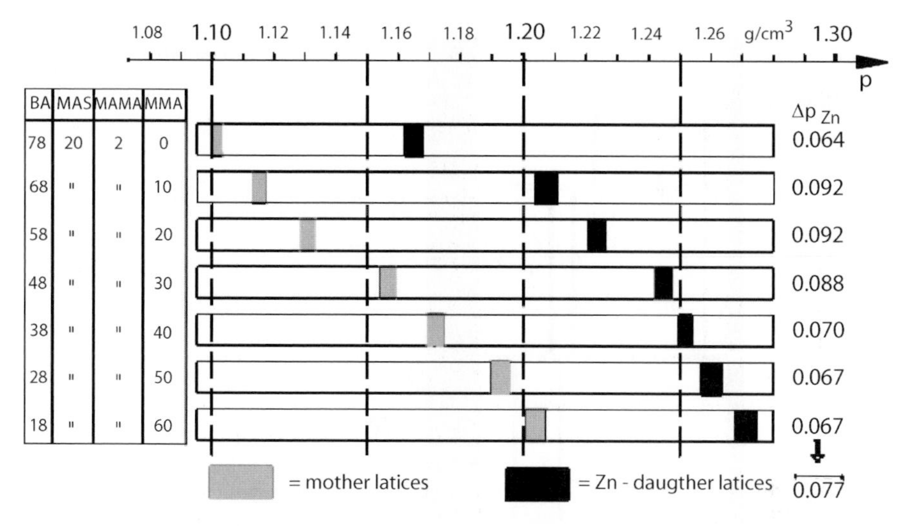

Fig. 6.22. Summarizing schematic diagram of the density gradient results on the seven (78–*p*) n-BA/20 MAA/2 MAMA/*p* MMA mother latices (*open bars*) and their Zn-metallized daughter latices (*closed bars*). *Inserts*: corresponding molecular composition (*left*); degree of density increase $\Delta\rho$ by metallization (*right*; reprinted from [4] with permission)

all cases a complete hydrogen–metal ion exchange took place, as was the case for the "softer" latices.

Kinetics of 2 $H^+ \leftrightarrow Zn^{2+}$ Ion Exchange

In order to now study 2 $H^+ \leftrightarrow Zn^{2+}$ ion exchange in mixtures, each daughter latex was mixed with its corresponding mother latex at a 1:1 mass ratio, whereas the original (high) latex concentration of 10 wt% was maintained. These mixtures were

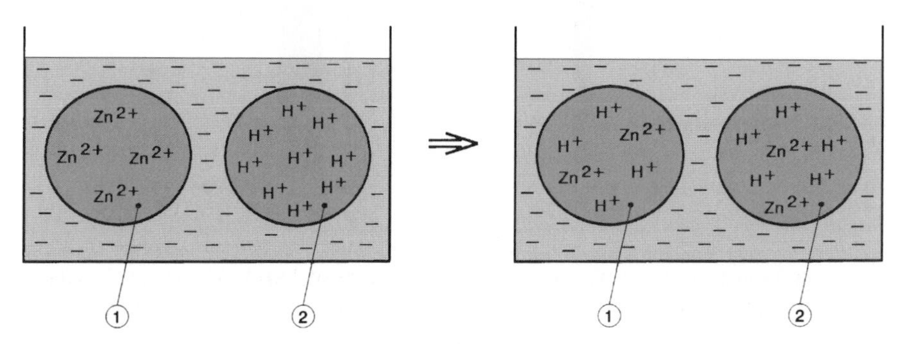

Fig. 6.23. Schematic diagram of latex particles before (*left*) and after (*right*) a complete 2 $H^+ \leftrightarrow Zn^{2+}$ ion exchange between a mother and a daughter latex (reprinted from [4] with permission)

Fig. 6.24. Density gradient Schlieren photographs of the seven (78–*p*) n-BA/20 MAA/2 MAMA/*p* MMA mother/daughter latex (1:1) mixtures after 1 h exchange time. These measurements are done simultaneously in one single run, using an eight-cell rotor and a multiplexer (reprinted from [4] with permission)

prepared at room temperature, i.e., below the T_g of all latex pairs, with exception of the first one.

In Fig. 6.23, a sketch of the inferred ion exchange process can be seen. The plot on the left-hand side shows the situation before the exchange, the plot on the right-hand side that at the end of a complete exchange. By this process, both kinds of particles, having two different densities, are transformed into one, single kind of particles having a uniform particle density at the mean value of the initial particle densities.

In order to study the *kinetics* of this process in detail, samples were drawn from the mixing vessel after different exchange times. The samples were diluted to a ratio of 1:100 (by this dilution, further ion exchange is stopped), and then analyzed in a density gradient measurement.

Figure 6.24 shows the original density gradient Schlieren optical photographs obtained after an exchange time of 1 h. In the photograph of the "softest" pair, containing 0 wt% MMA, only a single turbidity band is to be seen, whereas the photographs of all other mixtures exhibit two distinct, very narrow bands, indicating two kinds of particles of different densities, i.e., different Zn contents. For the four samples containing between 30 and 60 wt% MMA, these densities correspond approximately to those of the initial components of the respective mixture.

The situation after an exchange time of 3 days is schematically plotted in Fig. 6.25. Now also for the mixtures containing 10 and 20 wt% MMA, the initial bands of the mother latex (open bars) and the daughter latex (shaded bars) have merged into one, single respective band (closed bars), in the mean position. This means that in these samples, 2 $H^+ \leftrightarrow Zn^{2+}$ ion exchange is fully completed after

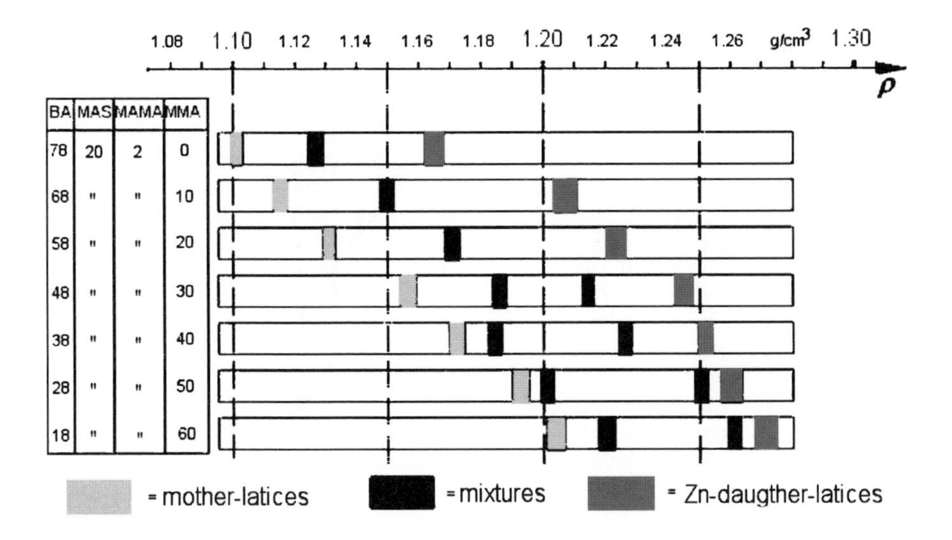

Fig. 6.25. Summarizing schematic diagram of the density gradient results of the seven (78–*p*) n-BA/20 MAA/2 MAMA/*p* MMA mother latices (*open bars*), their daughter latices (*shaded bars*), and respective (1:1) mixtures (*closed bars*) after 3 days exchange time (reprinted from [4] with permission)

3 days. For the four remaining "harder" samples, the distance between the mixture bands has decreased, compared to the 1-h situation of Fig. 6.24, and it will continue to decrease progressively with increasing exchange time.

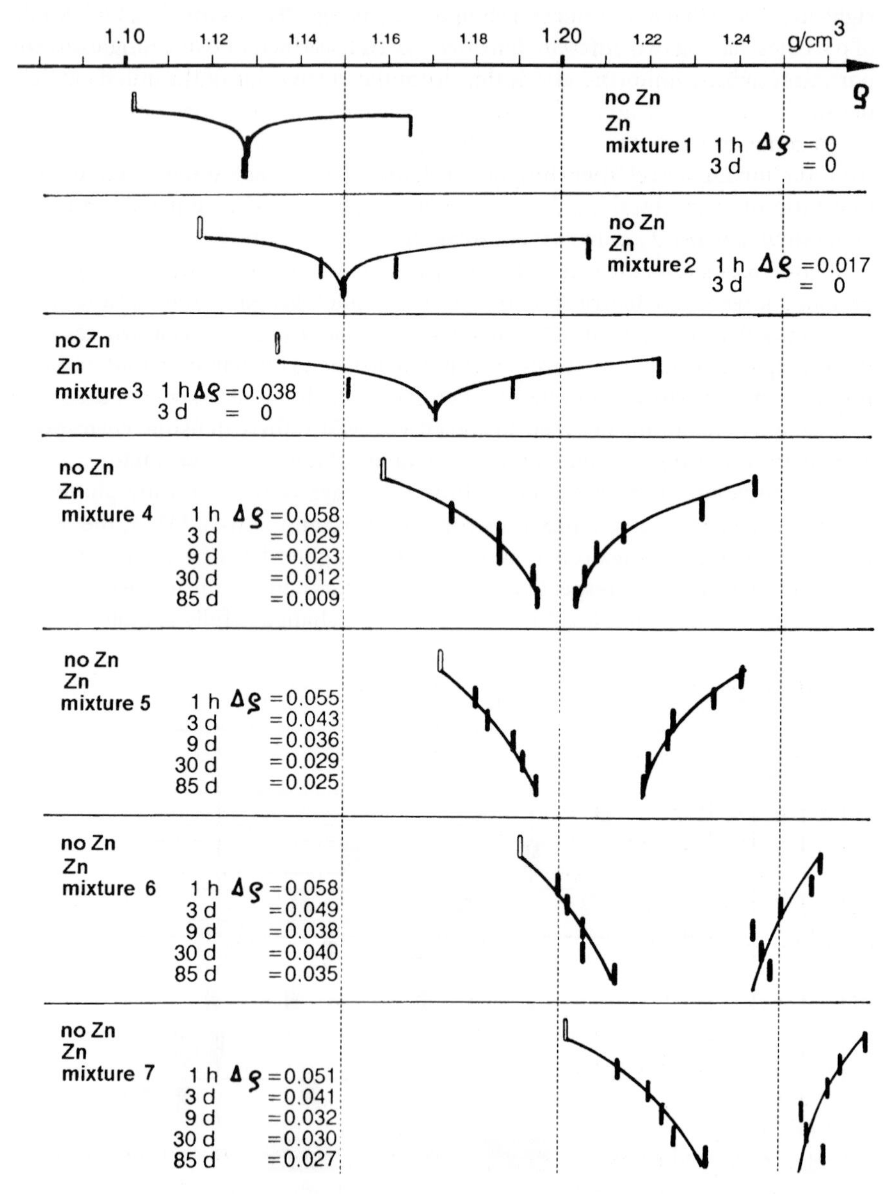

Fig. 6.26. Summarizing schematic diagram of the density gradient results of the seven (78–*p*) n-BA/20 MAA/2 MAMA/*p* MMA (1:1) mother/daughter latex mixtures after different exchange times (for details, see text) (reprinted from [4] with permission)

In Fig. 6.26, all our density gradient measurement results on these seven mixtures after exchange times of 1 h, and 3, 9, 30, and 85 days are summarized. For each mixture, the obtained density position of the turbidity bands within the density gradient is plotted along the ϱ axis as a function of the exchange time indicated. It is clearly shown that in all seven mixtures, complete 2 $H^+ \leftrightarrow Zn^{2+}$ ion exchange takes place even at room temperature, if the exchange process is maintained for a sufficiently long time. The rate of this process depends on the MMA content, i.e., on the "hardness" of the latices, and this is how it can be controlled: the higher the MMA content, the lower is the exchange rate.

This ion exchange is a strongly cooperative, quantized process. We do not understand it completely, but it is fascinating! During the whole exchange process, we always see two distinct, well-defined particle populations with different Zn contents. Within each of these two single populations, however, the corresponding Zn content is exactly uniform. Our expectation was that this exchange would be a statistical process, and we would obtain broader density gradient bands with increasing exchange time, which means a broad Zn distribution within every population. For which reason does this 2 $H^+ \leftrightarrow Zn^{2+}$ ion exchange occur in small, quantum-like portions? Why are all particles "watching" each other inside a population, controlling that each particle has exactly the same Zn content? Why is the process a cooperative one, and not a statistical one? With these open questions, we will close Sect. 6.1, dealing with the combination of different AUC methods, and open Sect. 6.2, in which other analytical techniques are combined with the AUC techniques.

6.2 Combination of AUC with Other Techniques

The application examples in Sects. 6.1.1 to 6.1.3 above are an impressive demonstration of the versatility and power of the different AUC methods, and their combination for the analysis of complex systems of synthetic polymers and colloids. However, sometimes the AUC alone cannot solve all analytical problems. In this case, we have to use additionally other analytical techniques, especially on samples having a particle density distribution (or other distributions) beside a size distribution. As mentioned in the introduction of this Chap. 6, the solution of such problems is a *"global analysis"* [12–15]. At present, we still are at the beginning of this approach, the rational being to use as many (fractionation) methods as possible for the same sample in order to gather information about the sample *distributions* in both size and density (and other parameters, too). Such methods are the AUC (s run, equilibrium run, DG run), EM, DLS, and FFF. The latter two methods also yield distributions of the diffusion coefficient. Recent examples of global analysis are (i) the combination of AUC (s runs) and flow-FFF (D distribution) for the analysis of the organic–inorganic hybrid colloid ferritin [12], and (ii) the combination of AUC and DLS in size and shape distribution analysis of macromolecules [13]. The author, P. Schuck, also presented a corresponding data evaluation program, called SEDPHAT (see [14, 15]).

In this Sect. 6.2, we will present some application examples of method combinations, in Sect. 6.2.1 the combination of AUC and sedimentation field flow fractionation (SFFF), and in Sect. 6.2.2 the combination of AUC and electron microscopy (EM). Simultaneously using different analytical techniques for the same sample is often a good method to disclose the special power, but also the typical limits of these different techniques. The following examples are not real global analysis examples, such as the one described in [12]. For a better understanding, we have chosen simple examples that have unambiguous results, and point out the advantageous AUC features.

6.2.1 AUC and SFFF

Sedimentation field flow fractionation (SFFF) is also a *centrifugation* technique (that is the reason why we present this example here in our AUC book), suitable for particle size distribution measurements in the nm range and, in contrast to the AUC, also for the *preparative* fractionation of such samples (albeit only in µg amounts, and in combination with a fraction sampler). In the following, we will analyze a ten-component mixture of nearly monodisperse polystyrene latex particles (the diameter $d_{p,i}$ varies in the range 67 – 1220 nm, the mass percentage m_i of every component is 10 wt%, and $i = 1, 2, ..., 9, 10$, the running number of the ten different components). Both techniques, the AUC and SFFF, will be applied, and some measurements using scanning electron microscopy (SEM) and DLS are included. All details of this analysis, especially literature concerning SFFF, can be found in [5]. A quite recent FFF review, especially for flow-FFF, is [16]. Note that the ten-component sample was already described in Sect. 3.5.2 of this book.

Figure 6.27 outlines the principles of SFFF. The general field flow fractionation (FFF), developed in 1965 by Giddings [6], is a chromatographic method. The "fractionation column" is a long and very thin channel, shaped like a belt (see Fig. 6.27a). Typical dimensions are $94 \times 2.0 \times 0.0254$ cm, resulting in a channel or void volume of $V_0 = 4.8\,\text{cm}^3$. A carrier medium (e.g., a solvent or water with 0.1% surfactant) is pumped continuously through the channel in the form of a laminar parabolic stream with typical flow rates of 1 – 3 cm^3/min. At the inlet of the channel, there is an injection port for the sample to be analyzed (the typical injected sample volume is 0.005 cm^3 for a sample concentration $c = 1$ wt%). A force F perpendicular to the belt-like channel (see Fig. 6.27a) interacts with the flowing sample particles, and pushes them toward the outer channel wall, i.e., away from the center of the parabolic carrier stream. The lateral movement of the sample particles depends on their size, density, electrical charge, etc., resulting in a fractionation of the sample into its different components, which can be monitored in a detector located behind the channel outlet. The various FFF methods are distinguished by the different forces F used, e.g., electrical, flow or *centrifugal* forces. If we use the latter, we speak of *sedimentation* field flow fractionation (SFFF). In this case (see Fig. 6.27a), the flat channel is bent in a circular way, like a real belt, and introduced into an (ultra)centrifuge. Also during rotation of this belt-like channel inside the SFFF rotor, a continuous parabolic laminar carrier flow through it is made possible by means of a rotating face seal. Generally, all FFF apparatus are fractionation devices according to their different parameters, and thus well suited for global analysis.

Figure 6.27b schematically shows the fractionation of a bimodal mixture of small and large particles inside the belt-like SFFF channel, in four steps. The resulting analytical curve (called the retention curve, or fractogram or elugram) of a UV detector (254 nm) is presented at the bottom of Fig. 6.27b. In the first step, shortly after the injection of the sample into the running rotor, the carrier flow is interrupted by stopping the pump for 10 min. Within this "relaxation" time, all particles sediment radially (i.e., perpendicularly to the two channel walls with their very small distance $w = 0.0254\,\mathrm{cm}$) toward the outer wall. Also within this time period, back-diffusion takes place, and sedimentation–diffusion equilibrium with an exponential radial concentration profile $c(r)$ is reached (as in an AUC equilibrium run). Small and large particles assemble superimposed in thin layers near the outer wall at the entrance of the channel.

In the second step, the pump is started, again creating the continuous laminar parabolic carrier flow through the channel. This flow transports the superimposed layers along the channel, thereby creating a *fractionation* of the two layers, because the small particles form a thicker layer moving faster in the faster stream lines of the parabolic profile, whereas the larger particles have a thinner, more compressed and slower moving layer, closer to the outer wall. Low molar mass material, such as UV-absorbing impurities, are not retained, moving through the channel with the highest possible velocity and reaching first the outlet of the channel and then the detector, thereby creating the first peak in the fractogram. This peak of unretained material marks the zero retention time t_0, or the void volume V_0 of the channel. At the beginning of the third step, the small, but fastest particles will pass the channel outlet and the detector, creating the second peak in the fractogram at the retention time t_e, or the corresponding retention (= elution) volume V_e. Similarly, the fourth step begins when the large, but slowest particles pass the channel outlet and the detector, creating the third peak of the fractogram. In contrast to size exclusion chromatography, in SFFF analysis the smaller particles exit the "column" before the larger particles.

The SFFF theory of Giddings (see [5]) relates the measured retention times, t_0, t_e, or retention volumes, V_0, V_e, to the desired particle diameter d_p:

$$\frac{V_0}{V_e} = R = 6\lambda_F \left[\coth \frac{1}{2\lambda_F} - 2\lambda_F \right] = f(d_p) \tag{6.2}$$

where the quotient $V_0/V_e = R$ is called the retention quotient, and the parameter λ_F is a dimensionless characteristic layer thickness related to the particle diameter d_p via

$$\lambda_F = \frac{kT}{Fw} = \frac{RT}{\omega^2 r w \left(\varrho_p - \varrho_s\right) N_A \left(\Pi \frac{d_p^3}{6} \right)} \tag{6.3}$$

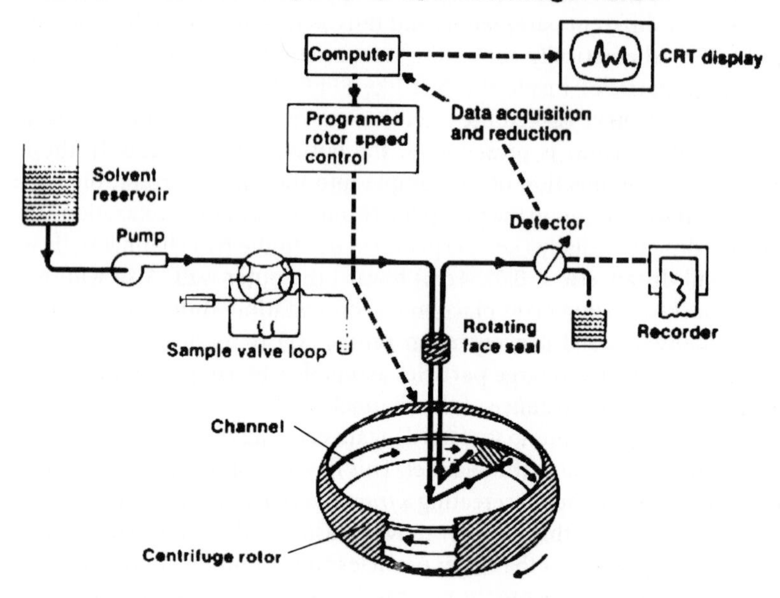

a

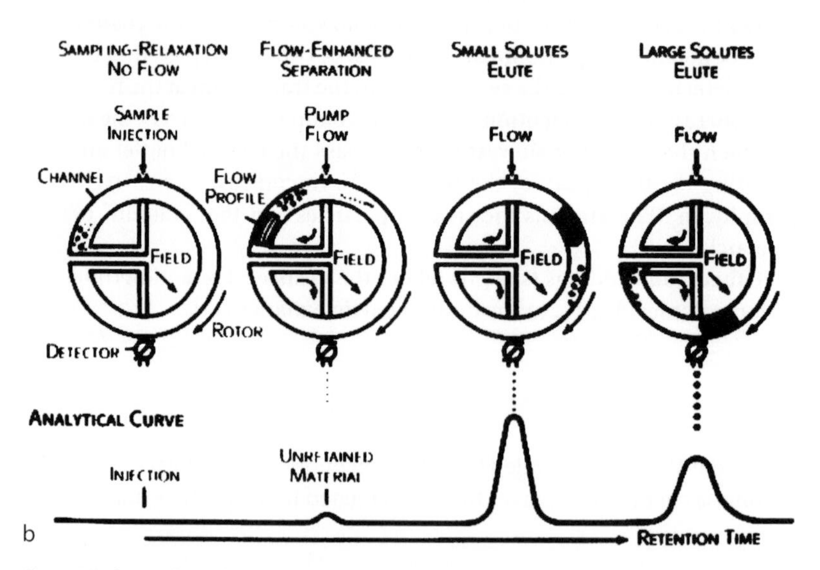

b

Fig. 6.27a,b. Outline of the chromatographic sedimentation field flow fractionation (SFFF) method. **a** General view of the SFFF arrangement, with its different elements. **b** Details and time sequence of the four-step fractionation process according to particle size inside the rotating belt-like channel, and the resulting analytical curve (= fractogram or elugram) of the UV detector (reprinted from [7] with permission)

where ω is the SFFF rotor speed, r the radial distance of the SFFF channel from the rotor axis (15.5 cm in the University of Utah SFFF apparatus), ϱ_p the particle density (1.055 g/cm^3 for polystyrene particles), ϱ_s the density of the carrier fluid, and N_A Avogadro's number.

After these introductory remarks concerning SFFF, we now start with comparative AUC and SFFF measurements for the determination of the particle size distribution of the ten-component mixture of nearly monodisperse polystyrene latex particles described above. The aim of these measurements is to reproduce not only the known diameter $d_{p,i}$ of the ten components but also their known percentages m_i. Beside AUC and SFFF, SEM and DLS will be used as well.

Figure 6.28 shows the AUC measurements for (Fig. 6.28a) each of the ten single components, and (Fig. 6.28b) the mixture of these ten components. The numerical $d_{p,i}$ and m_i values extracted from these AUC measurements, all done "field-programmed", i.e., with our usual, continuously increasing rotor speed ω from 0 to 40 000 rpm within 1.5 h, are summarized in Table 6.1 and in the insert of Fig. 6.28b. The comparison of these data shows that the AUC is able to reproduce all diameters within a measuring error of $\pm 5\%$, and all mass percentages within $\pm 15\%$.

Figure 6.29 shows the corresponding SFFF measurements: in Fig. 6.29a, a *field-programmed* $\omega(t)$ run, with an initial high field of 2000 rpm (692 g), followed by an exponentially decreasing $\omega(t)$ field with a time constant of 20 min (the flow rate in this case is 1.0 cm^3/min, the total running time 4 h), and in Fig. 6.29b, three measurements with $\omega = constant$ but different fields of 500, 111, and 28 g (in all three cases, the flow rate is 2.8 cm^3/min and the total running time 1.5 h). We start by discussing the field-programmed fractogram in Fig. 6.29a. Field-programmed $\omega(t)$ runs are the universal standard technique of SFFF (in contrast to PSD runs

Table 6.1. Particle diameters $d_{p,i}$ and mass percentages m_i for the ten-component mixture of nearly monodisperse polystyrene latex particles determined with different techniques (AUC, SFFF, SEM, and DLS)

Component no.	AUC single comp.		AUC ten-comp. mixt.		SFFF, DuPont		SFFF, Utah		DLS	SEM
	D (nm)	%	D (nm)	%	D (nm)	%	D (nm)	%	D (nm)	D (nm)
1	67	10	72	14	70	26	96	–	–	–
2	113	10	121	9	119	16	113	8	–	–
3	166	10	172	11	175	15	162	9	–	–
4	246	10	259	8	261	13	259	16	252	–
5	318	10	320	10	319	11	320	18	307	–
6	356	10	379	8	374	9	376	17	381	–
7	486	10	515	9	479	7	524	10	–	530
8	680	10	665	8	588	3	716	9	–	680
9	840	10	870	10	695	0.5	917	7	–	896
10	1220	10	1180	13	805	0.5	1221	6	–	1216

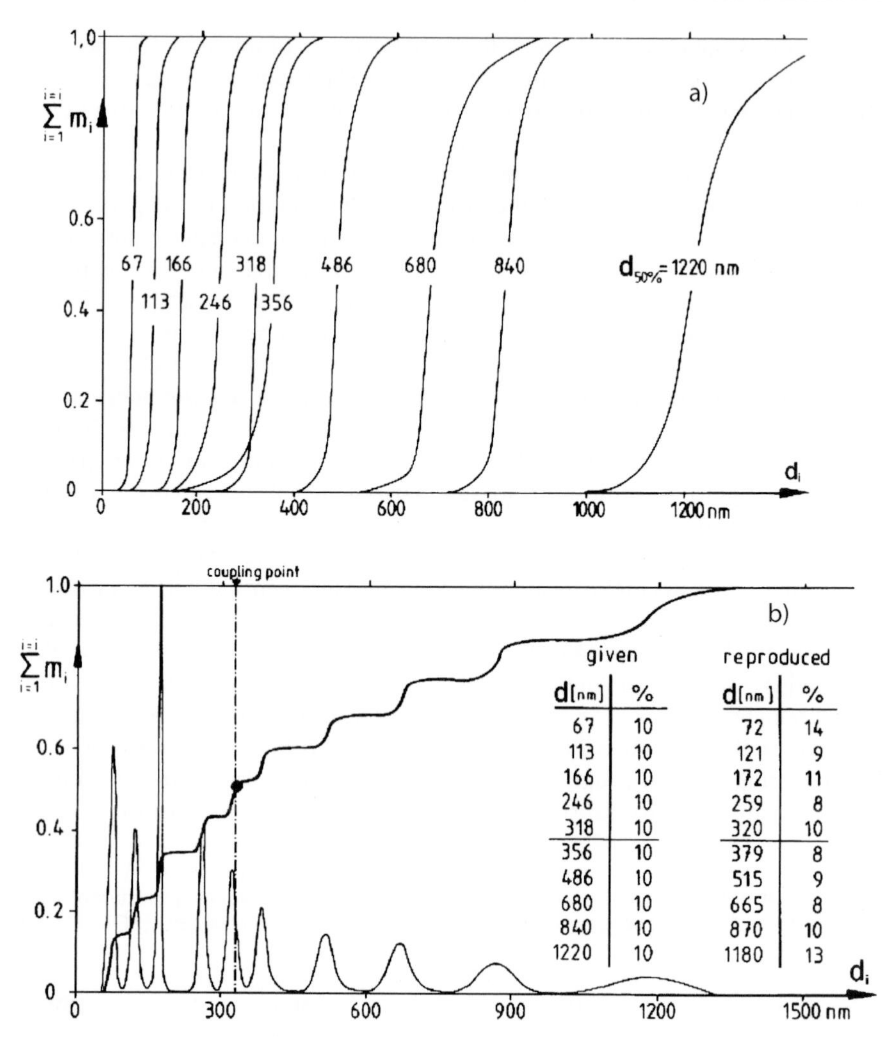

given d [nm]	%	reproduced d [nm]	%
67	10	72	14
113	10	121	9
166	10	172	11
246	10	259	8
318	10	320	10
356	10	379	8
486	10	515	9
680	10	665	8
840	10	870	10
1220	10	1180	13

Fig. 6.28a,b. AUC particle size distributions, **a** of the ten separate, single polystyrene latex components, and **b** of the ten-component mixture with an equal mass percentage of 10 wt% of each component (reprinted from [5] with permission)

in the AUC, where $\omega(t)$ *in*creases exponentially with time, in this SFFF mode $\omega(t)$ *de*creases exponentially with time). Unfortunately, this fractogram shows only eight resolved peaks, which belong to the eight largest particles (166–1220 nm). The two smallest components, numbers 1 and 2 (67 and 113 nm respectively), are not resolved, and close to the narrow impurity (or void volume) peak at low elution volumes. One reason is that the maximum field, 2000 rpm, of the University of Utah SFFF apparatus that was used for these measurements is not high enough. Therefore, the ten-component sample was sent to the DuPont laboratory for a routine analysis (analytical conditions not optimized) with the best SF^3 Particle Fractiona-

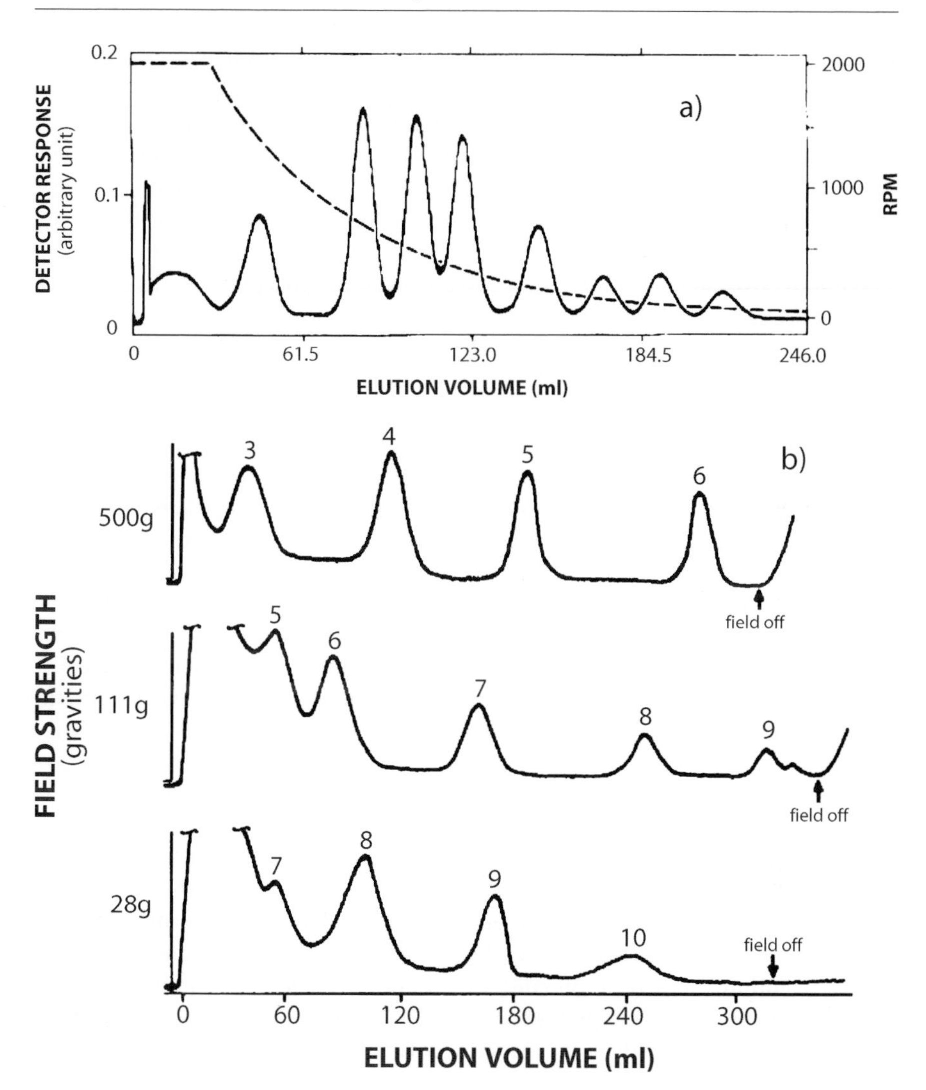

Fig. 6.29a,b. SFFF fractograms of the ten-component polystyrene latex mixture, using the University of Utah SFFF apparatus, **a** working with a programmed field $\omega(t)$, i.e., starting with the maximum field 2000 rpm (692 g) and then decreasing this exponentially within 4 h to zero, **b** working with three constant, but different fields of 500, 111, and 28 g, i.e., with ω = constant (reprinted from [5] with permission)

tor available at present, having a maximum field of 18 000 rpm (35 000 g). With this instrument, it was indeed possible to resolve all ten components. The resulting $d_{p,i}$ and m_i values are listed in Table 6.1. Although the diameter assignments made at DuPont for the smaller particles ($d_{p,i} < 500$ nm) were in good agreement with the AUC data, the table shows a systematically increasing deviation from the known diameters, with an increase in size. From Table 6.1, it is also clear that the quantification of the amount of the larger components is less accurate (and too low), as recoveries appear to decrease with increasing particle size. Systematic studies of this effect (see Fig. 6.30) show that excessive compression of the equilibrium layers of the largest particles at the beginning of the experiment with its high field will result in partial adsorption of the particles at the outer wall of the channel.

An analogous evaluation of the field-programmed fractogram of the University of Utah SFFF apparatus (Fig. 6.29a), i.e., calculating $d_{p,i}$ with (6.2) from the V_e-maximum position of the eluted peak, and m_i from the area of the peak (corrected for Mie scattering), yields a result analogous to the DuPont fractionator: (i) all SFFF diameters for particles larger than 500 nm are too low, (ii) also the SFFF m_i values for these particles are too low, and (iii) this deviation toward smaller SFFF m_i values increases systematically at higher $d_{p,i}$ values.

This effect of excessively low recoveries for large particles was systematically studied in Fig. 6.30 with the University of Utah SFFF apparatus as a function of (i) field strength (that means, also of retention time), and (ii) particle size, using three monodisperse polystyrene standard particles (Seragen) with known diameters of 394, 597, and 895 nm. Every measuring point in Fig. 6.30 is the result of a separate 1.5-h SFFF run at constant field. The recovery rates were calculated from the corresponding peak areas, corrected for Mie scattering. All calculated recovery rates at higher fields in Fig. 6.30 are normalized to zero field. Figure 6.30 shows a clear trend: the higher the diameter, and the higher the field strength, the lower is the recovery rate. Correct recovery rates of large particles are obtained only by an extrapolation procedure to zero field. This is a major disadvantage of the SFFF method, if one has to analyze broadly distributed particle mixtures with diameters larger than 400 nm.

An alternative way to the field-programmed SFFF technique of Fig. 6.29a, yielding perhaps more accurate $d_{p,i}$ and m_i values of larger particles, is to make several consecutive runs at constant but different fields, and to determine $d_{p,i}$ and m_i values only for particles of moderate layer compression, i.e., in the analytically useful retention range of 5–30 column volumes V_0. Such measurements of the ten-component sample are presented in Fig. 6.29b. Here, three fractograms were collected at constant fields in the "weak", "intermediate" and "strong" field range

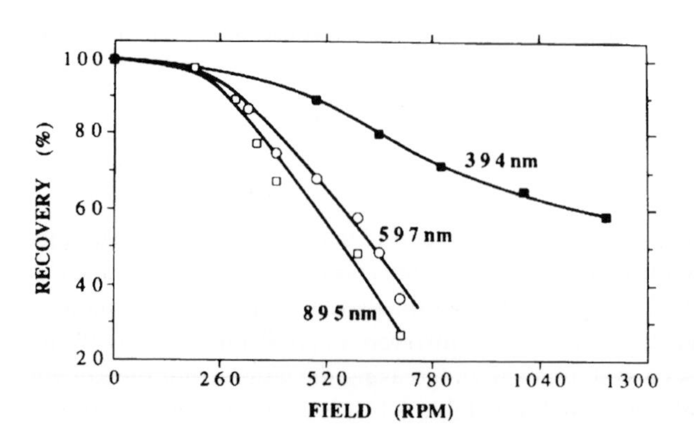

Fig. 6.30. Sample recovery measurements with the University of Utah SFFF apparatus as a function of field strength and particle size, using three monodisperse polystyrene standard particles (Seragen) of 394, 597 and 895 nm (reprinted from [5] with permission)

(28, 111, and 500 g) of the University of Utah SFFF apparatus. Each component peak in Fig. 6.29b is numbered from 1 to 10, corresponding to those in Table 6.1. The weak field (28 g) permitted detection and quantification of component 10 (1220 nm), which was otherwise undetectable, whereas it was clearly inadequate for resolution of the finer particles, which are resolved by the stronger fields of 111 and 500 g. Nevertheless, also at the high field of 500 g, the two smallest components 1 and 2 (67 and 113 nm respectively) are not resolved.

If we evaluate the fractograms of Fig. 6.29b by calculating $d_{p,i}$ for the different components at different field strength, using (6.2) and the measured V_0/V_e values, we obtain different and again excessively low numbers for the $d_{p,i}$ value for the same component. These systematic deviations from the known $d_{p,i}$ values increase with particle diameter and field strength, illustrated in Fig. 6.31. This figure shows not only the three fractograms of Fig. 6.29b, but also several other fractograms of the ten-component sample at additional constant field strengths.

These data demonstrate that there are no deviations for the 356- and 318-nm particles, neither for smaller ones. The reasons for the failure of (6.2) for particles larger than 356 nm are (i) steric exclusion of large particles from the outer wall (the theory of (6.2) uses point-like particles), and (ii) possible interactions of the particles and the outer wall in the form of velocity- and size-dependent lift forces directed toward the center of the channel. The present ten-component mixture of monodisperse particles proved to be an ideal sample for demonstrating the existence of such wall effects in SFFF.

The dataset in Fig. 6.31 reflects the ten-component sample SFFF behavior under a large number of different field strengths. These fields were chosen such that they allowed us to extrapolate the diameter data for each particle component to a value at zero field, presumably free from wall effects. For the four largest particles, a fourth-degree polynomial fit was shown to represent the data with a 99% confidence level, whereas sizes of smaller particles were determined by linear extrapolation to zero field. Extrapolated diameters obtained in this way are listed in Table 6.1, and are seen to compare well with those from AUC over the entire size range. The smallest particles (67 nm) were barely retained even at the maximum field strength of the

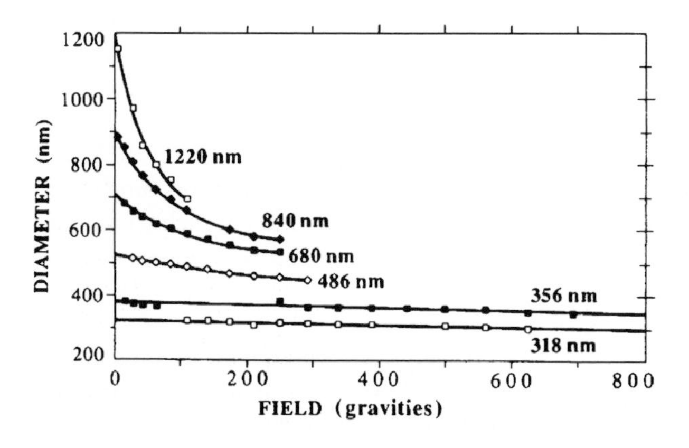

Fig. 6.31. Effect of field strength on particles' diameter d_p determined by SFFF using (6.2). Summary of constant field measurements, using the ten-component sample and the University of Utah SFFF apparatus (reprinted from [5] with permission)

University of Utah SFFF apparatus (2000 rpm, 692 g), and were therefore excluded from this extrapolation procedure.

Also listed in Table 6.1 are the mass percentages m_i of nine of the sample's ten components, calculated from the areas of the peaks (Mie scattering-corrected) in the different fractograms (the smallest particles, 67 nm, were not resolved). Although these constant-field m_i values are closer to the known values of 10 wt% than the field-programmed values, these SFFF values are not good enough. The comparison of all datasets in Table 6.1 shows clearly that the AUC is superior to SFFF in particle size distribution measurements of broadly distributed particle samples, especially if they contain components with diameters larger than 400 nm.

In the final part of this Sect. 6.2.1, we will describe a major advantage of the SFFF method, namely, the ability to separate a sample *preparatively* into fractions of uniform size (as mentioned above, only in µg amounts), which can be subjected to one or several secondary analysis steps. That is not possible with AUC. We will demonstrate this ability of SFFF with our ten-component sample, and the two secondary analysis methods SEM and DLS. During the constant-field measurements, presented in Fig. 6.29b, single-component peaks were "cut out" of the fractogram (via a fraction sampler positioned after the UV detector) for further analysis by SEM and DLS. The concentrations reached were above the detection limit of the DLS instrument used (Brookhaven BI-90, fixed angle 90°) for only three "cut" fractions (246, 318, and 356 nm). However, these three DLS diameters (252, 307, and 381 nm, see Table 6.1) are in good agreement with the AUC values.

For SEM measurements, it was possible to prepare nine "cuts" (again, the smallest component, 67 nm, could not be resolved and prepared). Prior to SEM analysis, the collected fractions were concentrated on Nucleopore filters with pore sizes of 100 or 200 nm. The filtered samples were mounted on copper stubs and gold-coated prior to imaging. The resulting SEM electron micrographs of the nine "cuts", and additionally that of the starting ten-component sample are compiled in Fig. 6.32.

These SEM micrographs give a clear indication that the SFFF fractionation is efficient, and produces highly monodisperse "cuts" without evidence of contamination by other components of the sample. The magnification is insufficient to permit accurate sizing of the smaller components, and Table 6.1 therefore contains only SEM-size assignments for the four largest SFFF fractions. These SEM sizes (530, 680, 896, and 1216 nm) are in good agreement with the AUC values.

In summary, the above Sect. 6.2.1 is an excellent demonstration of the synergy effects in combining different measuring techniques such as AUC, SFFF, SEM and DLS in analyzing complex colloidal systems. Using the same sample for comparison measurements manifests the strong points as well as the limits of each technique.

6.2.2 AUC and EM

AUC and electron microscope (EM) are two analytical techniques that complement each another in an excellent manner. AUC allows us to analyze very broadly distributed particulate systems, because millions of particles are "counted", and

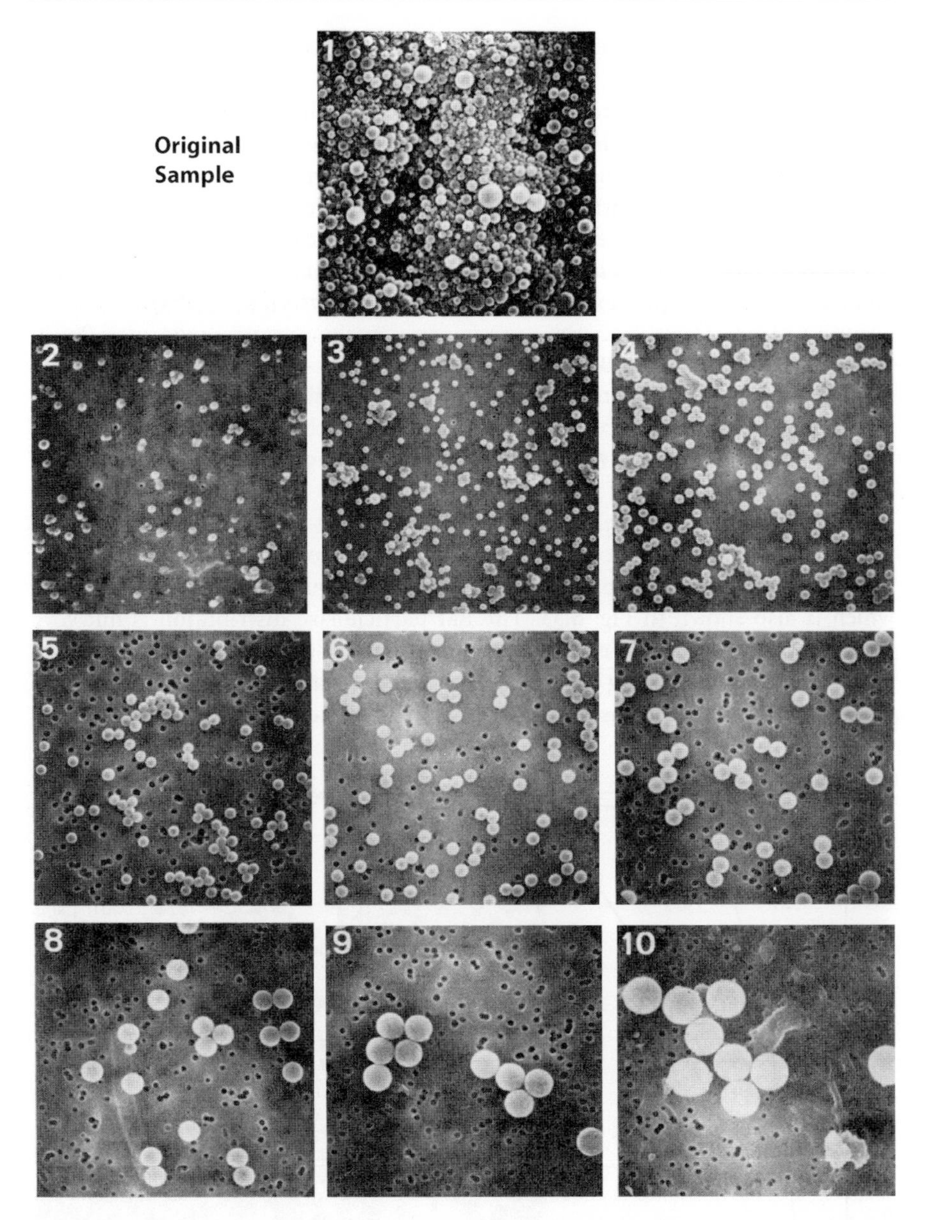

Fig. 6.32. SEM records of the starting ten-component sample (*top*), as well as of the single resolved SFFF fractions numbers *2* through *10*. Sizes measured for the four largest particle fractions from these micrographs are included in Table 6.1 (reprinted from [5] with permission)

it is possible to measure at low concentration in a "safe" manner, i.e., without appreciably disturbing the original system by a destructive preparation. However, the AUC always measures "Stokes-equivalent spheres", and does not deliver any information about the shape of the particles (spheres, rods, ellipsoids or coils) and their inner structure. That is possible with EM. Still, EM requires a complex preparation of the samples, and often only ten up to a few hundred particles can be counted on an EM micrograph. That is not enough for the determination of a very broad particle size distribution. In this case, several electron micrographs with different magnifications are necessary to obtain a representative picture of the whole sample. In the following, we present four combined AUC/EM examples: (i) copper phtalocyanine pigments, (ii) Mn/Zn ferrite particles in a magnetic fluid, (iii) colloid gold nanoparticles embedded in a transparent film of polyvinylpyrrolidone, and (iv) the very complex example of high-impact polystyrene (HIPS).

Copper Phtalocyanine Pigments

Figure 6.33 shows an electron micrograph of needle-shaped crystals of a copper phtalocyanine dye, used in printing inks, together with the particle size distribution curve that we obtained by AUC measurement, yielding Stokes-equivalent sphere diameters of 20 – 150 nm.

We also analyzed the electron micrograph of the needle-shaped crystals in Fig. 6.33 by means of an optical image processing computer program to measure the electron microscope (EM) particle size distribution of the needle lengths, which is also shown in Fig. 6.33. The EM particle size distribution is not very accurate, because we counted only 991 particles and it was very difficult to distinguish between neighboring particles. Thus, the AUC particle "size" distribution of Stokes-equivalent spheres is more representative for the PSD shape and broadness

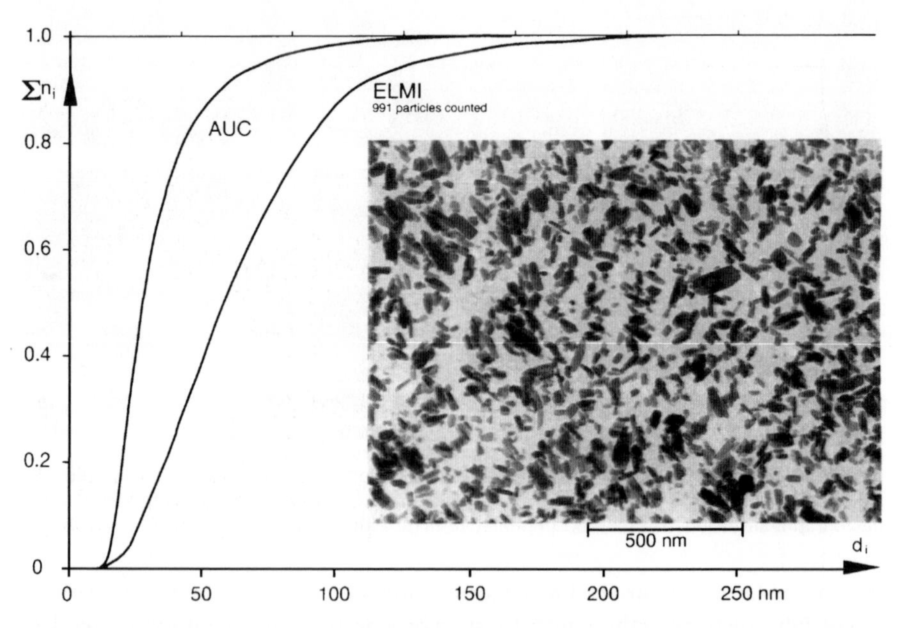

Fig. 6.33. Electron micrograph, AUC particle size distribution, and EM particle size distribution of a copper phtalocyanine pigment (reprinted from [8] with permission)

of the whole sample. Nevertheless, the (restricted) EM particle size distribution agreed reasonably well with the AUC particle size distribution. The EM particle sizes would be expected to be higher, because the needle length (EM) is being compared with the equivalent sphere diameter (AUC). In contrast to the AUC, only the EM is able to reveal that the copper phtalocyanine pigments are needle-shaped.

Mn/Zn Ferrite Particles of a Magnetic Fluid

Figure 6.34 shows another example with non-spherical particles. Rather than being dispersed in water, the heavy ($\varrho_P = 4.2\,g/cm^3$) Mn/Zn ferrite nanoparticles were dispersed in the organic solvent triethylene glycol (TEG). This demonstrates a great advantage of the AUC particle sizing method, namely, that it is possible to measure in every solvent or dispersion medium, not only in water, as is necessary for many other methods. The Mn/Zn ferrite dispersion is a magnetic fluid that can be switched in a magnetic field, and it may perhaps be used in the near future in coupling devices, seals, and dampers. Our task was to study the agglomeration of the primary 20-nm particles under different conditions. That is only possible with AUC, because every EM preparation will disturb the original agglomeration status considerably. Under the special conditions shown in Fig. 6.34, it was found that the Mn/Zn ferrite dispersion consisted of 20 wt% of non-agglomerated primary 20-nm particles, and 80 wt% of highly agglomerated particles with (sphere-equivalent) diameters of 40–400 nm. The micrograph in Fig. 6.34 shows, in agreement with AUC, correctly measured single 20-nm Mn/Zn ferrite particles, but the bigger agglomerates are not present in the original sample – rather, they are arbitrary, and created during the EM preparation.

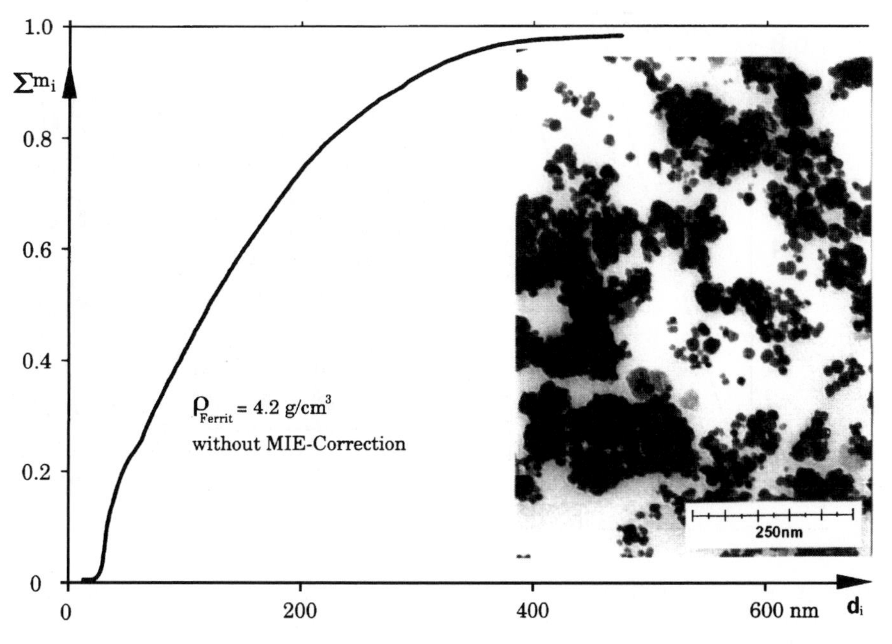

Fig. 6.34. Electron micrograph and AUC particle size distribution of Mn/Zn-ferrite particles in a magnetic fluid, dispersed in triethylene glycol, concentration 1 g/l (reprinted from [9] with permission)

Colloid Gold Nanoparticles Embedded in a Film of Polyvinyl Pyrrolidone

Figure 6.35 shows AUC-PSD measurements of gold colloids such as in the historical example of Rinde and Svedberg [10] in 1924. Colloidal gold nanoparticles have a very high density of $\varrho_p = 19.3\,\mathrm{g/cm^3}$ (and a complex refractive index $n(\mathrm{Au}) = 0.706 + i \cdot 2.42$ at $\lambda = 546\,\mathrm{nm}$ and $T = 25\,°\mathrm{C}$, which one has to know for the Mie theory).

In the introductory Chap. 1 of this book (see Figs. 1.1 and 1.2), we already described this historical example. Rinde and Svedberg created their colloid gold particles in pure water. By contrast, our sample was prepared in an aqueous solution of polyvinyl pyrrolidone, which was dried in order to obtain a transparent polymer film. The film itself has been tested as a fast nonlinear optical switch for optical computers.

The electron micrograph of a very thin 200-nm slice of this film in Fig. 6.35 (cut with a microtome) shows that single, non-agglomerated gold particles, with diameters of roughly 10–60 nm, are homogeneously distributed within this film. This information about the homogeneous distribution cannot be obtained by AUC. However, the AUC delivers a precise particle size distribution of these gold particles, also shown in Fig. 6.35, by again dissolving this film in water and carrying out a sedimentation run with the turbidity detector at $\lambda = 546\,\mathrm{nm}$. The result is a unimodal, nearly Gaussian particle size distribution with diameters of 5–50 nm.

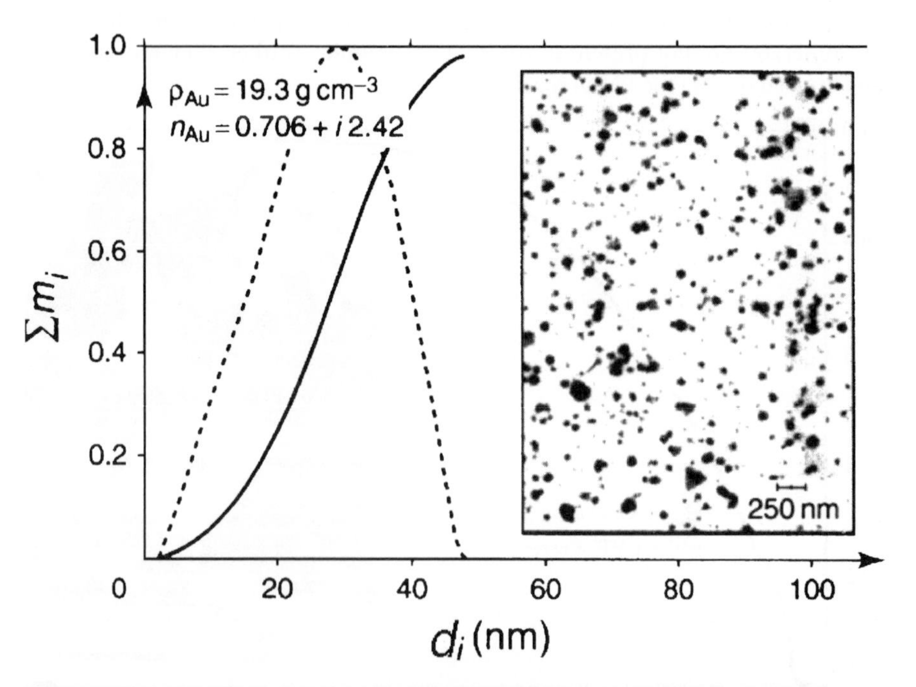

Fig. 6.35. Electron micrograph and AUC particle size distribution of colloid gold nanoparticles embedded in a film of polyvinyl pyrrolidone; dispersed in H_2O at a concentration of 0.1 g/l (reprinted from [9] with permission)

In contrast to the AUC particle size distribution in Fig. 6.35, the one in Fig. 1.2 is not entirely correct, because Rinde and Svedberg were not yet able to carry out a Mie correction. EM and AUC diameters in the example of Fig. 6.35 agree well. These gold particles have extremely small diameters. Therefore, we call them *nano*particles, and the particle size distribution of Fig. 6.35 demonstrates the lower measuring limit of the AUC particle sizing method with a light scattering/turbidity detector (with interference, Schlieren optics, or UV detectors, this lower limit is decreased to about 1 nm; see, for example, Figs. 1.2, 3.23 and 3.28).

High-Impact Polystyrene (HIPS)

In the following HIPS example, we combine only AUC and EM. Nevertheless, this is also a global analysis example. Indeed, only by a combination of *s* runs, DG runs, and PSD measurements with AUC and EM can we demonstrate the complex structure of this sample with its broad distributions in particle size and in particle density.

Figure 6.36 shows the transmission electron micrograph (TEM) of a very thin, 250-nm microtome slice (= cut) of high-impact polystyrene. This plastic engineering material is used worldwide for household appliances, automotive parts, toys, housings of telephones, radios, etc. Its special advantage is that nearly every shape of a part can easily be produced by injection molding, and that these parts show brightness, are rigid, but not as brittle as pure polystyrene, rather having a high-impact strength. The latter property is created by a special chemical synthesis: 8 wt% of (soft) polybutadiene is dissolved in styrene monomer, and the mixture is polymerized under intense stirring. By varying the stirring speed, the broadness of the particle size distribution of the resulting (soft) polybutadiene/polystyrene "rubber" particles contained in the (hard and brittle) polystyrene matrix can be

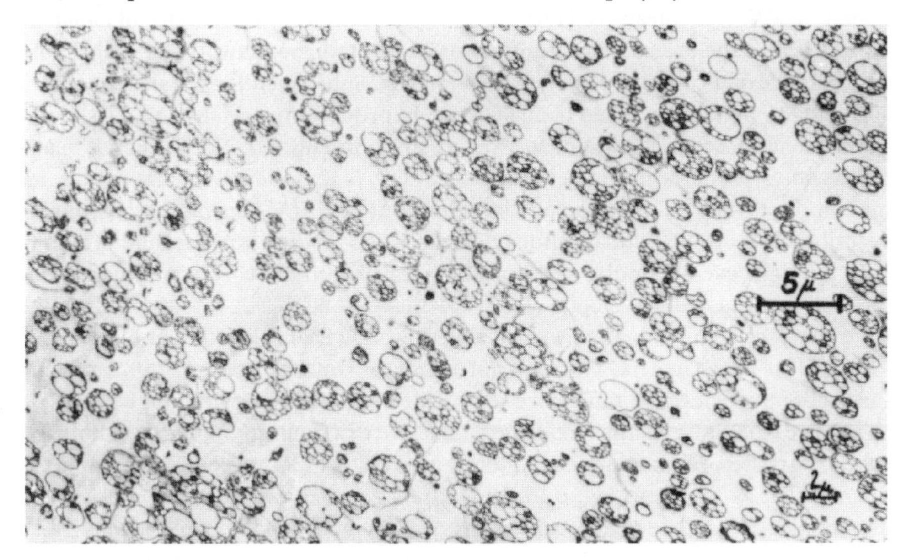

Fig. 6.36. TEM electron micrograph of a microtome-cut thin slice of high-impact polystyrene (HIPS), consisting of 92 wt% polystyrene (*white*) and 8 wt% polybutadiene (*stained black* by osmium tetraoxid)

controlled, and so the high-impact strength of this material, too. A similar material, soft PBA particles inside of a hard PSAN matrix, has been discussed in Sect. 6.1.1.

Only electron micrographs, such as shown in Fig. 6.36, and not the AUC (!), can reveal the complex inner structure of this high-impact polystyrene: The white areas in Fig. 6.36 consist of polystyrene, whereas the black areas consist of polybutadiene stained by osmium tetroxide. Bizarrely formed (nevertheless, nearly sphere-like) embedded polybutadiene rubber particles with diameters of 1000–5000 nm can be seen within the continuous (is it really continuous?) white polystyrene matrix, which again in its inside shows included white polystyrene sub-particles.

The following questions arising from Fig. 6.36 can be answered only by AUC, but not by EM. (i) Are the rubber particles inside the polystyrene matrix isolated particles, or are they connected by a polybutadiene network, not visible in Fig. 6.36, i.e., is this matrix really a continuous one? (ii) If there are single, isolated rubber particles, what is their S/Bu composition (is it constant, or variable?), and what is their particle size distribution? (iii) What is the molar mass of the matrix polystyrene, and of the included polystyrene? (iv) Is polystyrene grafted onto the (crosslinked) polybutadiene rubber particles, inside and outside, and what is the grafting degree? This questionnaire is typical for a *global analysis*. In the following, we will answer these questions by a combination of different AUC techniques: sedimentation runs, particle size distribution runs, density gradient runs, and preparative ultracentrifugation.

Figure 6.37 shows a sedimentation run of HIPS, dissolved in methylethyl ketone at a concentration of $c = 10$ g/l. Starting with a very low rotor speed of 1000 rpm, we see a continuous sedimentation of the (crosslinked!) rubber particles, as a fast turbidity front to the cell bottom within 15 min at a resulting velocity of $s = 20\,000$ S. This allows us to roughly estimate the lowest diameter (to approximately 1000 nm) by means of (1.9). This is a first hint for the existence of isolated particles, and the absence of a continuous polybutadiene network, as well as for the presence of a continuous polystyrene matrix. Subsequently, we raise the rotor speed to the maximum speed of 60 000 rpm, in order to for the 3000-times slower polystyrene macromolecules of the matrix to sediment, too. The Schlieren peak of this polystyrene, seen in the lower part of Fig. 6.37, yields two pieces of information: first, the complete molar mass distribution $W(M)$, using the technique described in Sect. 3.4.2 and a scaling relation $s = KM^{a}$ (we do not present this here), with a weight-average molar mass of $M_w = 240\,000$ g/mol, corresponding to the (average) sedimentation velocity of the peak maximum $s_0 = 12$ S; second, following from the Schlieren peak area, we calculate that 84 wt% of the whole polystyrene molecules is in the matrix. The remainder, 16 wt%, must be grafted or included in the rubber particles, and must have sedimented with them very rapidly.

Figure 6.38 presents the (rough) particle size distribution of the (crosslinked) rubber particles, estimated independently by AUC and EM, using electron micrographs such as shown in Fig. 6.36. Only a few hundred particles are counted to obtain this EM particle size distribution, and thus it is not a very precise one. This is also due to the fact that we are not sure whether we have properly corrected

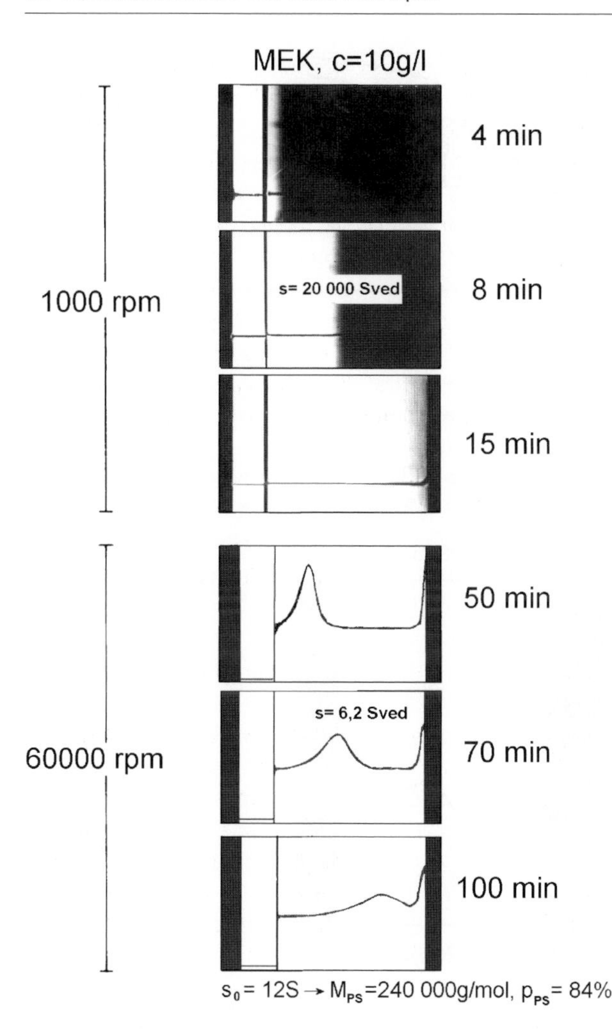

Fig. 6.37. Sedimentation run of high-impact polystyrene in methylethylketone at $c = 10$ g/l. In the first part of the run at 1000 rpm (*top*), we see the fast rubber particles sedimenting as a turbidity band, and in the second part at 60 000 rpm (*bottom*), we see the slow polystyrene matrix macromolecules sedimenting as a Schlieren peak

the so-called "tomato salad effect", resulting from cutting thin slices with the microtome. The AUC particle size distribution was measured with our particle sizer, described in Sect. 3.5.1, using turbidity optics and a very low HIPS concentration of 0.4 g/l, again in methylethyl ketone. This is also not very precise, because the particle density ϱ_p and refractive index n_p are not exactly known. Further ϱ_p and n_p are not really uniform, because of possible variations in the S/Bu composition (see the following density gradient in Fig. 6.39), and in the degree of crosslinking/swelling. Therefore, we assumed reasonable, average values to calculate the PSD. Indeed, we found that the two particle size distributions in Fig. 6.38 correspond reasonably well. Both are very broad, and the average diameter of 2500 nm is very high, which, it should be noted, is significant for the excellent high-impact strength of this material. It is also easy to explain why the EM distribution of 1000–6000 nm is somewhat smaller than the AUC distribution of 500–8000 nm:

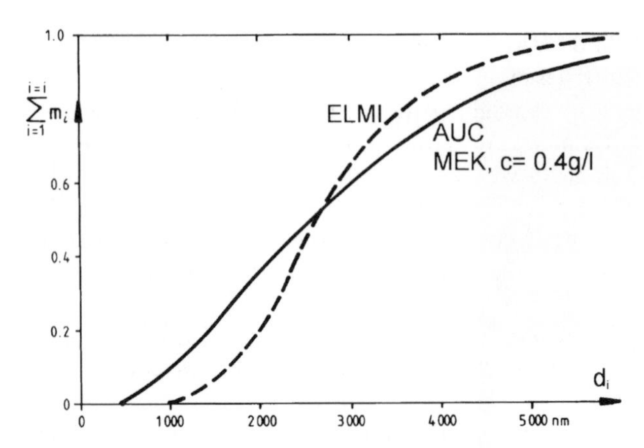

Fig. 6.38. Particle size distribution of the rubber particles inside a high-impact polystyrene sample, measured independently with AUC and with electron microscopy (TEM)

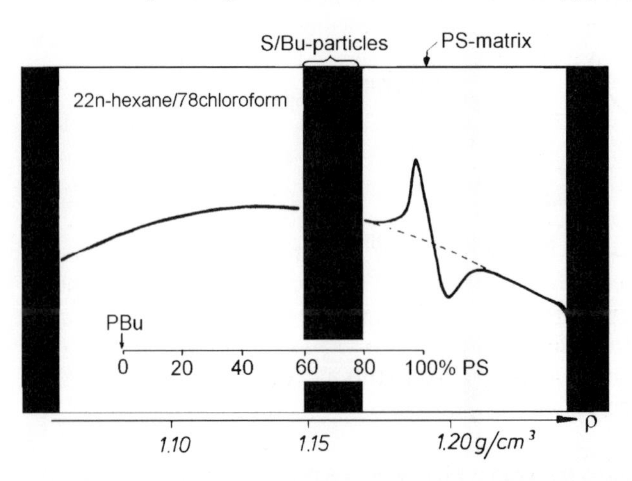

Fig. 6.39. Schlieren photograph of a static 22 n-hexane/78 chloroform AUC density gradient of a high-impact polystyrene sample, showing the very broad turbidity band of the dispersed rubber particles, and the double Schlieren peak of the dissolved polystyrene matrix macromolecules. A S/Bu composition axis is drawn into the photograph

the AUC recognizes *all* particles, and counts millions of them, whereas the EM, at only one magnification, does not detect either the smallest, or the biggest particles in this very broad distribution.

Figure 6.39 shows the static 22 n-hexane/78 chloroform density gradient run of the HIPS sample. Since the exact radial density positions of pure polybutadiene, $\rho = 1.08\,g/cm^3$, and of pure polystyrene, $\rho = 1.19\,g/cm^3$, are well-known in this density gradient, we can draw a S/Bu composition axis into the corresponding Schlieren photograph, as shown in Fig. 6.39 (although the absolute particle density values of PS and PBu, 1.055 and $0.899\,g/cm^3$, are falsified due to preferential solvation of chloroform, this composition axis remains valid). The large density difference between polybutadiene and polystyrene allows us to estimate rather precisely the S/Bu composition of S/Bu particles or macromolecules within this gradient. In this Schlieren photograph, we see on the right-hand side the double Schlieren peak of the (pure) polystyrene matrix macromolecules, whereas on the left-hand side, at the position of pure polybutadiene, we do not see anything, which proves that no free pure polybutadiene macromolecules are present in this sample. Therefore, the whole amount of the starting polybutadiene macromolecules (8 wt%) must have been incorporated completely and covalently into the crosslinked S/Bu rubber particles. In Fig. 6.39, these particles show a broad turbidity band, i.e., a high degree of chemical heterogeneity in the form of a strong, continuous S/Bu composition variation between 60/40 and 80/20 wt%. The average S/Bu composition of the rubber particles is 70/30 wt%.

Now, we come to the decisive and most difficult question (cf. above): which part of the 70 wt% polystyrene is covalently grafted onto the rubber particles, and which part is only included (cf. not able to permeate through the crosslinked polybutadiene membranes surrounding the included polystyrene)? An answer to this question will be given by a sophisticated experiment using EM techniques and AUC density gradients.

Figure 6.40 explains this experiment. The electron micrograph on the left-hand side of this figure, and the density gradient below have already been shown in Figs. 6.36 and 6.39. From the original HIPS material, we extracted the continuous polystyrene matrix completely with metylethyl ketone in a Soxhlett extractor. As a result, we obtained a gel consisting exclusively of rubber particles. The evidence is given by the density gradient of this gel, as can be seen in the middle of Fig. 6.40 (on the right-hand side). The density gradient shows the turbidity band of rubber particles, but no double Schlieren peak of polystyrene macromolecules from the matrix. Above this density gradient, in the upper right-hand corner of Fig. 6.40, again a transmission electron micrograph of a thin slice of this gel, cut with a microtome, is shown. One can see that the rubber particles are fully packed and deformed within this gel. Every complete particle is surrounded by thin black, parallel double lines. These lines represent the two closed polybutadiene membranes surrounding the neighboring rubber particles.

We now cut the original material (shown on the left in Fig. 6.40), as well as the rubber particle gel (shown on the right), into very thin, successive 250-nm slices by means of a microtome. Parallel lines in both micrographs of Fig. 6.40 visualize the thickness of these slices. Both assemblages of successive slices were analyzed again in two density gradients. The result is presented in the lower two Schlieren photographs of Fig. 6.40. All polystyrene macromolecules within these slices, which are covalently grafted onto the crosslinked polybutadiene, cannot diffuse away from the rubber particles, whereas all polystyrene macromolecules, which are only included, will do exactly this. In fact, in the sliced gel density gradient we now see a double Schlieren peak of the formerly included polystyrene macromolecules. We determined its molar mass to be $M = 170\,000$ g/mol. From the area of this double Schlieren peak, and the finding that, due to the loss of the

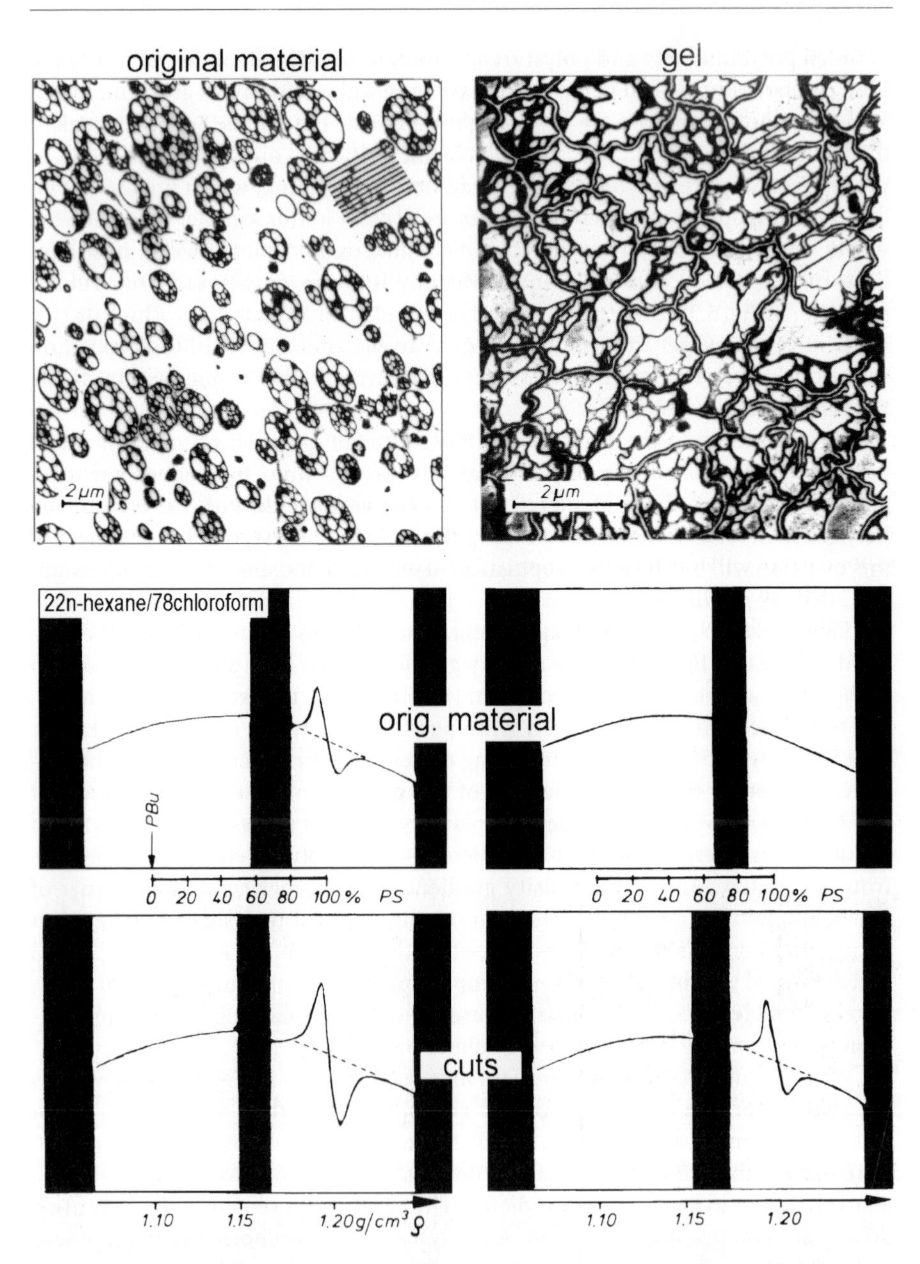

Fig. 6.40. TEM electron micrographs (*above*) and static AUC density gradients (*middle*) of original high-impact polystyrene (*left*), and of a gel of the extracted isolated rubber particles (*right*). Additionally, two density gradients (*below*) of very thin, 250-nm slices (microtome cuts) of both materials are shown

included heavy polystyrene, the radial position of the middle of the gel turbidity band has shifted by only 10 wt% to the left on the composition axis (from 70/30 to 60/40 S/Bu), it follows that the dominating part of the polystyrene within the rubber particles is covalently grafted onto the polybutadiene membranes, inside and outside. This is, beside the broad rubber particle size distribution, another reason for the good high-impact strength of HIPS.

Figure 6.41 is another electron micrograph of the isolated rubber particles, done with a special EM technique, namely, the *scanning* electron microscope (SEM). The preparation of these particles was not done in a Soxhlett extractor, but rather in a preparative ultracentrifuge (PUC). Original HIPS was dissolved in toluene and centrifuged in a PUC beaker, and all fast-sedimenting rubber particles were assembled in a thin layer at the bottom. The supernatant solution of matrix polystyrene was decanted, the beaker filled again with pure toluene, and the rubber particles redispersed. Then, the PUC run was repeated, and the resulting second thin rubber particle layer air-dried. This dried layer was used for the SEM electron micrograph in Fig. 6.41. The electron micrograph shows impressive *isolated, single* rubber particles of a bizarre, moon-like shape, with numerous deep craters or "footballs" turned upside down. Nevertheless, the three-dimensional SEM images show that most of the rubber particles are sphere-like, with SEM diameters of 1000–7000 nm, which is in good agreement with the AUC results.

We consider that this series of experiments with high-impact polystyrene is a good example for a *global analysis.* It demonstrates the strong power and the synergy effects of the combination of different analytical techniques. It also illustrates the outstanding and unique ability of the AUC to analyze systems in which both species are simultaneously present, i.e., macromolecules *and* microparticles.

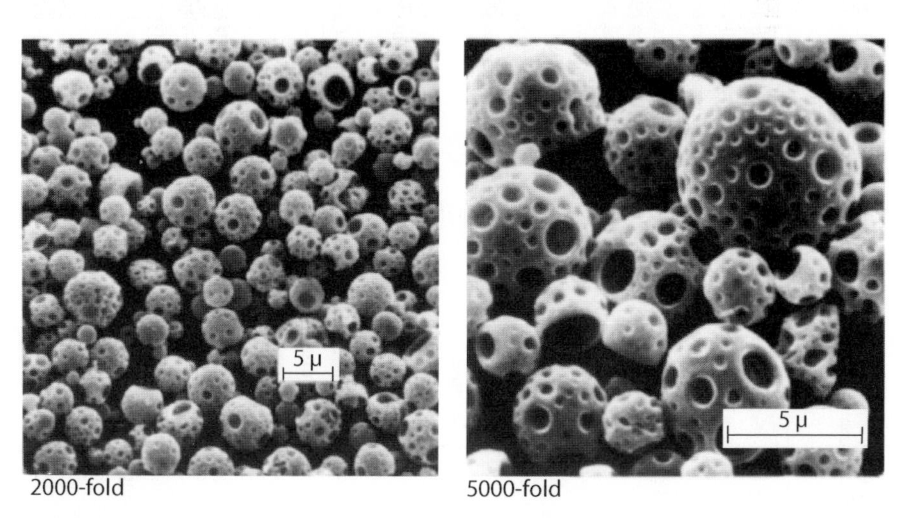

Fig. 6.41. Scanning electron micrograph (SEM) of isolated rubber particle from high-impact polystyrene

Table 6.2. Application of AUC to different nanoparticulate systems (reprinted from [17] with permission)

System	Experiment type/quantity	References
Latex dispersions	Sedimentation velocity/Particle size distribution; Static and dynamic density gradient/particle density	Polystyrene [18], Polystyrene, Polybutylacrylatestyrene, Polybutadiene & Polyethylacrylate [19], Polystyrene [20], Polystyrene, Polybutylacrylate (also styrene & acrylonitrile grafted), Polybutadiene, acrylic homopolymer & copolymer dispersions [1], Polychloropropene [21], Polystyrene, Polystyrene/Polybutadiene, Polystyrene-co-butadiene, Polychloropropene [22], Polystyrene, Polystyrene-co-butadiene, styrene & acrylonitrile grafted polybutadiene [23], Polystyrene [9], Polyurethane [24], Polystyrene [25], Polystyrene, Polybutadiene, Polybutylacrylate-co-butadiene [26]
Inorganic nanoparticles	Sedimentation velocity/Particle size distribution	Au [10], Au [27], ZrO_2 [28], Au_{55}-cluster [29], Pt & ZnO [30], ZnO, Au, CdS, Pt [31], CdS [24], Silica [25], β-FeOOH [26], CdS [32], CdS [33], iron oxide [34], Au, Pd, surfactant encapsulated $(NH_4)[H_3Mo_{57}V_6(NO)(6)O\text{-}183\ H_2O(18)$ [35]
Inorganic complexes	Sedimentation equilibrium/molar mass	Zr-complexes [36], Zr-complexes [37]
Magnetic fluids	Sedimentation velocity	Fe_3O_4 [38], Fe_3O_4 [39]
Microgels	Sedimentation velocity/composition & density	Styrene-Butadiene [40], Acrylic acid [3], Polystyrene-Poly-4-vinylpyridine microgels [41]
Gels	Sedimentation velocity & equilibrium/Swelling degree and pressure, amount of soluble compounds	Gelatin, κ-Carrageenan, Agar, Casein [42], Gelatin, κ-Carrageenan, Agar, Casein [43] and other systems (Reviews)
Emulsions	Sedimentation equilibrium	SDS stabilized n-decane in water [44], Food emulsions [45], Legumin stabilized n-decane in water [46], Legumin stabilized n-decane in water [47]
Supramolecular assemblies and polymers	Sedimentation equilibrium/molar mass, sedimentation velocity/particle size	Co-coordination arrays [48], Co-coordination arrays [49], Fe-coordination polymers [50], Co-coordination arrays [51]

Table 6.2. (continued)

System	Experiment type/quantity	References
Polyelectrolyte complexes	Sedimentation velocity/particle size & composition, Sedimentation equilibrium/Molar mass & interaction & membrane characteristics; synthetic boundary/ membrane formation	Alginate/chitosan [52], Polysyrenesulfonate/Polydiallyldimethylaminchlorid–acrylamide [53]
Micelles	Sedimentation velocity/composition; Dynamic density gradient/composition	Polymampholyte/florinated & hydrogenated dodecanoic acid [54], Lipid and detergent micelles [55], Review, different systems [56], Polystyrene-b-polyisoprene, Polystyrene-b-poly(ethylene-co-propylene) [57], Polylactide-co-polyethyleneglycol [58], Enzymes in reverse micelles [59]
Dendrimers	Sedimentation velocity & Sedimentation equilibrium/Hydrodynamics	Carbohydrate coated polypropyleneimine dendrimers [60], Lactosylated polyamidoamine dendrimers [61]
Hybrid colloids	Sedimentation velocity/composition/particle size	Au/Polystyrenesulfonate microgels, CdS or Au in polystyrene-b-poly-4-vinylpyridin micelles, Pt in Polyethyleneoxide-Polymethacrylic acid [31], Calciumphosphates in alkylated Polyethyleneoxide-b-polymethacrylic acid [62], FeOOH, NiOOH & CoOOH in κ-carrageenan microgels [71], Review of various systems [63], CdS in reverse micelles [64], Pt and Pd in Polystyrene-b-polyethyleneoxide/cetylpyridiniumchloride mixed micelles [65], BSA coated Au,Au, Pt & Pd in Poly-2-vinylpyridine-b-polyethyleneoxide micelles [66], Pt, Pd, Rh & Cu in Polyethyleneoxide-b-polyethyleneimine [67]
Organic colloids	Sedimentation equilibrium; Sedimentation velocity/composition, molar mass	Cu-phtalocyanine [9], κ-Casein particles [68], κ-Casein particles [69]
Nanocapsules	Sedimentation velocity, density gradient/composition	Oil filled Polybutylcyanoacrylate nanocapsules [70]

6.3 Literature Examples: AUC and Nanoparticles

The AUC application examples of the above Sects. 6.1 and 6.2 are all extracted from the authors' works. However, there are many other examples in the literature. In an article recently published by Cölfen [17], such papers were summarized in the form of a table, and grouped according to the nanoparticle systems investigated. This table is reproduced in Table 6.2. This compilation is certainly not exhaustive or representative. It is only meant to guide the reader to the original literature, should solutions for the analysis of a particular nanoparticulate system be sought.

References

1. Mächtle W (1992) In: Harding SE, Rowe AJ, Horton JC (eds) Analytical ultracentrifugation in biochemistry and polymer science. The Royal Society of Chemistry, Cambridge, p. 147
2. Kirsch S, Doerk A, Bartsch E, Sillescu H, Landfester K, Spiess HW, Mächtle W (1999) Macromolecules 32:4508
3. Mächtle W, Ley G, Streib J (1995) Prog Colloid Polym Sci 99:144
4. Mächtle W, Ley G, Rieger J (1995) Colloid Polym Sci 273:708
5. Li J, Caldwell KD, Mächtle W (1990) J Chromatogr 517:361
6. Giddings JC (1966) Sep Sci 1:123
7. Kirkland II, Yau WW (1982) Science 218:121
8. Mächtle W (2000) In: Meyers RA (ed) Encyclopedia of Analytical Chemistry. Wiley, Chichester, p. 5337
9. Mächtle W (1999) Biophys J 76:1080
10. Svedberg T, Rinde H (1924) J Am Chem Soc 46:2677
11. Müller HG, Schmidt A, Kranz D (1991) Prog Colloid Polym Sci 86:70
12. Cölfen H, Völkel A (2003) Eur Biophys J 32:432
13. Schuck P (2002) Size and shape distributions of macromolecules in solution by global analysis of sedimentation and dynamic light scattering. Lecture Advances in Analytical Ultracentrifugation and Hydrodynamics, 8–11 June 2002, Autrans, France
14. Schuck P (2003) Anal Biochem 320:104
15. http://www.analyticalultracentrifugation.com/sedphat.htm
16. Cölfen H, Antonietti M (2000) Adv Polym Sci 150:67
17. Cölfen H (2004) In: Nalwa HS (ed) Encyclopedia of Nanoscience and Nanotechnology, vol I. American Scientific Publishers, Stevenson Ranch, CA, p. 67
18. Cantow HJ (1964) Makromol Chem 70:130
19. Scholtan W, Lange H (1972) Kolloid Z Z Polym 250:782
20. Müller HG (1989) Colloid Polym Sci 267:1113
21. Nichols JB, Kramer EO, Bailey ED (1932) J Phys Chem 36:326
22. Lange H (1995) Part Part Syst Charact 12:148
23. Müller HG, Herrmann F (1995) Prog Colloid Polym Sci 99:114
24. Müller HG (1997) Prog Colloid Polym Sci 107:180
25. Lechner MD, Mächtle W (1999) Prog Colloid Polym Sci 113:37
26. Mächtle W (1984) Makromol Chem 185:1025
27. Rinde H (1928) The Distribution of the Sizes of Particles in Gold Sols. PhD Thesis, University of Uppsala
28. Cölfen H, Schnablegger H, Fischer A, Jentoft FC, Weinberg G, Schlögl R (2002) Langmuir 18:3500
29. Rapoport DH, Vogel W, Cölfen H, Schlögl R (1997) J Phys Chem B 101:4175
30. Cölfen H, Pauck T (1997) Colloid Polym Sci 275:175
31. Cölfen H, Pauck T, Antonietti M (1997) Prog Colloid Polym Sci 107:136
32. Börger L, Cölfen H (1999) Prog Colloid Polym Sci 113:23
33. Börger L, Cölfen H, Antonietti M (2000) Colloids Surf A 163:29
34. Nichols JB (1931) Physics 1:254, and [21]
35. Kurth DG, Lehmann P, Volkmer D, Cölfen H, Koop MJ, Müller A, Du Chesne A (2000) Chem Eur J 6:385

36. Kraus KA, Johnson JS (1953) J Am Chem Soc 75:5769
37. Johnson JS, Kraus KA (1956) J Am Chem Soc 78:3937
38. Seifert A, Buske N, Strenge K (1991) Colloids Surf 57:267
39. Seifert A, Buske N (1993) J Magn Magn Mater 122:115
40. Müller HG, Schmidt A, Kranz D (1991) Prog Colloid Polym Sci 86:70
41. Remsen EE, Thurmond KB, Wooley KL (1999) Macromolecules 32:3685
42. Cölfen H (1999) Biotechnol Genet Eng 16:87
43. Cölfen H (1995) Colloid Polym Sci 273:1101
44. Strenge K, Seifert A (1991) Prog Colloid Polym Sci 86:76
45. Seifert A, Strenge K, Schultz M, Schmandtke H (1991) Nahrung 9:989
46. Seifert A, Schwenke KD (1995) Prog Colloid Polym Sci 99:31
47. Krause JP, Wustneck R, Seifert A, Schwenke KD (1998) Colloids Surf B 1:119
48. Schubert D, van den Broek JA, Sell B, Durchschlag H, Mächtle W, Schubert US, Lehn JM (1997) Prog Colloid Polym Sci 107:166
49. Tziatzios C, Durchschlag H, Sell B, van den Broek JA, Mächtle W, Haase W, Lehn JM, Weidl CH, Eschbaumer C, Schubert D, Schubert US (1999) Prog Colloid Polym Sci 113:114
50. Schütte M, Kurth DG, Linford MR, Cölfen H, Möhwald H (1998) Angew Chem Int Ed Engl 37:2891
51. Schubert D, Tziatzios C, Schuck P, Schubert US (1999) Chem Eur J 5:1377
52. Wandrey C, Bartkowiak A (2001) Colloids Surf A 180:141
53. Karibyants N, Dautzenberg H, Cölfen H (1997) Macromolecules 30:7803
54. Thünemann AF, Sander K, Jaeger W, Dimowa R (2002) Langmuir 18:5099
55. Lustig A, Engel A, Zulauf M (1991) Biochim Biophys Acta 1115:89
56. Roxby RW (1992) In: Harding SE, Rowe AJ, Horton JC (eds) Analytical ultracentrifugation in biochemistry and polymer science. The Royal Society of Chemistry, Cambridge, UK, p. 609
57. Pacovsca M, Prochazka K, Tuzar Z, Munk P (1993) Polymer 34:4585
58. Hagan SA, Coombes AGA, Garnett MC, Dunn SE, Davies MC, Illum L, Davis SS (1996) Langmuir 12:2153
59. Chebotareva NA, Kurganov BI, Burlakova AA (1999) Prog Colloid Polym Sci 113:129
60. Pavlov GM, Korneeva EV, Jumel K, Harding SE, Meijer EW, Peerlings HWI, Stoddart JF, Nepogodiev SA (1999) Carbohydrate Polym 38:195
61. Pavlov GM, Korneeva EV, Roy R, Michailova NA, Ortega PC, Perez MA (1999) Prog Colloid Polym Sci 113:150
62. Antonietti M, Breulmann M, Göltner CG, Cölfen H, Wong KKW, Walsh D, Mann S (1998) Chem Eur J 4:2493
63. Cölfen H (2001) Habilitation Thesis, Potsdam
64. Robinson BH, Towey TF, Zourab S, Visser AJWG, Vanhoek A (1991) Colloids Surf 62:175
65. Bronstein LM, Chernychov DM, Timofeeva GI, Dubrovina LV, Valetsky PM, Obolonkova ES, Khokhlov AR (2000) Langmuir 16:3626
66. Bronstein LM, Sidorov SN, Valetsky PM, Hartmann J, Cölfen H, Antonietti M (1999) Langmuir 15:6256
67. Sidorov SN, Bronstein LM, Valetzky PM, Hartmann J, Cölfen H, Schnablegger H, Antonietti M (1999) J Colloid Interface Sci 212:197
68. Farell HM, Wickham ED, Dower HJ, Piotrowski EG, Hoagland PD, Cooke PH, Groves ML (1999) J Protein Chem 18:637
69. Farell HM, Kumosinski TF, Cooke PH, Hoagland PD, Wickham ED, Unruh JJ, Groves ML (1999) Int Dairy J 9:193
70. Wohlgemuth M, Mächtle W, Mayer C (2000) J Microencaps 17:437
71. Jones F, Cölfen H, Antonietti M (2000) Colloid Polym Sci 278:491

7 Recent Developments and Future Outlook

In 1991, one of the authors of this book wrote a paper about the future requirements for modern analytical ultracentrifuges [1]. The four most important requirements proposed in this paper were:

(i) multi-hole rotors,
(ii) multiple detecting systems,
(iii) automatic online data analysis, and
(iv) a new running technique with a variable rotor speed $\omega(t)$ during a run.

In a recent feature article about analytical ultracentrifugation of nanoparticles, Cölfen [2] wrote "it is amazing that small parts of these requirements were understood in only a few specialized laboratories, although their potential benefit is obvious". Nevertheless, some of these proposed requirements are realized: eight-hole rotors are standard, the Optima XL-A/I now has two simultaneous online detectors (absorption and interference), Laue [3] recently presented a powerful new fluorescence detector inside an XL-A/I, and the authors of this book together with coworkers presented a Schlieren optics detector (needed for density gradients; [4]). Additionally, there has been a dramatic improvement of data analysis due to new, fast and powerful computer programs described by Philo [5], Demeler [6], Schuck [7,8], Stafford [9], Lechner [10], and Behlke [25], which are partly mentioned in the foregoing chapters. To date, however, no 50-hole rotors, nor any triple or quadruple detectors have been developed. Nearly all detectors are still not fast enough, and the quality of their primary measuring data should be higher. Thus, the full power of the new data analysis cannot be used fully today. Furthermore, the benefits of the variable rotor speed $\omega(t)$ technique are used only rarely, although recent efforts are dedicated to using variable rotor speeds on the Optima XL-A/I [26,27].

The authors of this book are optimistic that in the next two decades, the requirements proposed in 1991, but not yet developed, will be realized. Especially, we appeal to the manufacturers of analytical ultracentrifuges to attend to this business. Each laboratory working with GPC, liquid and size exclusion chromatography or field flow fractionation, i.e., each laboratory analyzing polymers, polyelectrolytes, colloids, emulsions or nanoparticles (and there are a lot of them!) is a potential user of AUC technology. It has to be stressed here that a more user-friendly design of the AUC, especially of the measuring cells, is believed to directly increase the number of potential users. It is one major drawback of AUC technology that a relatively high expertise is needed to run an apparatus.

In the following three sections, some details of the requirements remaining to be realized are repeated briefly. Additionally, some new ideas and possible improvements of the AUC technique are proposed. In Sect. 7.1, new instrumentation and detectors will be discussed, whereas Sect. 7.2 will deal with new AUC methods, and Sect. 7.3 with new data analysis.

7.1 New AUC Instrumentation and Detectors

The present standard AUC, the Optima XL-A/I of Beckman-Coulter, is a good basis for further improvements of the AUC technique. We hope Beckman-Coulter, or other manufacturers will carry this out, but it could also be a task for specialized scientific laboratories, because the XL-A/I is a compact modular instrument that has enough space inside the rotor chamber to introduce new detectors and other modifications.

However, it is not easy to make use of this type of modularity, as well as of the space inside and below the rotor chamber, because of the associated safety and warranty loss aspects. Therefore, a wish to be addressed to the manufacturing company arises from the work as developer of new AUC techniques: the modularity should be increased, e.g., by combining the XL-A/I with detectors from other manufacturers, thus turning the XL-A/I more into a development platform, rather than leaving it as a closed system. The advantage for the manufacturing company is obvious: the development work would be distributed onto more shoulders (consider the LINUX example from the world of computer software), and the attractiveness of the AUC technology would be increased and remain at a high level.

We will discuss in this Sect. 7.1 four possible and hypothetical improvements, concerning (i) rotors, (ii) variable speed technique $\omega(t)$, (iii) measuring cells, and (iv) detectors. Although this is yet "science fiction", we hereby intend to catalyze further methodological and hardware development.

Rotors

Four-hole rotors with a maximum speed of 60 000 rpm, and eight-hole rotors with 50 000 rpm are the standard at present. However, in order to measure simultaneously as many samples as possible, 16-, 50-, or 100-hole rotors are desirable, and would substantially increase the AUC efficiency. Naturally, that is not possible for a maximum speed of 60 000 rpm, but this maximum speed is needed only rarely. In most cases, 30 000 or 40 000 rpm are enough. For the industrially important particle size distribution measurement of colloids and polymer dispersions, a maximum rotor speed of 20 000 rpm, or lower is sufficient, and for equilibrium runs to measure molar masses, 10 000 rpm is often enough. A slower rotor speed strongly reduces the cell tightening and leakage problems, as, of course, will do short running times. Figure 7.1 shows how such a 16- (or 50- or 100-) hole rotor could look like (with nine different circularly arranged detectors).

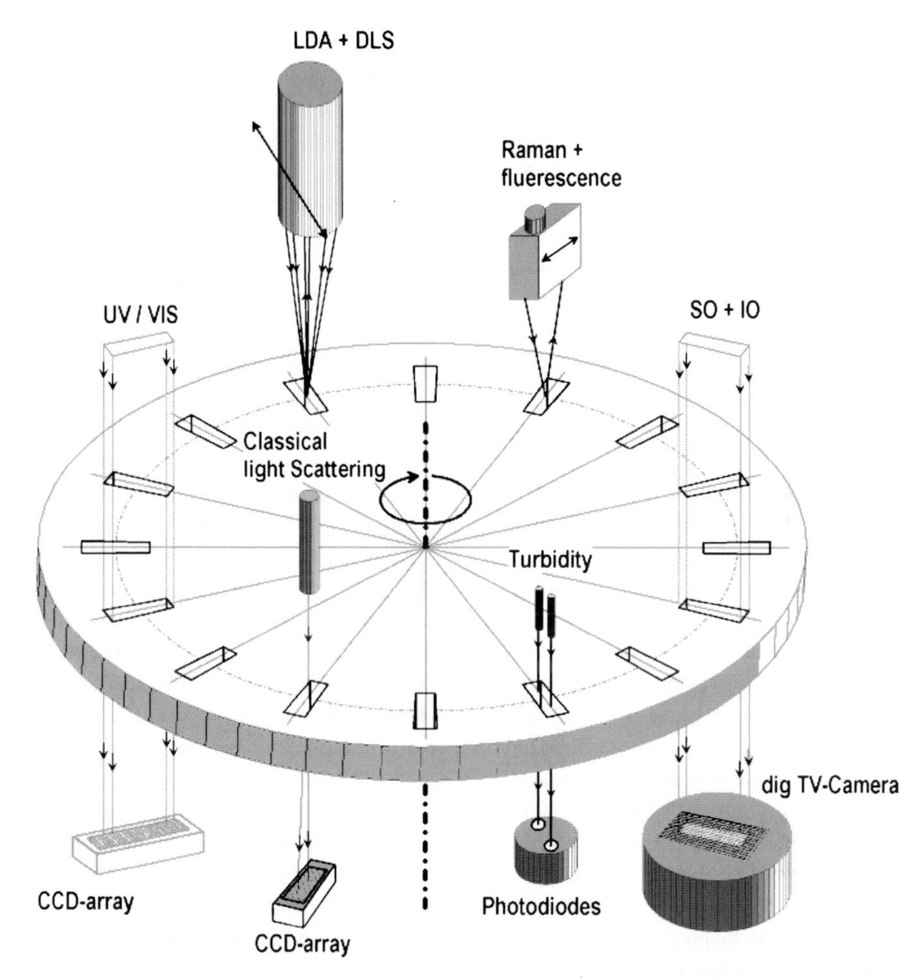

Fig. 7.1. Sketch of an AUC device with a 16-hole rotor and nine circularly arranged detectors for simultaneous multi-detection (Schlieren+interference, turbidity, classical light scattering, UV/Vis, laser Doppler anemometer+DLS, and Raman+fluorescence)

The sector-shaped holes are cut directly into the rotor core. One of these holes can serve as a reference cell (= counterbalance), which (i) bears defined radius reference marks for the radial calibration, and (ii) serves also as cell number 0 in the multiplexer mode. Rectangular "holes" with sector-shaped inserts of metal or Epon are hypothetical. Two circular glass plates (not shown), with a hole in the center for the rotor axle, have to be clamped onto the upper and the lower side of this rotor to tighten the "cells" and to serve as cell windows. Such a device could be part of an efficient particle sizer. For some detection techniques (fluorescence, Raman, and laser Doppler anemometry), where it is possible to work with back-scattered measuring light, one needs only one circular glass plate, and the sector-shaped holes are cut only into the upper side of the rotor core.

Variable Rotor Speed (or Sweeping) Technique $\omega(t)$

Nearly all AUC runs are done with a constant rotor speed ω during the run. The reason is that the standard XL-A/I instrument requires this constant rotor speed. However, the XL-A/I vacuum-included, electrically controllable induction motor allows also a rotor speed *variation* during a run, for example, an exponentially increasing one up to the chosen maximum speed. Still, to do this one needs a special electronic chip. It is then possible to choose different $\omega(t)$ speed profiles via an additional computer. Such an exponential speed profile is presented in Sect. 3.5.1 and Figs. 3.17 and 3.19a for particle size distribution measurements. All AUC runs in which we find the term $\omega^2 t$ in the evaluation equations (these are all kinds of sedimentation velocity runs, see (1.5), (1.9), (3.2), (3.10), and (3.24)) are practicable with a variable rotor speed, rather than a constant one. In this case, however, we have to replace $\omega^2 t$ in these equations by the running time integral $\int \omega^2 \, \mathrm{d}t$. This running time integral has to be measured continuously and very precisely during the run. Usually, this is done with the same computer, i.e., the one that controls the rotor drive and the speed profile $\omega(t)$.

The variable rotor speed technique $\omega(t)$ offers some interesting advantages. Often, it shortens the total running time of a run. Some measurements are possible only with this technique, such as the measurement of extremely broad particle size distributions, and the H_2O/D_2O analysis (see Sect. 3.5.3). Already in 1984, we demonstrated this [11], and proved experimentally that the two different running techniques, ω_{constant} and ω_{variable}, are completely equivalent (see also [12]). Since that time, we perform all particle size distribution measurements with the same speed profile: $\omega(t)$ is always increased exponentially from 0 to a maximum speed of 40 000 rpm within 1 h. Thus, we never have to ask, before a run, what the most suitable constant rotor speed is (difficult to answer for unknown samples!). Rather, in every run, we measure the correct particle size distribution with this exponential or sweeping technique $\omega(t)$, for very small particles of only 30 nm, as well as for very large particles of 3000 nm, showing a sedimentation velocity that is 10 000 times higher. Simultaneous measurement of seven different unknown samples, such as in Fig. 3.17 (or of 15 samples, as in Fig. 7.1), is possible only with the variable rotor speed technique. A sweeping technique $\omega(t)$ is also used in sedimentation field flow fractionation (SFFF) in order to measure very broad particle size distributions. However, in this case, one starts with a high rotor speed, and decrease it exponentially to zero (see Fig. 6.29 and Sect. 6.2.1).

A method suitable for broadly distributed particles should also be applicable to broadly distributed, dissolved *macromolecules*. However, to date the variable rotor speed technique has not been put into practice for measurements of molar mass distributions. Thus, we propose to do this in the future. If this can be realized, it could result in a renaissance of molar mass distribution measurements via AUC, and become a hard competitor to size exclusion chromatography (SEC), the leading method at present in this field. At the beginning of such an "exponential" sedimentation velocity run for the measurement of very broad molar mass distributions, i.e., at low rotor speeds, we "see" and detect only the fast, large macromolecules.

Later, at high rotor speeds (when the large molecules have sedimented completely to the cell bottom), we detect only the slower, smaller macromolecules. During the run, we will never see all macromolecules simultaneously. However, if we record during the run, at many different experimental times $t_1, t_2, t_3, ...$, many momentary partial (!) radial concentration distributions, $c(r, t_1), c(r, t_2), c(r, t_3), ...$, it should be possible to subsequently transform and combine these, by means of a powerful data analysis, into a "master curve" representing the time-independent sedimentation coefficient distribution $g(s) = \mathrm{d}G(s)/\mathrm{d}s$ of the total sample, and resulting in the desired molar mass distribution (see Sect. 3.4.2). As mentioned in the introduction, there are recent, first attempts to use the sweeping technique $\omega(t)$ for *MMD* measurements [27].

Especially in combination with synthetic boundary techniques (see Sect. 3.6) and very fast measuring data detection, we see interesting new analytical possibilities by using the proposed variable rotor speed technique $\omega(t)$. One field is the (fast) measurement of average molar masses M and molar mass distributions (*MMD*) via sedimentation velocity runs, using Svedberg's equation (1.8), and scaling laws such as (3.11) (see Sect. 3.4). The idea behind this is that at the beginning of such an "exponential" synthetic boundary run, at low rotor speeds and shortly after the superimposition of the solvent on top of the solution, the spreading of the Gaussian radial concentration distribution around the superimposition radius inside the measuring cell, $c(r, t)$, is controlled mainly by diffusion spreading, and thus the average diffusion coefficient D can be measured. Perhaps even the complete diffusion coefficient distribution $g(D) = \mathrm{d}G(D)/\mathrm{d}D$ could be reasonably well approximated, too. In the course of the experiment, at high rotor speeds, the spreading and especially the migration of the concentration boundary is controlled mainly by sedimentation, and thus the average sedimentation coefficient s as well as the complete sedimentation coefficient distribution $g(s) = \mathrm{d}G(s)/\mathrm{d}s$ can also be measured. Determining $g(s)$ is certainly much easier than determining $g(D)$. Nevertheless, the combination of variable rotor speed technique, synthetic boundary runs, and fast online data analysis (including fast detectors) has the potential to solve the old problem of separation between diffusion and sedimentation spreading within a sedimenting concentration boundary of broadly distributed samples, implemented into the two terms of Lamm's differential equation ((1.10); see also Sect. 3.3.6).

Measuring Cells

All cells, and the variety of different 3-, 12-, and 30-mm centerpieces of the older Model E centrifuges can be used in the newly designed rotors of the XL-A/I, with one exception – the 30-mm cells/centerpieces, which cannot be used due to the slightly reduced height of the new rotors. For the 12-mm standard cell available at present, this requires a 30/12 = 2.5-fold higher concentration in order to obtain a detector signal comparable to the one obtained with a 30-mm cell. Since it is possible to shorten the 30-mm cells and centerpieces to 25 mm, the required concentration increase is very low. AUC manufacturers should offer such 25-mm cells for runs with very low concentrations, where one is near the ideal case $c \rightarrow 0$.

Rather than increasing the number of rotor holes, as shown in Fig. 7.1, in order to measure as many samples as possible simultaneously, it is a simpler alternative to increase the number of holes, or the number of sector-shaped compartments in the *centerpieces*. Already at present, it is possible to measure simultaneously the particle size distributions of 16 different samples with the existing fast turbidity detectors, if one uses an eight-hole rotor and standard 12-mm, double 2° sector centerpieces. It is conceivable to cut four parallel (long), 1° sector-shaped compartments into such a centerpiece, or four or six shorter 2° sectors (then in a parallel and radial arrangement). In this case, a reduction of the maximum rotor speed, and more sensitive and faster detectors are required.

We expect a great future potential and new analytical possibilities for newly designed synthetic boundary cells, or centerpieces, in the field of fast dynamic density gradients (see Sect. 4.3). The same applies to the study of surface reactions inside an AUC cell, e.g., the formation of polyelectrolyte complexes and membranes, or studies of the first steps in crystal growth and metal cluster formation (see also [2]). For this purpose, we need synthetic boundary centerpieces (see Sect. 2.3) with valves or capillaries, where the momentum of superimposition is better defined and controllable. This momentum is mostly correlated with a special rotor speed. In the case of the capillary-type cell, it depends additionally on the capillary diameter, and the surface tension between the solvent, cell window and centerpiece material, and in the case of the valve-type cell, on the elasticity of a rubber or mechanical spring. Often, this momentum is arbitrary and changes with time. In particular, we consider that the valve-type cell offers the possibility of better-defined momentums by construction of more sophisticated springs/valves with well-defined and controllable superimposition rotor speeds, for example, 5000, 10 000 or 20 000 rpm. It would be desirable to have a two-step synthetic boundary cell with two storage bins and two valves, which open at different rotor speeds, for example, at 5000 and 30 000 rpm. This cell could be used for studies of chemical two-step surface reactions inside the AUC cell. An urgent appeal is made to the AUC manufacturers to develop and to supply such synthetic boundary centerpieces!

Detectors

The lack of appropriate detectors is surely the most important disadvantage of present analytical ultracentrifugation. The existing detecting systems are not very precise, not fast enough, and the data acquisition often happens to be not really "online". Only a precise and fast online data acquisition with large datasets would allow us the entire use of the already existing, powerful data analysis software. Thus, detector improvements, and the development of new detectors are the most important tasks for the future of AUC.

At present, beside the turbidity detector, the Optima XL-A/I interference optics detector is the best one, because it is fast, online, and does not need a (slow!) mechanical radial scanning device. On the other hand, it lacks in data precision, because of sapphire window distortions. Perhaps this can be eliminated by software improvement and additional blind measurements.

The present XL-A/I UV/Vis detector, which is very important for chemical differentiation, lacks also precision and, above all, its mechanical radial scanning device is too slow. This important online detector cannot be used for fast sedimentation velocity runs. In particular, it is not suited for synthetic boundary runs and sedimentation runs with eight-cell rotors. A German group led by H. Cölfen at the Max-Planck Institute of Colloids and Interfaces in Potsdam (Germany), and a group of specialists working at BASF (represented by the authors) are currently trying to replace the slow scanning device by a fast diode array.

Unfortunately, no fast commercial online Schlieren optics detector for the XL-A/I is available on the market at present, only a homemade detector [4]. This versatile Schlieren optics detector is required for the chemical analysis in AUC density gradients, with their steep radial refractive index gradients inside the measuring cells (partly compensated by wedge windows; see Sect. 4.2.3).

The improvement of the existing detectors is possible by (i) faster and more extensive data acquisition, (ii) replacement of slow mechanical scanning devices by faster ones, or better, by linear diode arrays, (iii) increasing of data precision by more sensitive and higher-resolution photomultipliers, avalanche photodiodes, and higher-resolution CCD cameras, and (iv) reduction of data noise by additional automatic blind measurements and mathematical subtraction.

Beside the existing detectors, there is a demand for new detecting systems, particularly for more real multi-detecting facilities suitable for the analysis of complex samples. The standard online multi-detector should be a combination of at least four detectors: interference, Schlieren, UV/Vis, and turbidity. It is desirable to add even more detectors, as proposed in Fig. 7.1, such as fluorescence, infrared, Raman, laser Doppler anemometry (LDA), static, and dynamic light scattering detectors. Some of these proposed new detectors could be realized in the near future; for some others, we will have to wait until faster and more sensitive diodes, more powerful light sources, faster flashlights, and a better optical glass fiber technique have been developed. A prototype of the fluorescence detector exists already [3]. An AUC-Raman detector (see Fig. 7.2), having a similar setup and also using confocal optical technique and back-scattering of the measuring light, was proposed by Schrof [13], but it is not yet realized.

Details of this micro-focus Raman detector are described in [13]. Similarly to the UV/Vis, and especially the fluorescence detector, a Raman detector would be a specific detector, able to "see" and to identify special Raman-active components in a complex sample mixture. This facilitates chemical analyses. Because a Raman and a fluorescence detector have a similar optical arrangement, it will perhaps be possible to combine these into one optical and mechanical unit, as shown in Fig. 7.1. Also the very similar interference and Schlieren optics detectors are united into one optical path in Fig. 7.1.

A completely new detector would be a micro-focus LDA detector (= laser Doppler anemometer), mounted on a radial scanning device and working with back-scattered measuring light [14]. The advantage of such an LDA detector would be the possibility to measure radial local sedimentation coefficients online within

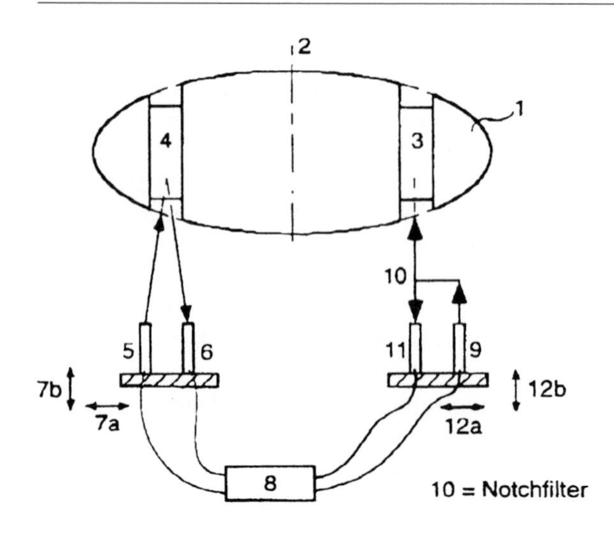

Fig. 7.2. Sketch of a Raman detector inside an analytical ultracentrifuge (1 rotor, 2 rotor axle, 3 + 4 measuring cells, 5 + 9 laser light sources, 6 + 11 detection systems, 7 + 12 radial and axial scanning devices, 8 evaluation unit, 10 notch filter) (reprinted from [13] with permission)

an AUC cell, and thus particle sizes, without any knowledge of the radius positions of the meniscus and bottom. Further advantages of the LDA are that (i) it is a single-particle detector requiring only low concentrations near the ideal case $c \to 0$, and (ii) it offers the possibility to analyze the scattered light in a dynamic way (DSL), thereby allowing us to measure D. The simultaneous measurement of D and s on single particles could yield M via the Svedberg equation (1.8).

7.2 New AUC Methods

A variety of existing AUC methods has been described in the foregoing Chaps. 1–6 of this book: sedimentation runs, equilibrium runs, M^* runs, synthetic boundary runs, static and dynamic density gradient runs, Archibald runs, meniscus depletion runs, and others. Nevertheless, the improvement of the existing, and the development of completely new methods are desirable and possible. Thinking about new methods often creates new instrumentation and new data analysis – and vice versa. In the following, we will discuss some new ideas and developments: (i) radial pH gradients within an AUC cell, (ii) a preparative dynamic density gradient method, and (iii) a few other new AUC methods.

Radial pH Gradient Within an AUC Cell

If a binary mixture, e.g., a light and a heavy component, a light and a heavy solvent, or a light solvent and a heavy solute, is centrifuged up to equilibrium, a radial density gradient (see Chap. 4), and simultaneously a radial concentration gradient (see Chap. 5) will be established within the AUC cell. Additionally, other radial gradients are established simultaneously, such as pH gradients or solubility gradients, depending on the nature of the two components. This was the idea of Lucas and Cölfen [15], and of one of the authors. Other radial gradients should be possible if physicochemical parameters other than pH or solubility are applied.

Introducing a small amount of a third component into such a two-component gradient mixture offers new analytical possibilities for this third component, for example, the preparative isolation of critical crystal nuclei in solutions.

To create a radial pH gradient means to create a radial H^+ ion gradient inside the AUC cell. This is not trivial, because these H^+ ions are the most rapidly diffusing species existing. Drastic conditions have to be applied in order to induce a radial accumulation of H^+ ions. The authors of [15] found two ways to do this. The first option was to add a substance of very high density, I^- ions (in the form of iodine acid, H^+I^-) to water. The second way was to dissolve heavy (poly)anions of high molar mass, such as poly(sodium 4-styrene sulfonate), NaPSS (see also Sects. 3.3.5 and 3.4.1), or poly(acrylic acid) in water, so that H^+ ions would accumulate by electrostatic interaction toward the bottom of the ultracentrifuge cell. Indeed, when they used a maximum rotor speed of 60 000 rpm and waited 2–3 days, this led to the formation of a radial pH gradient, as shown in Fig. 7.3 (left-hand side) for the case of NaPSS/water, visualized by the indicator Bromocresol purple, which changes its color from purple (at the top) to yellow (at the bottom) between pH 6.8 and 5.2. The same color change can be observed for HI/water (see Fig. 7.3, right-hand side), where the colored iodine (at the bottom) is indicative for the position of I^-, and thus for the accumulation of H^+. The pH varies between 1.0 and 0.6 in this case.

In the experiment shown in Fig. 7.3, the authors [15] used, only for demonstration purpose, a preparative ultracentrifuge. Furthermore, they performed this experiment also in an analytical ultracentrifuge using the same pH gradient (and others, too), and used the simultaneous detector (interference and UV/Vis) of the XL-A/I to measure the radial concentration of the different components and (indirectly) the radial pH gradient. Figure 7.4 shows such an example. The heavy-density gradient component, NaPSS in this case, is measured via the shift of the interference fringes, and the third component to be analyzed, $BaCrO_4$, is measured via its UV absorption at 375 nm.

Fig. 7.3. Demonstration of PUC pH gradient formation by bromocresol purple as pH indicator, (*left*) in a water/NaPSS (140 000 g/mol) gradient, centrifuged for 2 days at 60 000 rpm at 25 °C, and (*right*) in a water/HI gradient centrifuged at 60 000 rpm and 25 °C for 3 days (reprinted from [15] with permission)

The dissolution of small inorganic $BaCrO_4$ crystals in this pH gradient was assessed as follows (this pH gradient is simultaneously a solubility gradient). $BaCrO_4$ ($K_p = 1.17 \times 10^{-10}$ at 25°C) is reversibly transformed into the more

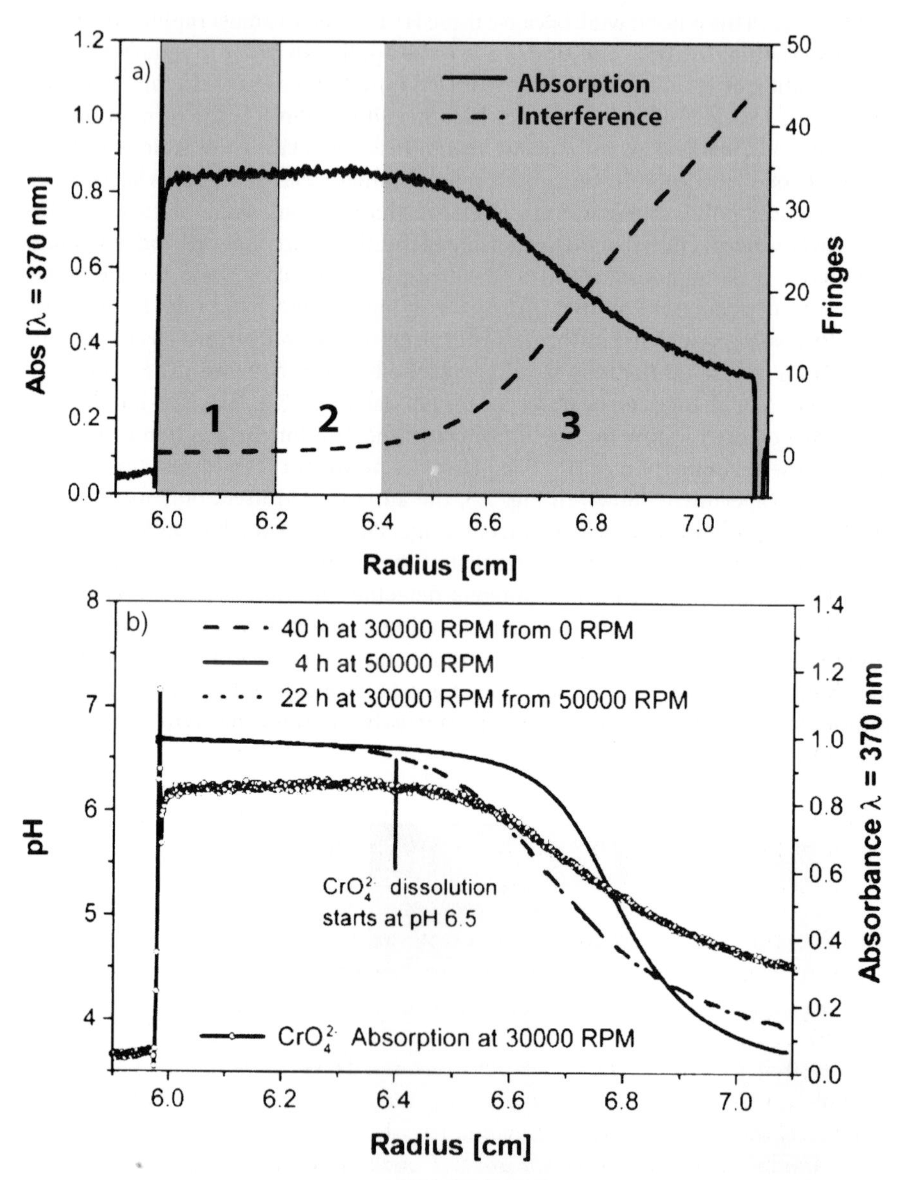

Fig. 7.4. **a** Different regions in an AUC pH gradient as monitored by UV absorption at 370 nm to follow the $BaCrO_4$ concentration, and by interference optics to determine the NaPSS (140 000 g/mol) concentration. **b** Local radial pH within the AUC cell, and the response from the $BaCrO_4$ monitored by UV absorption at 370 nm (reprinted from [15] with permission)

soluble $BaCr_2O_7$ upon a decrease in pH, according to (7.1):

$$2CrO_4^{2-} + 2H_3O^+ \leftrightarrows Cr_2O_7^{2-} + 3H_2O \tag{7.1}$$

This transformation is also discernible as a color change from yellow to orange, and can thus be followed via the XL-A/I UV/Vis detector (see Fig. 7.4). For details of this interesting AUC experiment in a radial pH/solubility gradient, the reader is referred to [15]. Indeed, it was possible to monitor the transition of $BaCrO_4$ to the more soluble $BaCr_2O_7$ with decreasing pH. This demonstrates the possibility to perform chemical reactions within a pH gradient, which was also shown to be in thermodynamic equilibrium by the path independence proof (Fig. 7.4).

Lucas, Cölfen and one of the authors [15] studied another radial solubility gradient, which was not connected with a pH gradient: a 30:70 vol% mixture of tetrahydrofuran (THF) and water, which are fully miscible. They analyzed the resulting radial gradient (60 000 rpm, 14 h) via the XL-A/I UV/Vis detector at 275 nm, and found that the THF:water relation was 36:64 vol% at the meniscus, and 23:77 vol% at the bottom. The idea behind this approach was to introduce small sedimenting inorganic particle crystals into such solubility gradients, which are insoluble/soluble (e.g., copper sulfate pentahydrate) or slightly/readily soluble in the two different gradient solvents. These (heavy) crystals should sediment within this radial gradient with increasing solubility up to that radial position where they would be dissolved completely into small ions, which are no longer able to sediment. This radial point is their dissolution point. Shortly before their complete dissolution, critical crystal nuclei should exist, which would perhaps be isolable on a preparative scale.

Absorption scans monitoring the copper sulfate pentahydrate showed [15] that, in fact, the concentration increased toward the bottom of the cell, where the water fraction was higher. To date, however, no evidence of critical crystal nuclei of copper sulfate has been observed in the AUC. Their formation at a critical THF fraction would have become apparent in a discontinuous slope of the concentration gradient. Perhaps the radial THF/water solubility gradient was not steep enough in this experiment.

The Preparative Dynamic Density Gradient Method (PDDG)

In the foregoing pH gradient section, we demonstrated (compare Figs. 7.3 and 7.4) that an experiment done in an analytical ultracentrifuge (using only 0.5 ml solution) can be repeated in a preparative ultracentrifuge (using about 50-ml beakers/tubes). This is often done for the isolation/preparation of a component in a mixture. In the following, we present a recent example, namely, the transformation of the analytical (fast!) dynamic H_2O/D_2O density gradient (described in Sect. 4.3) into a preparative one, realized by one of the authors and Lechner [16]. This new type of dynamic density gradient has been named "preparative dynamic density gradient method", abbreviated PDDG.

Rather than synthetic boundary cells in an AUC, standard preparative centrifugation tubes of 38 ml in a PUC are used, as shown in Fig. 7.5. Unfortunately,

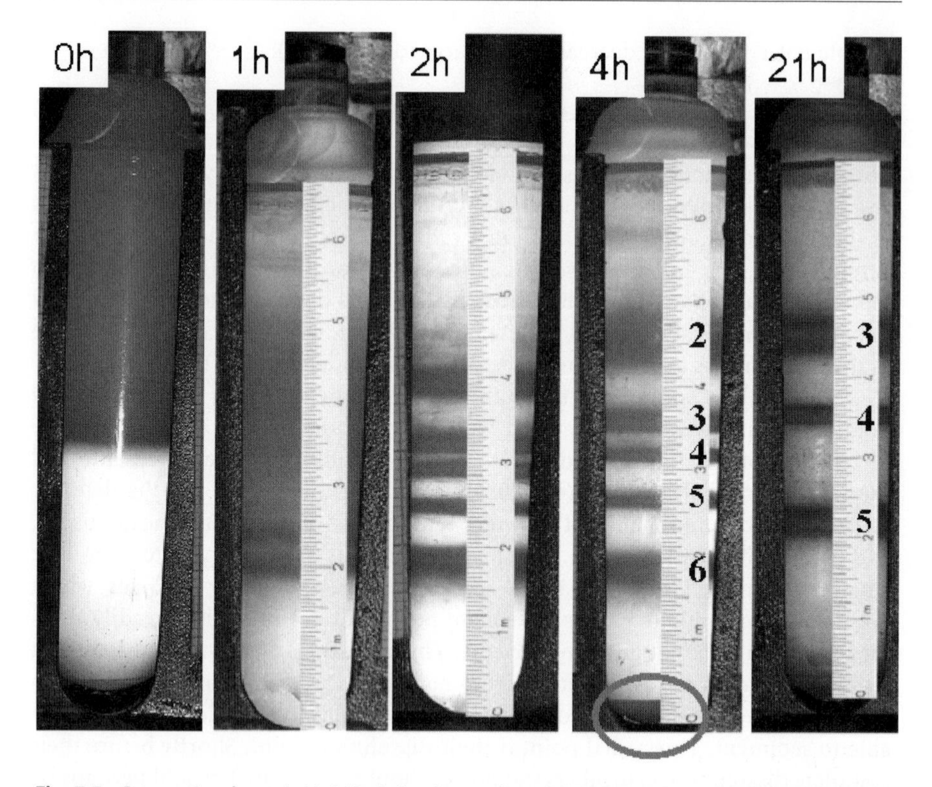

Fig. 7.5. Preparative dynamic H_2O/D_2O density gradient (PDDG) experiment at different times (at 90 000 *g*). An 11-component latex mixture with 11 different particle densities was analyzed. As indicated, five of these components are separated after 2 h in this gradient. The others are on the meniscus, or on the bottom of the tube (reprinted from [16] with permission)

these tubes have no gravity valve, and no separate storage bin, unlike the synthetic boundary cells. Thus, the superimposition of light H_2O ($\rho = 0.997\,\mathrm{g/cm^3}$) onto heavy D_2O ($1.105\,\mathrm{g/cm^3}$) can not be done during the PUC run. Therefore, this has to be carried out before the run, in the following manner.

For a start, the (upright-standing) tube is half filled with water, which contains the sample to be analyzed or fractionated. Then, the pure D_2O is layered smoothly *under* the H_2O by means of a syringe connected with a thin hose. The free end of the hose is introduced through the H_2O down to the bottom of the tube, and the D_2O is pressed in softly until the tube is filled completely. Subsequently, the tube is introduced into the swing-out rotor, and the centrifugation is started. Now, as in the analytical dynamic H_2O/D_2O density gradient (see Sect. 4.3), H_2O and D_2O molecules undergo an inter-diffusion process that eventually leads to a homogeneous H_2O/D_2O mixture (after a long time of some days). While this equilibrium is being reached, the H_2O/D_2O concentration varies over the cell radius, resulting in a radial density gradient varying from $0.997\,\mathrm{g/cm^3}$ near the meniscus to $1.105\,\mathrm{g/cm^3}$ near the bottom. This density gradient changes with time – at the beginning it is very steep, and later it becomes increasingly flat. A theory based on both Fick's laws was developed by Börger et al. [16] to calculate this radial density gradient in a PDDG tube, and its time dependence, i.e., $\rho(r, t)$.

During the experiment, fractions of the (fast) sample, dispersed in the H_2O section of the tube, will sediment to that radius position in the tube where particle density and gradient density are matching, as can be seen in Fig. 7.5. The 11-component latex mixture (see Table 4.1) with 11 different particle densities is used as a sample in the same way as in the analytical H_2O/D_2O density gradient of Fig. 4.12. The result is, of course, the same. We see five of these components gathered in small turbidity bands inside this tube, component numbers 2, 3, 4, 5, and 6, because they match the PDDG density range of 0.997 – 1.105 g/cm^3. The lighter component 1 has floated to the meniscus, whereas the heavier components 7, 8, 9, 10, and 11 have sedimented completely to the bottom of the tube. However, there is a very important difference between the preparative and the analytical dynamic density gradient (nomen est omen!). In contrast to the analytical gradient, the PDDG delivers *preparative* sample fractions, yielding enough substance to be used for further analytical purposes. For the isolation of these fractions, the ultracentrifuge is stopped, and (i) the preparative tube is transferred into a fractionation device where it is cut into slices containing the different fractions, or (ii) the content of the tube is pumped through a fraction sampler. Figure 7.5 demonstrates that the PDDG is a new, fast preparative fractionation method. In most cases, the fractionation is completed within 1 or 2 h. Similarly to the analytical dynamic (H_2O/D_2O) density gradient, also the preparative dynamic density gradient is restricted to fast-moving particles with diameters $d_p > 20$ nm.

Some Other New AUC Methods

A drawback of static density gradients (see Chap. 4) are the long measuring times of 1 – 5 days necessary to reach equilibrium. However, *during* this process, every static density gradient is in principle also a dynamic one, $\rho(r, t)$, i.e., it changes with time: the change is fast at the beginning of the experiment, and becomes progressively slower at the end. Basically, there are two possibilities to use these faster "pseudo-static" density gradients for analytical purposes. The first one is to calculate $\rho(r, t)$ theoretically, as described above for the PDDG, but this can only be done when a new theory describing the attainment of static density gradient equilibrium will have been developed. The second option is to measure $\rho(r, t)$ as a function of time: (i) by recording continuously the concentration $c(r, t)$ of one of the two main density gradient components, which has already been realized by using metrizamide (or Nycodenz) in static H_2O/metrizamide density gradients (see [17]), or (ii) by adding sets of calibrated density markers to the density gradient (as for the 11-component latex particle mixture mentioned above), also recently realized in [18] (also see Sect. 4.2.1).

As mentioned in Sect. 7.1 on the topic of "measuring cells", new synthetic boundary cells offer the possibility of new analytical methods in the fields of studying surface reactions inside the AUC cell, e.g., the formation of polyelectrolyte complexes and thin membranes, or studies of the first steps in crystal growth and metal cluster formation (see also [2] and [17] in Chap. 6).

Above, we discussed different radial gradients within an AUC cell, in particular density, concentration, pH, and solubility gradients. However, we did not discuss the radial pressure gradient. Such a radial pressure gradient (see Sect. 3.3.3) is established within every AUC cell in every kind of AUC run, also in a pure solvent. It is independent of time, and the function $p(r)$ can be calculated if rotor speed, filling height, solvent density, and solvent compressibility coefficient κ are known. At maximum rotor speed (60 000 rpm), the radial pressure gradient varies from 1 bar at the meniscus to 250 bar at the bottom. It is conceivable to use this pressure gradient for some analytical purposes. For example, one could measure compressibility coefficients κ of solvents and solutions, or of single macroscopic particles introduced into such pressure gradients. It should also be possible to study the compression of gels as a function of pressure (see [29] and [30]).

In 95% of all AUC experiments performed worldwide, the sedimentation velocity method or the sedimentation equilibrium method are used, where fractionation is achieved according to size/molar mass. Only in 5% of the cases is the density gradient method used, mainly because of its high potential to fractionate according to particle density – in other words, according to chemical composition. Even today, the density gradient method is a nearly unknown continent in the AUC world. We recommend that the scientific community should explore it more intensively in the near future.

The AUC particle sizing method, using a turbidity detector (see Sect. 3.5.1), is the most powerful and most perfect AUC method at present, especially in combination with the H_2O/D_2O analysis method (see Sect. 3.5.3). Nevertheless, it can still be improved. One improvement was proposed by us already [12], namely, the introduction of a second, parallel measuring laser beam into the *PSD* device (see Fig. 3.13). We also presented [12] initial successful measurements with a device having the first laser beam in the first third, and the second laser beam in the last third of the cell sector length ($r_b - r_m$). However, the full potential of this new two-beam *PSD* method (or three, or even more beams!) has not been explored until now. We are confident that this modification will lead to improvements in *PSD* data quality and, in particular, allow us to carry out more precise H_2O/D_2O analyses. If the precision of measurement is high enough, then it will be possible to obtain the complete particle *density* distribution (in addition to the *PSD*), i.e., the chemical heterogeneity of an unknown latex sample. A higher precision could also lead to a higher density range of the H_2O/D_2O analysis, perhaps up to about 0.5 – 20 g/cm^3, which is substantially wider than the range accessible by static density gradients (0.85 – 1.9 g/cm^3).

7.3 New AUC Data Analysis

High-precision multi-detection and powerful, fast online data analysis are the keys for a more frequent use of the AUC in the future, in particular for the investigation of complex inhomogeneous mixtures showing distributions in size, molar mass, density/chemical composition, electrical charge, shape, degree of crosslinking, etc. The aim is a global AUC analysis (perhaps in combination with other methods), and the simultaneous determination of all these different distributions. There is interdependence between multi-detection and data analysis, but at present data analysis is more developed as multi-detection. Nevertheless, also the data analysis should be improved. In the following, we will discuss three ways to improve the existing, and to develop new data analyses: (i) online controlling of the AUC instrument, (ii) improvement of measuring data quality via data analysis, and (iii) improvement of data evaluation of data analysis.

Online Controlling of the AUC Instrument

The registration of measuring data during AUC runs is not always really online (as in the case of turbidity, interference, and UV/Vis detectors). For example, data registration by Schlieren optics (mostly used for density gradients and some fast synthetic boundary runs) is still done via photographs on photo-plates, or by means of digital CCD cameras. When the AUC run is finished, these "photos" are evaluated offline. At present, no sophisticated PC programs are available to do this offline photograph evaluation fully automatically. Usually, manual help with a cursor on a PC TV screen is required, for example, to define r_m and r_b, or to follow the Schlieren curve. So far, there exists only one attempt in the literature [19] to replace this manual help by an automatic image-processing program. Thus, for the future of Schlieren optics, it is desirable to combine CCD photograph registration and image processing to obtain a real online data registration during the AUC run. Furthermore, it is desirable to have this automatic detection (and also controlling!) of r_m and r_b within the existing XL-A/I online data registration programs of the interference and the UV/Vis detector.

Internally, the Optima XL-A/I is equipped with a computer that controls the machine and all relevant parameters, i.e., rotor speed, temperature, flash light, laser diodes, detectors, repetition rates of the scanning devices, etc. For safety reasons, this internal computer is not accessible for the operator of the XL-A/I, or only in a very restricted manner. However, it would be advantageous to have an additional external computer for the control of the XL-A/I by the operator in special cases. One option was already mentioned above, namely, selecting and controlling different rotor speed profiles $\omega(t)$. Another idea is to check the meniscus position r_m continuously via continuous radial scans, and to detect in this way, for instance, any slow disturbing leakage of the cell. Continuous scans (together with continuous data analysis) can also be used to check reaching the equilibrium state in equilibrium and density gradient runs, and to stop the run automatically when there is no difference anymore (within the errors of measurement) between the last two successive scans, or when a 99% equilibrium is reached. In this way, it should also be possible to check the solvent plateau and the solution plateau continuously during sedimentation velocity runs, in order to make sure that the law of conservation of mass and the radial dilution rule (3.5) are fulfilled at all times. We hope somebody will develop such intelligent data analysis programs in the near future to improve the control of the machine itself, and of the whole AUC run.

Improvement of Measuring Data Quality via Data Analysis

The best way to obtain high-quality measurement data is to use very sensitive detectors, with a signal-to-noise ratio as high as possible. However, physics determines the limits here. Also the data of the best detector can be improved by data analysis. A striking example is the time derivative method to evaluate sedimentation velocity runs (see Sect. 3.3.6 and [9]), a data analysis in which a significant improvement of the signal-to-noise ratio of the experimental data is reached by subtracting successive scans from each other, resulting in the elimination of systematic errors in the optical patterns, and a decrease of random noise.

Imperfections and distortions of sapphire windows are a major problem for interference optics detection. They create time-independent distortion of the detector signal. Often these distortions increase with rotor speed. We propose to develop a data analysis that eliminates these distortions in the following manner. One uses always the same cells with the same windows, placed always in the same position within the cells, measuring these cells, filled only with solvent, at the usual standard rotor speeds and storing the resulting "blind" signal data in the evaluation computer. If real samples are measured, these stored blind signals are subtracted from the sample signal. Another possible way to eliminate window distortions is to record at the beginning of every run, immediately after the final rotor speed is reached, a first scan and to subtract this from all following measuring scans. The blank correction routine of the XL-A/I, and the two programs SEDFIT [7] and dc/dt [9] solve parts of this window problem for constant rotor speed, but not for the variable rotor speed technique $\omega(t)$.

One way to reduce statistical noise is to use as many scans as possible, followed by averaging. This is valid for each kind of evaluation. This is a need especially for the present XL-A/I UV/Vis detector. To realize this, we need, as mentioned above, a radial scanning device much faster than the existing one. This would allow us to multiply the number of scans per run. Another disturbing problem is the presence of UV-active impurities in solvents and samples. It is conceivable to eliminate this problem by improving the measuring signal, as proposed above, via "blind" scans, or scans at the beginning or at the end of a run, which are subtracted subsequently from the measuring scans. Scans at the end of the run are preferred, but only if the sample has sedimented completely to the cell bottom. Beside a faster scanning device, new data analysis programs that subtract this impurity noise, if possible automatically and online, are required to carry out such corrections. There are first programs [28] that are indeed able to do this, but they should be improved and work really online.

Schlieren optics data can be improved in two ways via data analysis. The first, also by additional "blind" photographs, is to correct for (i) inhomogeneous illumination of the cells, and (ii) pixel heterogeneities, and possible dark currents of the CCD camera (see [19]). The second way is by developing a powerful image processing program to analyze such corrected CCD Schlieren photographs (see, for example, Figs. 3.24, 4.1, 4.7, and 4.13), involving (i) the radial calibration and automatic detection of radial positions of the meniscus, bottom and turbidity bands in density gradients, (ii) determining the coordinates of Schlieren curves and the areas of Schlieren peaks, and (iii) extracting from a series of Schlieren photographs, taken during a sedimentation velocity run, the complete distribution of the sedimentation coefficient $g(s) = \mathrm{d}G(s)/\mathrm{d}s$. Indeed, A. Rowe and coworkers, University Nottingham, England, are developing a Schlieren online evaluation.

The extraction/construction of $g(s)$ "master curves", as mentioned in Sect. 7.1, is important for very broadly distributed samples, and is a difficult task for all optical detection systems. Unfortunately, the time derivative method [9] fails for such broadly distributed samples, where all components of the sample can not be seen simultaneously on single or two successive scans/photographs. In the first scans of such a sedimentation velocity run, we see only the fast, large components, and in the last scans we see only the slow, small components. However, if it would be possible to combine all these different time-dependent scans into a single time-independent $g(s)$ "master curve" by means of a new data analysis program, then this would be the key to a renaissance of molar mass distribution measurements via AUC. As mentioned above, there are first attempts to realize this [26,27].

Improvement of Data Evaluation

Once we have the best possible primary measuring data, these data have to be transformed subsequently via evaluation programs into sedimentation and diffusion coefficients, sizes, molar masses, particle densities, or into the corresponding distributions. For this transformation, we need physical theories and a good data evaluation/analysis.

At present, there are physical theories for most of the evaluation problems, but some of these should be improved. One missing theory is, as mentioned above, the time dependence of $c_S(r,t)$ during sedimentation equilibrium and static density gradient runs to reach final equilibrium. There is also a need for a theory to calculate more precisely the radial density function, $\rho(r)$, within static density gradients (see Chap. 4 and [20–22]).

Good evaluation theories are already existing, too. Nevertheless, improvements and new developments are desirable. A first step to make the fast Archibald runs (see Sect. 5.5) attractive again for molar mass measurements is the recent evaluation program of Schuck and Millar [23]. For calibration of static density gradients with marker particles of known densities [18], one needs better interpolation programs to calculate $\rho(r)$ for every rotor speed, filling height, and mixing ratio of the two gradient components. Some evaluation programs do not offer possibilities to correct for effects of, for example, pressure, concentration and speed dependence, or Johnston–Ogston and non-ideality (see Sects. 3.2 and 3.3). However, these correction possibilities should be part of every program. Also, the correction of diffusion broadening during sedimentation velocity runs of very broadly distributed samples should be improved.

A *global* AUC data analysis of the future could be a combination of all single data analysis programs: Archibald, time derivative, approximate, and finite element solutions of Lamm's equation and equilibrium programs. In every AUC run, one should use Archibald programs at the beginning, sedimentation velocity programs in the middle part, and equilibrium programs at the end of the run.

In the introduction of this Chap. 7, we mentioned that the most powerful and most sophisticated data analysis programs at present are those extracting $g(s) = dG(s)/ds$ from sedimentation velocity runs [5–9], using the time derivative method or, more recently, fits of a series of concentration profiles $c(r, t)$ to approximate or finite element solutions of the Lamm differential equation (1.10). The finite element program SEDFIT of Schuck [7] is able to determine correct molar mass distributions of polydisperse samples. SEDFIT is also able to evaluate the particle size distribution of very small polydisperse nanoparticles with $d_p < 10\,\text{nm}$, where the diffusion correction is of great importance. We demonstrate this in Fig. 7.6 with a recent example, given by Cölfen [2], relating to AUC particle sizing of gold colloids.

Figure 7.6 is the last figure of this book. However, also the first two figures of this book, Figs. 1.1 and 1.2, are concerned with *gold colloids* and with an AUC particle

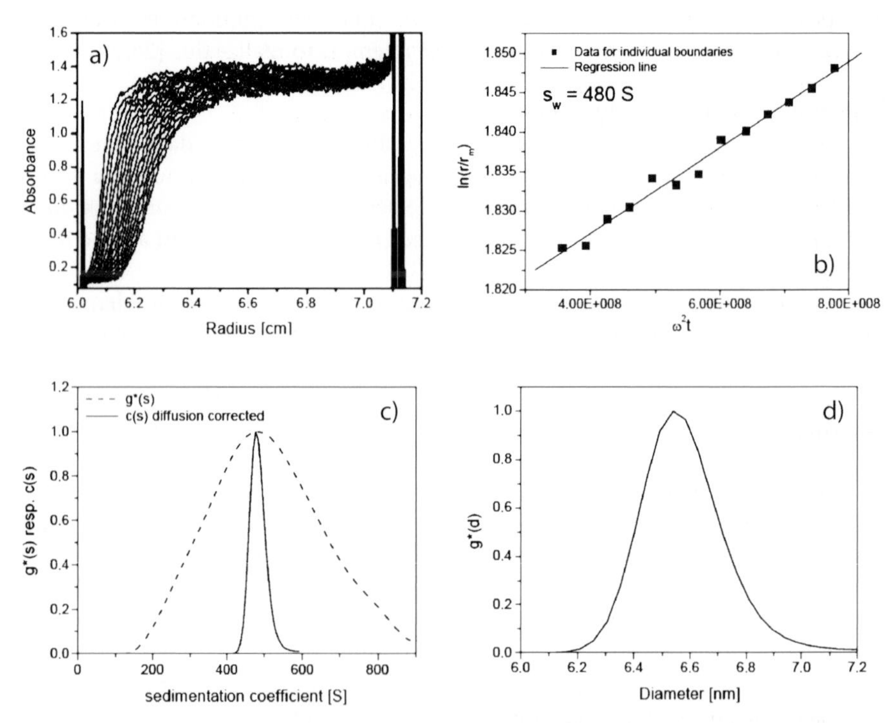

Fig. 7.6a–d. Sedimentation velocity experiment on gold colloids in water at 5000 rpm and 25 °C, illustrating three different evaluation methods. **a** Experimental XL-A/I raw data acquired with an UV/Vis detector at 575 nm. The interval between two radial scans is 2 min. **b** (Average) sedimentation coefficient calculated with (3.2). **c** Apparent sedimentation coefficient distribution $g^*(s)$ from the time derivative method ((3.10) and [9]), as well as diffusion-corrected sedimentation coefficient distribution $c(s) = g(s)$ according to Schuck [7]. **d** Final resulting diffusion-corrected particle size distribution calculated with Stokes' equation (1.9) (reprinted from [2] with permission)

size distribution measurement. They show the very first AUC experiment done by the great The(odor) Svedberg [24] in 1924. It is intriguing to compare Svedberg's measurements on gold colloids with those done 80 years later, and to look for similarities and differences. The most important difference is that Svedberg did not carry out any diffusion correction, and therefore his particle size distribution is falsified, i.e., it is too broad.

Figure 7.6a shows the (very noisy!) raw data of the sedimentation velocity experiment on gold colloids, done with the modern XL-A/I UV/Vis scanner. In the following, three different evaluation methods of these data are demonstrated by Cölfen. Figure 7.6b presents the simplest evaluation of these scans, according to (3.2), yielding an average sedimentation coefficient of 480 S. Figure 7.6c shows two, more complex evaluations, first, an evaluation using the time derivative method ((3.10), and [9]), yielding the apparent sedimentation coefficient distribution $g^*(s)$ (* signifies without diffusion correction), and second, an evaluation using a finite element solution of the Lamm equation according to Schuck [7], yielding the diffusion-corrected (real!) sedimentation coefficient distribution $g(s) = c(s)$. The comparison of $g^*(s)$ and $g(s) = c(s)$ in Fig. 7.6c demonstrates impressively how important this correction of the diffusion broadening is. Figure 7.6d shows the transformation of $g(s) = c(s)$ into the final diffusion-corrected particle size distribution of the gold colloid particles. Their average diameter, 6.5 nm, is bigger than that of Svedberg's gold particles, 1.5 nm (Fig. 1.2).

With these two gold colloid examples, a historical one and a recent one, we close our book and Chap. 7 dealing with recent developments and future outlook. Both authors are looking with interest and expectation into the future of the AUC. Will the optimistic expectations of an AUC renaissance be fulfilled? Will the proposed improvements and requirements in AUC instrumentation and data analysis be realized? The future will give us the answer.

References

1. Mächtle W (1991) Prog Colloid Polym Sci 86:111
2. Cölfen H (2004) Polymer News 29:101
3. MacGregor IK, Anderson AL, Laue TM (2004) Biophys Chem 108:165
4. Mächtle W (1999) Prog Colloid Polym Sci 113:1, and Börger L, Lechner MD, Stadler M (2004) Prog Colloid Polym Sci 127:19
5. Philo JS (2000) Anal Biochem 279:151
6. Demeler B, Saber H (1998) Biophys J 74:444
7. Schuck P (2000) Biophys J 78:1606
8. Schuck P (2003) Anal Biochem 320:104
9. Stafford WF (1992) Anal Biochem 203:295
10. Lechner MD, Mächtle W (1991) Makromol Chem 192:1183
11. Mächtle W (1984) Makromol Chem 185:1025
12. Mächtle W (1999) Biophys J 76:1080
13. Schrof W, Rossmanith P, Mächtle W (1999) European Patent EP 0 893 682 A2
14. Cölfen H, personal communication
15. Lucas G, Börger L, Cölfen H (2002) Prog Colloid Polym Sci 119:11
16. Börger L, Lechner MD (2005) Colloid Polym Sci, online published

17. Rossmanith P, Mächtle W (1997) Prog Colloid Polym Sci 107:159
18. Mächtle W, Lechner MD (2002) Prog Colloid Polym Sci 119:1
19. Klodwig U, Mächtle W (1989) Colloid Polym Sci 267:1117
20. Lechner MD, Mächtle W, Sedlack U (1997) Prog Colloid Polym Sci 107:148
21. Lechner MD, Mächtle W, Sedlack U (1997) Prog Colloid Polym Sci 107:154
22. Lechner MD, Borchard W (1999) Eur Polymer J 35:371
23. Schuck P, Millar DB (1998) Anal Biochem 259:48
24. Svedberg T, Rinde H (1924) J Am Chem Soc 46:2677, and Svedberg T (1927) Nobel lecture
25. Behlke J, Ristau O (1997) Biophys J 72:428
26. Müller HG (2004) Prog Colloid Polym Sci 127:9
27. Stafford WF, Braswell EH (2004) Biophys Chem 108:273
28. Philo JS (1997) Biophys J 72:435, and http://www.jphilo.mailway.com/download.htm
29. Cölfen H (1995) Colloid Polym Sci 273:1101
30. Cölfen H (1999) Biotechnol Genetic Eng Rev 16:87

8 Subject Index

Printing: Krips bv, Meppel
Binding: Stürtz, Würzburg